Global Implications of Emerging Technology Trends

Francisco José García-Peñalvo
University of Salamanca, Spain

A volume in the Advances in IT Standards and Standardization Research (AITSSR) Book Series

IGI Global
DISSEMINATOR OF KNOWLEDGE

Published in the United States of America by
IGI Global
Information Science Reference (an imprint of IGI Global)
701 E. Chocolate Avenue
Hershey PA, USA 17033
Tel: 717-533-8845
Fax: 717-533-8661
E-mail: cust@igi-global.com
Web site: http://www.igi-global.com

Library of Congress Cataloging-in-Publication Data

Names: García Peñalvo, Francisco José, 1971- editor.
Title: Global implications of emerging technology trends / Francisco José García Peñalvo, editor.
Description: Hershey, PA : Information Science Reference, [2018] | Includes bibliographical references.
Identifiers: LCCN 2017033305| ISBN 9781522549444 (h/c) | ISBN 9781522549451 (eISBN)
Subjects: LCSH: Technological innovations.
Classification: LCC T173.8 .G585 2018 | DDC 303.48/33--dc23
LC record available at https://lccn.loc.gov/2017033305

This book is published in the IGI Global book series Advances in IT Standards and Standardization Research (AITSSR) (ISSN: 1935-3391; eISSN: 1935-3405)

British Cataloguing in Publication Data
A Cataloguing in Publication record for this book is available from the British Library.

For electronic access to this publication, please contact: eresources@igi-global.com.

Advances in IT Standards and Standardization Research (AITSSR) Book Series

Kai Jakobs
RWTH Aachen University, Germany

ISSN:1935-3391
EISSN:1935-3405

Mission

IT standards and standardization are a necessary part of effectively delivering IT and IT services to organizations and individuals, as well as streamlining IT processes and minimizing organizational cost. In implementing IT standards, it is necessary to take into account not only the technical aspects, but also the characteristics of the specific environment where these standards will have to function.

The **Advances in IT Standards and Standardization Research (AITSSR) Book Series** seeks to advance the available literature on the use and value of IT standards and standardization. This research provides insight into the use of standards for the improvement of organizational processes and development in both private and public sectors.

Coverage

- User-Related Issues
- Economics of Standardization
- Case Studies on Standardization
- Standards in the Public Sector
- Standards and Technology Transfer
- Standardization for Organizational Development
- Open source and standardization
- Risks of Standardization
- Technological innovation and standardization
- Managing Standardization Implementation

IGI Global is currently accepting manuscripts for publication within this series. To submit a proposal for a volume in this series, please contact our Acquisition Editors at Acquisitions@igi-global.com or visit: http://www.igi-global.com/publish/.

The Advances in IT Standards and Standardization Research (AITSSR) Book Series (ISSN 1935-3391) is published by IGI Global, 701 E. Chocolate Avenue, Hershey, PA 17033-1240, USA, www.igi-global.com. This series is composed of titles available for purchase individually; each title is edited to be contextually exclusive from any other title within the series. For pricing and ordering information please visit http://www.igi-global.com/book-series/advances-standards-standardization-research/37142. Postmaster: Send all address changes to above address.

Titles in this Series

For a list of additional titles in this series, please visit: www.igi-global.com/book-series

Effective Standardization Management in Corporate Settings
Kai Jakobs (RWTH Aachen University, Germany)
Business Science Reference • copyright 2016 • 418pp • H/C (ISBN: 9781466697379) • US $205.00 (our price)

Business Process Standardization A Multi-Methodological Analysis of Drivers and Consequences
Björn Münstermann (University of Bamberg, Germany)
Business Science Reference • copyright 2015 • 448pp • H/C (ISBN: 9781466672369) • US $245.00 (our price)

Modern Trends Surrounding Information Technology Standards and Standardization Within Organizations
Kai Jakobs (RWTH Aachen University, Germany)
Information Science Reference • copyright 2015 • 339pp • H/C (ISBN: 9781466663329) • US $225.00 (our price)

Evolution and Standardization of Mobile Communications Technology
DongBack Seo (University of Groningen, The Netherlands and Hansung University, South Korea)
Information Science Reference • copyright 2013 • 328pp • H/C (ISBN: 9781466640740) • US $195.00 (our price)

Information Technology for Intellectual Property Protection Interdisciplinary Advancements
Hideyasu Sasaki (Ritsumeikan University, Japan)
Information Science Reference • copyright 2012 • 367pp • H/C (ISBN: 9781613501351) • US $195.00 (our price)

Frameworks for ICT Policy Government, Social and Legal Issues
Esharenana E. Adomi (Delta State University, Nigeria)
Information Science Reference • copyright 2011 • 352pp • H/C (ISBN: 9781616920128) • US $180.00 (our price)

Toward Corporate IT Standardization Management Frameworks and Solutions
Robert van Wessel (Tilburg University, Netherlands)
Information Science Reference • copyright 2010 • 307pp • H/C (ISBN: 9781615207596) • US $180.00 (our price)

New Applications in IT Standards Developments and Progress
Kai Jakobs (Aachen University, Germany)
Information Science Reference • copyright 2010 • 368pp • H/C (ISBN: 9781605669465) • US $180.00 (our price)

Data-Exchange Standards and International Organizations Adoption and Diffusion
Josephine Wapakabulo Thomas (Rolls-Royce, UK)
Information Science Reference • copyright 2010 • 337pp • H/C (ISBN: 9781605668321) • US $180.00 (our price)

IGI Global
DISSEMINATOR OF KNOWLEDGE

701 East Chocolate Avenue, Hershey, PA 17033, USA
Tel: 717-533-8845 x100 • Fax: 717-533-8661
E-Mail: cust@igi-global.com • www.igi-global.com

Table of Contents

Educational and Learning Technologies

Detailed Table of Contents

The global integration of competence-based education and training systems and the search for a generalized common framework for the incorporation of key competences in the curriculums of national education systems have generated a growing need for information literacy as a way of advancing to the awaited knowledge society. Large-scale assessments of student performance present criterion variables such as language, mathematics, or science, but it is noticeable how these assessments leave aside contents from other key competences such as information literacy. This chapter shows a theoretical approach to the subject and an example of an empirical study that aims to shed some light to the topic of information literacy by analysing the relationship between the level of information literacy shown by a student and their academic performance in subjects such as language and mathematics. The results suggest that it is possible to develop an instrument for the assessment of the complex information literacy competence, and which is also easy to administer in the classroom.

This chapter presents a new approach of a quantitative analysis used to research the understanding of visual literacy issues. The objective of the research is to find common patterns, opinions, and behaviors between different people regarding the use of visual communication and people's state of visual literacy, while also considering the possible cultural differences related. To explain visual literacy and its

The confluence of thousands of students in a MOOC is an opportunity to manage all the knowledge generated through the creation of open educational resources (OER), especially when a connectivist approach is applied and the MOOC makes use of virtual learning communities. The challenge is transferring the flow of knowledge, activity, and interactions of the course to the community and making that transference sustainable and ongoing over time. For this purpose, the use of elements of gamification to train and retain the knowledge creators of the community along with the use of social networking platforms is proposed. This chapter analyses several editions of a MOOC and the opportunity offered by the use of different types of learning (formal, non-formal, and informal) that occur in them, thus characterizing patterns to train the open content and knowledge generation through gamification. From the results, indicators for managing successful and sustainable knowledge communities are proposed along with indicators for persistence and interaction between participants.

There are numerous examples from recent years of the incorporation of all types of applications and technological systems into the classroom at all educational levels, with the aim to improve both student motivation and academic performance; we can group these initiatives under the term technology-enhanced learning, TEL. The TEL research field has been profoundly involved with the development and application of collaboration apps. Computers, mobile devices, and applications play diverse roles at different times along the project lifecycle. The most common lifecycle comprises four distinct phases: design, implementation, approval, and final assessment. In this chapter, the authors discuss key concepts of these TEL phases as well as some different approaches that can be defined as "good technological practices" and their main results in order to implement technologies in the formative process.

Mobile devices and apps are placed in a prominent position in the daily routine of all people. The fast evolution of these devices and their main advantages have caused a real impact in the society. Currently, they are essential tools to be totally connected anywhere and to consult and access information of any field. The use of these devices has been increasing since the last decade. The emergence of new and sophisticated devices and new services has contributed to this sparkling uptrend. One of the fields that society is using these mobile technologies is in learning. The aim of this chapter is to describe the current situation of these technologies and to make an approach of the future tendencies of these tools. To achieve this goal, it was necessary to conduct a survey and involve different undergraduate students of the university and different professionals. Results reveal that students are using more and more apps and mobile devices but there is an important gap between students and professionals so it is still necessary to boost their relevance to improve their potential use.

Preface

This book, *Global Implications of Emerging Technology Trends*, is focused on the Information Technology Research as a fully multicultural, multidisciplinary and interdisciplinary research filed, which has a broad scope of application areas, with the aim to face up the complex Knowledge Society problems and challenges we currently have to solve (García-Peñalvo, 2014, 2015b, 2015c, 2013).

The book comprises fourteen chapters that are organized in three main sections: information, media and coding literacy (Lee & So, 2014; National Research Council Committee on Information Technology Literacy, 1999; Vee, 2013), educational and learning technologies (Herold, 2016; Robinson, Molenda, & Rezabek, 2008; Spector, 2015) and data-driven intelligent ecosystems (Bogdanova & Ackovska, 2010; Cruz-Benito, Therón, & García-Peñalvo, 2016).

1. INFORMATION, MEDIA, AND CODING LITERACY

Information literacy may be defined as the capacity of people to recognize their information needs; locate and evaluate que quality of information; store and retrieve information; make effective and ethical use of information; and apply information to create and communicate knowledge (Catts & Lau, 2008), or the ability to recognize the need for information and knowing how to access, evaluate, synthesize and communicate it (Moeller, Joseph, Lau, & Carbo, 2011).

The development of the knowledge society means social transformations in which citizens need new competences and skills to tackle with the new technological ecosystems (Rodríguez-de-Dios & Igartua, 2016). These competences are based on the combination of both information literacy and media literacy set of skills that are necessary for life and work nowadays.

This situation presents an increasing approach for introduce digital or information technology (IT) literacy from the early beginning of the individual development (Bers, Flannery, Kazakoff, & Sullivan, 2014; Kazakoff & Bers, 2012; Pinto-Llorente, Casillas-Martín, Cabezas-González, & García-Peñalvo, 2017). The most frequent approach to teaching digital literacy has been to gradually encourage the learning of programming, and the term code-literacy (diSessa, 2000; Prensky, 2008) has been coined to referrer the process of teaching children programming tasks, from the simplest and most entertaining to the most complex.

Consequently, at the same time that children learn human languages, both for speaking and writing, natural languages, encompassing all matters related with the experimental sciences (physics, chemistry, biology, etc.), and humanity languages, involving social sciences and humanities, it is also necessary they learn digital languages, in which ones of the competences to be success in the digital world are included, using coding as the way to solve problems and computational thinking as working paradigm (Llorens-Largo, 2015).

A code-literate person means that can read and write in programming languages (Román-González, 2014), computational thinking is referred to the underlying problem-solving cognitive process that allows it. Thus, coding is a key way to enable computational thinking (Lye & Koh, 2014) and computational thinking may be applied to various kinds of problems that do not directly involve coding tasks (García-Peñalvo, 2016c; García-Peñalvo & Cruz-Benito, 2016; García-Peñalvo, Reimann, Tuul, Rees, & Jormanainen, 2016; Wing, 2006). An example of this is TACCLE 3 – Coding European Project that is devoted to promote computational thinking and coding in pre-university studies, specially in primary schools, all around Europe (García-Peñalvo, 2016a; García-Peñalvo, Hughes, et al., 2016; TACCLE 3 Consortium, 2017).

The information and media literate section of the book is composed by four chapters.

Martínez-Abad et al. expose that large-scale assessments of student's performance present criterion variables such as language, mathematics, or science, but it is noticeable how these assessments leave aside contents from other key competences such as information literacy. Authors, extending a previous work (Martínez-Abad, Torrijos-Fincias, & Rodríguez-Conde, 2016), present a theoretical approach to the subject and an example of an empirical study that aims to shed some light to the topic of information literacy by analyzing the relationship between the level of information literacy shown by a student and their academic performance in subjects such as language and mathematics. The results suggest that it is possible to develop an instrument for the assessment of the complex information literacy competence, and which is also easy to administer in the classroom.

García-Sánchez et al. focus their work on visual literacy (Fransecky & Debes, 1972). They present a new approach of a quantitative analysis used to research about the understanding of visual literacy issues, with the aim of finding common patterns, opinions and behaviors between different people regarding the usage of visual communication and people's state of visual literacy, while also considering the possible cultural differences related. In order to conduct these cross-cultural analyses, authors propose a new kind of quantitative questionnaire-based instrument that includes a section to measure the cultural characteristics of the individual and their level of literacy. This instrument proposal is the main result of these chapter, since the research field of visual literacy lacks this kind of quantitative approaches.

Adrián R. Vila presents a study of the action performed by the publishing industry in the context of the transposition into digital format of printed books comprising a Latin American and Caribbean literary corpus. The designed corpus includes works and authors labelled as a Latin American segment of the Western Canon, in addition to those segments provided by feminist, queer, postcolonial, and/or decolonization critical theories. It is described/defined the digital ecosystem to which the corpus is transposed as well as some of the strategies implemented by the major e-book trade platforms and the main digital libraries to offer Latin American and Caribbean literature transposed into digital format are described/defined. Likewise, a reading extended to the influence of the postcolonial turn is proposed by considering cartonera publishing as a device of the postcolonial turn but, this time, widening the range of typographic forms as well as the postcolonial effect on the publishing field. This is the evolution of his previous work (Vila, 2016).

Klinge O. Villalba Condori defends that in order to introduce computational thinking in the classrooms, it is essential to modify the initial training of teachers restructuring the curricula of the Faculties or Professional Schools of Education, in any field or context it is going to be possible to identify areas of basic or general training and areas of specialized training, as well as reading or writing, computer thinking is currently essential because its application in any professional context is necessary. This research is very related to TACCLE 3 – Coding principles (García-Peñalvo, 2017; Reimann & Maday, 2017) and closed to experiences developed in different countries (Balanskat & Engelhardt, 2015; Brackmann, Couto Barone, Casali, & Hernández, 2016; García-Peñalvo, Llorens Largo, Molero Prieto, & Vendrell Vidal, 2017; Llorens Largo, García-Peñalvo, Molero Prieto, & Vendrell Vidal, 2017).

2. EDUCATIONAL AND LEARNING TECHNOLOGIES

Educational technology is defined as the study and ethical practice of facilitating learning and improving performance by creating, using and managing appropriate technological processes and resources (Richey, 2008).

The Association for Educational Communications and Technology (AECT) denoted instructional technology as the theory and practice of design, development, utilization, management, and evaluation of processes and resources for learning (Januszewski & Molenda, 2008).

Educational technology and instructional technology are terms that are often used interchangeably, and there are a growing number of people who recommend adopting a label that includes the word "learning" (Lowenthal & Wilson, 2010).

Taking this into account, educational technology is an inclusive term for both the material tools and the theoretical foundations for supporting learning and teaching, thus it refers to all valid and reliable applied education sciences, such as equipment, as well as processes and procedures that are derived from scientific research, and in a given context may refer to theoretical, algorithmic or heuristic processes. This means that educational technology is not restricted to high technology and is anything that enhances learning in a blended or online context (García-Peñalvo, 2015a; Herold, 2016).

Nevertheless, a modern motion of technology education means electronic and plays and important role in current society (Selwyn, 2011). Education technology or EdTech, refers to an area of technology devoted to the development and application of tools (including software, hardware, and processes) intended to promote education (Lazaro, 2014).

Educational technology encompasses different approaches in the literature: eLearning (García-Peñalvo & Seoane-Pardo, 2015; Gros & García-Peñalvo, 2016), instructional technology (Molenda, 1997), information and communication technology (ICT) in education (García-Peñalvo, 2008a), EdTech (Lazaro, 2014), learning technology (Berlanga & García-Peñalvo, 2005a, 2005b), multimedia learning (García-Peñalvo & García Carrasco, 2005), technology-enhanced learning (TEL) (Kirkwood & Price, 2014), computer-based instruction (CBI) (Kulik & Kulik, 1991), computer managed instruction (Day & Payne, 1987), computer-based training (CBT) (Williams & Zahed, 1996), computer-assisted instruction or computer-aided instruction (CAI) (Suppes & Morningstar, 1969), Internet-based training (IBT) or Web-based training (WBT) (Driscoll, 1997), flexible learning (Hill, 2006), virtual education, online education or digital education (García-Peñalvo, 2008b; Seoane Pardo & García-Peñalvo, 2014), collaborative learning (Dillenbourg, 1999a, 1999b), distributed learning (Oblinger & Maruyama, 1996), computer-mediated communication (Walther, 1996), cyberlearning (Frechette, 2006), multi-modal instruction (Steil, Röthling, Haschke, & Ritter, 2004), personal learning environments (Wilson et al., 2007), networked learning (Goodyear, 2005), virtual learning environments (VLE) or learning platforms (García-Peñalvo & García Carrasco, 2002), m-learning (Casany et al., 2012; Ramírez-Montoya & García-Peñalvo, 2017; Sánchez-Prieto, Olmos-Migueláñez, & García-Peñalvo, 2014), ubiquitous learning (Conde

González, Muñoz Martín, & García-Peñalvo, 2008; Joo-Nagata, Martínez Abad, García-Bermejo Giner, & García-Peñalvo, 2017; Yang, 2006) and Massive Open On-line Courses (MOOC) (García-Peñalvo, 2015d; García-Peñalvo, Fidalgo-Blanco, & Sein-Echaluce, 2018; López Meneses, Vázquez-Cano, & Román Graván, 2015; Martínez Abad, Rodríguez Conde, & García-Peñalvo, 2014; Martínez-Núñez, Borrás-Gene, & Fidalgo-Blanco, 2016).

This book comprises four chapters related to the educational and learning technologies.

Martín-Fernández et al. analyze several editions of a MOOC and the opportunity offered by the use of different types of learning (formal, non-formal and informal) that occur in them, thus characterizing patterns to train the open content and knowledge generation through gamification (Minović, García-Peñalvo, & Kearney, 2016; Sánchez i Peris, 2015). From results, indicators for managing successful and sustainable knowledge communities are proposed along with indicator for persistence and interaction between participants.

Fonseca et al. discuss about key concepts of the Technology Enhanced Learning phases as well as some different approaches that can be defined as "Good Technological Practices" and their main results in order to implement technologies in the formative process.

Briz-Ponce et al. describe the current situation of mobile devices and apps to make an approach of the future tendencies of these tools. To achieve this goal, it was necessary to conduct a survey and involve different undergraduate students of the University and different professionals. Results reveal that students are using more and more apps and mobile devices but there is an important gap between students and professionals so it is still necessary to boost their relevance to improve their potential use. This is study is an evolution of a previous one (Briz-Ponce & Juanes-Méndez, 2015).

A. García-Valcárcel and J. J. Mena Marcos aim at determining what in-service teachers think (teachers' opinion), know (technical knowledge) and do (tactual use) about ICT to promote professional collaborative learning. This is an updated version of a previous work (García-Valcárcel Muñoz-Repiso & Mena Marcos, 2016).

3. DATA-DRIVEN INTELLIGENT ECOSYSTEMS

A technological ecosystem is a metaphor to express a needed evolution of the traditional information systems (García-Peñalvo, 2016b, 2018). These are solutions based on the composition of different software components and services that share a set of semantically defined data flows. The result is a complex ecosystem that provides a set of services that each component separately does not offer and is able to evolve as a whole in a better way when its components does or when some components are dropped out or when new components are included. Moreover, the technological ecosystem is thought to offer a better user experience in the way that users are also part or components of the ecosystem.

The internal structure of the technological ecosystems is more complex than a traditional information system (García-Holgado & García-Peñalvo, 2017a, 2017b), this implies that these solutions should be taken into account in those cases in which the knowledge management (Fidalgo-Blanco, Sein-Echaluce, & García-Peñalvo, 2014, 2015) and solution-making processes are based on heterogeneous and complex data-driven architectures (García-Peñalvo et al., 2015).

The technological ecosystem metaphor comes from the Biology field and it has been transferred into software development because it reflects so well the evolutionary nature of software. There are several authors that use the definition of natural ecosystem to support their own technological ecosystem definition systems (Chen & Chang, 2007; Dhungana, Groher, Schludermann, & Biffl, 2010; Mens, Claes, Grosjean, & Serebrenik, 2014; Yu & Deng, 2011). This way, a technological ecosystem may be defined through a mapping with the main elements that appear in every natural ecosystem (García-Holgado & García-Peñalvo, 2014, 2016), i.e., the organisms or biotic factors, the physical environment in which they inhabit or abiotic factors and the relationships between organisms and organisms with the environment. Specifically, within a technological ecosystem there are a set of persons and software components that represent the role of the biotic factors; a set of elements that allow that ecosystem runs (hardware, communications, etc.), these are the abiotic factors; and a set of data flows that mean the relationships among the software components and among these components and the involved users.

This section of the book presents six chapters.

Murteira Mendes et al. present a Question Answering system to help clinical practitioners in a cardiovascular healthcare environment to interface Clinical Decision Support systems. This can be built by using an extended discourse representation structure, CIDERS and an ontology framework, Ontology for General Clinical Practice. CIDERS is an extension of the well-known DRT (Discourse Representation Theory) structures, intending to go beyond single text representation to embrace the general clinical history of a given patient represented in an ontology. The Ontology for General Clinical Practice improves the currently available state-of-the-art ontologies for medical science and for the cardiovascular specialty. It is shown the scientific and philosophical reasons of its present dual structure with a deeply expressive (SHOIN) terminological base (TBox) and a highly computable (EL++) assertions knowledge base (ABox). To be able to use the current reasoning techniques and methodologies authors made a thorough inventory of biomedical ontologies currently available in OWL2 format. This work is based on a previous work (Murteira Mendes, Pimenta Rodrigues, Rodríguez-Solano, & Fernandes Baeta, 2014).

Benaouda Rashid and García-Peñalvo present a study that concerns the conceptualization of an intelligent ecosystem (García-Peñalvo & García-Holgado, 2017) for the territorial planning, taking as an example the agriculture case, as a tool for decision making.

Bouarara deals on the unveiling of fresh bio-inspired techniques, such as, artificial social cockroaches (ASC), artificial haemostasis system (AHS) and artificial heart lungs system (AHLS) and their application for SPAM detection. For the experimentation, the author has used the benchmark SMS Spam corpus V.0.1 and the validation measures (recall, precision, f-measure, entropy, accuracy and error). He has optimizing the sensitive parameters of each algorithm (text representation technique, distance measure, weightings, and threshold). The results are positive compared to the result of artificial social bees and machine learning algorithms (decision tree C4.5 and K-means). This chapter is based on previous works of the author (Bouarara, Hamou, & Amine, 2016a, 2016b).

Poggi and Tomaiuolo review some typical social attacks that are conducted on social networking systems, describing real-world examples of such violations and analyzing in particular the weakness of password mechanisms. They also present some solutions that could improve the overall security of the systems. This is an updated version of this work (Franchi, Poggi, & Tomaiuolo, 2015).

Vázquez-Ingelmo et al. outline the technological evolution experimented by the Observatory for University Employability and Employment information system (Michavila, Martín-González, Martínez, García-Peñalvo, & Cruz-Benito, 2015) to become a data-driven technological ecosystem. This Observatory collects data from more than 50 Spanish universities and their graduate students (bachelor's degree, master's degree) with the goal of measuring the factors that lead to students' employability and employment (Michavila, Martínez, Martín-González, García-Peñalvo, & Cruz-Benito, 2016). The goals pursued by the Observatory need a strong technological support to gather, process and disseminate the related data. The system that support these tasks, has evolved from a standard (traditional) information system to a data-driven ecosystem, which provide remarkable benefits covering the Observatory requirements (Vázquez-Ingelmo, Cruz-Benito, & García-Peñalvo, 2017).

Yaokumah and Kumah propose a theoretical model that integrates security policy, monitoring, security operations, and security roles to examine employees' security compliance. Data were collected from two hundred and thirty-three IT security and management professionals. Using Partial Least Square Structural Equation Modelling and testing hypotheses, the study finds that information security policy has significant indirect influence on information security compliance. The effect of security policy is fully mediated by security roles, operations security activities, and security monitoring activities. Security policy strongly influences operations security activities and has the greatest effect on security roles and responsibilities. Among the three mediating variables, monitoring has the most significant influence on security compliance. This work is an extension of this previous research (Winfred Yaokumah, Brown, & Dawson, 2016).

Francisco José García-Peñalvo
University of Salamanca, Spain

ACKNOWLEDGMENT

This book has been partially done within the Spanish Government Ministry of Economy and Competitiveness DEFINES project (Ref. TIN2016-80172-R). I would like to thank the authors and the reviewers their effort to make this book a reality.

REFERENCES

Balanskat, A., & Engelhardt, K. (2015). *Computing our future. Computer programming and coding Priorities, school curricula and initiatives across Europe*. Brussels, Belgium: European Schoolnet. Retrieved from https://goo.gl/i5aQiv

Berlanga, A. J., & García-Peñalvo, F. J. (2005a). IMS LD reusable elements for adaptive learning designs. *Journal of Interactive Media in Education*, 11.

Berlanga, A. J., & García-Peñalvo, F. J. (2005b). Learning Technology Specifications: Semantic Objects for Adaptive Learning Environments. *International Journal of Learning Technology*, *1*(4), 458–472. doi:10.1504/IJLT.2005.007155

Bers, M. U., Flannery, L., Kazakoff, E. R., & Sullivan, A. (2014). Computational thinking and tinkering: Exploration of an early childhood robotics curriculum. *Computers & Education*, *72*, 145–157. doi:10.1016/j.compedu.2013.10.020

Bogdanova, A. M., & Ackovska, N. (2010). Data Driven Intelligent Systems. In M. Gusev (Ed.), *Proceedings of the ICT Innovations 2010* (pp. 1-9). ICT ACT.

Bouarara, H. A., Hamou, R. M., & Amine, A. (2016a). Artificial haemostasis system for modern information retrieval with 3D result-mining. *Journal of Information Technology Research*, *9*(2), 17–46. doi:10.4018/JITR.2016040102

Bouarara, H. A., Hamou, R. M., & Amine, A. (2016b). New Bio Inspired techniques in the filtering of spam. *Journal of Information Technology Research*, *9*(2), 44–77. doi:10.4018/JITR.2016040103

Brackmann, C. P., Couto Barone, D. A., Casali, A., & Hernández, S. (2016). Pensamento Computacional: Panorama nas Américas. In F. J. García-Peñalvo & J. A. Mendes (Eds.), *XVIII Simposio Internacional de Informática Educativa, SIIE 2016* (pp. 197–202). Salamanca: Ediciones Universidad de Salamanca.

Briz-Ponce, L., & Juanes-Méndez, J. A. (2015). Mobile Devices and Apps, Characteristics and Current Potential on Learning. *Journal of Information Technology Research*, *8*(4), 26–37. doi:10.4018/JITR.2015100102

Casany, M. J., Alier, M., Mayol, E., Piguillem, J., Galanis, N., García-Peñalvo, F. J., & Conde, M. Á. (2012). Moodbile: A Framework to Integrate m-Learning Applications with the LMS. *Journal of Research and Practice in Information Technology*, *44*(2), 129–149.

Catts, R., & Lau, J. (2008). *Towards Information Literacy Indicators*. Paris: UNESCO.

Chen, W., & Chang, E. (2007). Exploring a Digital Ecosystem Conceptual Model and Its Simulation Prototype. In *Proceedings of IEEE International Symposium on Industrial Electronics, 2007 (ISIE 2007)* (pp. 2933 - 2938). IEEE.

Conde González, M. Á., Muñoz Martín, C., & García-Peñalvo, F. J. (2008). M-learning, towards U-learning. In I. Arnedillo Sánchez, & P. Isaías (Eds.), *Proceedings of the IADIS International Conference Mobile Learning 2008* (pp. 196-200). IADIS Press.

Cruz-Benito, J., Therón, R., & García-Peñalvo, F. J. (2016). Software Architectures Supporting Human-Computer Interaction Analysis: A Literature Review. In P. Zaphiris & I. Ioannou (Eds.), *Learning and Collaboration Technologies. Third International Conference, LCT 2016, Held as Part of HCI International 2016, Toronto, ON, Canada, July 17-22, 2016, Proceedings* (pp. 125-136). Switzerland: Springer International Publishing. doi:10.1007/978-3-319-39483-1_12

Day, R., & Payne, L. (1987). Computer-managed instruction: An alternative teaching strategy. *The Journal of Nursing Education*, *26*(1), 30–36. doi:10.3928/0148-4834-19870101-08 PMID:3029349

Dhungana, D., Groher, I., Schludermann, E., & Biffl, S. (2010). Software ecosystems vs. natural ecosystems: learning from the ingenious mind of nature. In *ECSA '10 Proceedings of the Fourth European Conference on Software Architecture: Companion Volume* (pp. 96-102). New York, NY: ACM.

Dillenbourg, P. (1999a). *Collaborative Learning: Cognitive and Computational Approaches* (2nd ed.). New York, NY: Elsevier Science.

Dillenbourg, P. (1999b). What do you mean by collaborative learning? In P. Dillenbourg (Ed.), *Collaborative-learning: Cognitive and Computational Approaches*. Oxford, UK: Elsevier.

diSessa, A. A. (2000). *Changing minds: Computers, learning, and literacy*. Cambridge, MA: MIT Press.

Driscoll, M. (1997). Defining Internet-based and Web-based training. *Performance Improvement*, *36*(4), 5–9. doi:10.1002/pfi.4140360403

Fidalgo-Blanco, Á., Sein-Echaluce, M. L., & García-Peñalvo, F. J. (2014). Knowledge Spirals in Higher Education Teaching Innovation. *International Journal of Knowledge Management*, *10*(4), 16–37. doi:10.4018/ijkm.2014100102

Fidalgo-Blanco, Á., Sein-Echaluce, M. L., & García-Peñalvo, F. J. (2015). Epistemological and ontological spirals: From individual experience in educational innovation to the organisational knowledge in the university sector. *Program: Electronic Library and Information Systems, 49*(3), 266-288. doi:10.1108/PROG-06-2014-0033

Franchi, E., Poggi, A., & Tomaiuolo, M. (2015). Information and Password Attacks on Social Networks: An Argument for Cryptography. *Journal of Information Technology Research*, *8*(1), 25–42. doi:10.4018/JITR.2015010103

Fransecky, R. B., & Debes, J. L. (1972). *Visual literacy: A way to learn--a way to teach.* Washington, DC: Association for Educational Communications and Technology.

Frechette, J. (2006). Cyber-censorship or cyber-literacy? Envisioning cyber-learning through media education. In D. Buckingham & R. Willett (Eds.), *Digital Generations: Children, Young People, and the New Media*. New York, NY: Routledge.

García-Holgado, A., & García-Peñalvo, F. J. (2014). Architectural pattern for the definition of eLearning ecosystems based on Open Source developments. In J. L. Sierra-Rodríguez, J. M. Dodero-Beardo, & D. Burgos (Eds.), *Proceedings of 2014 International Symposium on Computers in Education (SIIE), Logrono, La Rioja, Spain, 12-14 Nov. 2014* (pp. 93-98). Institute of Electrical and Electronics Engineers. doi:10.1109/SIIE.2014.7017711

García-Holgado, A., & García-Peñalvo, F. J. (2016). Architectural pattern to improve the definition and implementation of eLearning ecosystems. *Science of Computer Programming*, *129*, 20–34. doi:10.1016/j.scico.2016.03.010

García-Holgado, A., & García-Peñalvo, F. J. (2017a). A metamodel proposal for developing learning ecosystems. In P. Zaphiris & A. Ioannou (Eds.), *Learning and Collaboration Technologies. Novel Learning Ecosystems. 4th International Conference, LCT 2017. Held as Part of HCI International 2017, Vancouver, BC, Canada, July 9–14, 2017. Proceedings, Part I* (pp. 100-109). Springer International Publishing. doi:10.1007/978-3-319-58509-3_10

García-Holgado, A., & García-Peñalvo, F. J. (2017b). Preliminary validation of the metamodel for developing learning ecosystems. In J. M. Dodero, M. S. Ibarra Sáiz, & I. Ruiz Rube (Eds.), *Fifth International Conference on Technological Ecosystems for Enhancing Multiculturality (TEEM'17)*. New York, NY: ACM.

García-Peñalvo, F. J. (2008a). Docencia. In J. Laviña Orueta & L. Mengual Pavón (Eds.), *Libro Blanco de la Universidad Digital 2010* (pp. 29–61). Barcelona: Ariel.

García-Peñalvo, F. J. (Ed.). (2008b). Advances in E-Learning: Experiences and Methodologies. Hershey, PA: Information Science Reference (formerly Idea Group Reference). doi:10.4018/978-1-59904-756-0

García-Peñalvo, F. J. (Ed.). (2013). *Multiculturalism in Technology-Based Education: Case Studies on ICT-Supported Approaches*. Hershey, PA: IGI Global. doi:10.4018/978-1-4666-2101-5

García-Peñalvo, F. J. (2014). Managing the Knowledge Society Construction. *International Journal of Knowledge Management*, *10*(4), iv–vii.

García-Peñalvo, F. J. (2015a). Cómo entender el concepto de presencialidad en los procesos educativos en el siglo XXI. *Education in the Knowledge Society*, *16*(2), 6–12. doi:10.14201/eks2015162612

García-Peñalvo, F. J. (2015b). Engineering contributions to a Knowledge Society multicultural perspective. *IEEE Revista Iberoamericana de Tecnologías del Aprendizaje*, *10*(1), 17–18. doi:10.1109/RITA.2015.2391371

García-Peñalvo, F. J. (2015c). Information Technology Research. *Journal of Information Technology Research*, *8*(1), iv–v.

García-Peñalvo, F. J. (2015d). Massive Open Online Courses as Data Sources for Making Decisions in Learning Processes. *Journal of Information Technology Research*, *8*(4), iv–vii.

García-Peñalvo, F. J. (2016a). A brief introduction to TACCLE 3 – Coding European Project. In F. J. García-Peñalvo, & J. A. Mendes (Eds.), *2016 International Symposium on Computers in Education (SIIE 16)*. IEEE. doi:10.1109/SIIE.2016.7751876

García-Peñalvo, F. J. (2016b). Technological Ecosystems. *IEEE Revista Iberoamericana de Tecnologias del Aprendizaje*, *11*(1), 31–32. doi:10.1109/RITA.2016.2518458

García-Peñalvo, F. J. (2016c). What Computational Thinking Is. *Journal of Information Technology Research*, *9*(3), v–viii.

García-Peñalvo, F. J. (2017). Pensamiento computacional en los estudios preuniversitarios. *El enfoque de TACCLE3*. Retrieved from https://repositorio.grial.eu/handle/grial/801

García-Peñalvo, F. J. (2018a). Technological ecosystems for enhancing the interoperability and data flows. *Journal of Information Technology Research*, *11*(1).

García-Peñalvo, F. J., & Cruz-Benito, J. (2016). Computational thinking in pre-university education. In F. J. García-Peñalvo (Ed.), *Proceedings of the Fourth International Conference on Technological Ecosystems for Enhancing Multiculturality (TEEM'16)* (pp. 13-17). New York, NY: ACM. doi:10.1145/3012430.3012490

García-Peñalvo, F. J., Fidalgo-Blanco, Á., & Sein-Echaluce, M. L. (2018). An adaptive hybrid MOOC model: Disrupting the MOOC concept in higher education. *Telematics and Informatics*. doi:10.1016/j.tele.2017.09.012

García-Peñalvo, F. J., & García Carrasco, J. (2002). Los espacios virtuales educativos en el ámbito de Internet: Un refuerzo a la formación tradicional. Education in the Knowledge Society, 3.

García-Peñalvo, F. J., & García Carrasco, J. (2005). Educational hypermedia resources facilitator. *Computers & Education*, *44*(3), 301–325. doi:10.1016/j.compedu.2004.02.004

García-Peñalvo, F. J., & García-Holgado, A. (Eds.). (2017). *Open Source Solutions for Knowledge Management and Technological Ecosystems*. Hershey, PA: IGI Global. doi:10.4018/978-1-5225-0905-9

García-Peñalvo, F. J., Hernández-García, Á., Conde-González, M. Á., Fidalgo-Blanco, Á., Sein-Echaluce Lacleta, M. L., Alier-Forment, M., & Iglesias-Pradas, S. et al. (2015). Learning services-based technological ecosystems. In G. R. Alves, & M. C. Felgueiras (Eds.), *Proceedings of the Third International Conference on Technological Ecosystems for Enhancing Multiculturality (TEEM'15)* (pp. 467-472). New York: ACM. doi:10.1145/2808580.2808650

García-Peñalvo, F. J., Hughes, J., Rees, A., Jormanainen, I., Toivonen, T., Reimann, D., . . . Virnes, M. (2016). *Evaluation of existing resources (study/analysis)*. Belgium: TACCLE3 Consortium. doi: 10.5281/zenodo.163112

García-Peñalvo, F. J., Llorens Largo, F., Molero Prieto, X., & Vendrell Vidal, E. (2017). Educación en Informática sub 18 (EI<18). *ReVisión*, *10*(2), 13–18.

García-Peñalvo, F. J., Reimann, D., Tuul, M., Rees, A., & Jormanainen, I. (2016). *An overview of the most relevant literature on coding and computational thinking with emphasis on the relevant issues for teachers*. Belgium: TACCLE3 Consortium. doi:10.5281/zenodo.165123

García-Peñalvo, F. J., & Seoane-Pardo, A. M. (2015). An updated review of the concept of eLearning. Tenth anniversary. *Education in the Knowledge Society*, *16*(1), 119–144. doi:10.14201/eks2015161119144

García-Valcárcel Muñoz-Repiso, A., & Mena Marcos, J. J. (2016). Information Technology As a Way To Support Collaborative Learning: What In-service Teachers Think, Know and Do. *Journal of Information Technology Research*, *9*(1), 1–17. doi:10.4018/JITR.2016010101

Goodyear, P. (2005). Educational design and networked learning: Patterns, pattern languages and design practice. *Australasian Journal of Educational Technology*, *21*(1), 82–101. doi:10.14742/ajet.1344

Gros, B., & García-Peñalvo, F. J. (2016). Future trends in the design strategies and technological affordances of e-learning. In M. Spector, B. B. Lockee, & M. D. Childress (Eds.), *Learning, Design, and Technology. An International Compendium of Theory, Research, Practice, and Policy* (pp. 1–23). Springer International Publishing. doi:10.1007/978-3-319-17727-4_67-1

Herold, B. (2016). Technology in Education: An Overview. *Education Week*. Retrieved from https://goo.gl/9CxpnL

Hill, J. R. (2006). Flexible Learning Environments: Leveraging the Affordances of Flexible Delivery and Flexible Learning. *Innovative Higher Education*, *31*(3), 187–197. doi:10.1007/s10755-006-9016-6

Januszewski, A., & Molenda, M. (2008). *Educational technology: A definition with commentary*. Mahwah, NJ: Lawrence Erlbaum Associates.

Joo-Nagata, J., Martínez Abad, F., García-Bermejo Giner, J., & García-Peñalvo, F. J. (2017). Augmented reality and pedestrian navigation through its implementation in m-learning and e-learning: Evaluation of an educational program in Chile. *Computers & Education*, *111*, 1–17. doi:10.1016/j.compedu.2017.04.003

Kazakoff, E., & Bers, M. (2012). Programming in a robotics context in the kindergarten classroom: The impact on sequencing skills. *Journal of Educational Multimedia and Hypermedia*, *21*(4), 371–391.

Kirkwood, A., & Price, L. (2014). Technology-enhanced learning and teaching in higher education: What is ‘enhanced’ and how do we know? A critical literature review. *Learning, Media and Technology*, *39*(1), 6–36. doi:10.1080/17439884.2013.770404

Kulik, C. C., & Kulik, J. A. (1991). Effectiveness of computer-based instruction: An updated analysis. *Computers in Human Behavior*, *7*(1-2), 75–94. doi:10.1016/0747-5632(91)90030-5

Lazaro, H. (2014). *What is EdTech an why should it matter to you?* Retrieved from https://goo.gl/7sPm62

Lee, A. Y. L., & So, C. Y. K. (2014). Media Literacy and Information Literacy: Similarities and Differences. *Comunicar*, *21*(42), 137–145. doi:10.3916/C42-2014-13

Llorens-Largo, F. (2015). Dicen por ahí. . . . que la nueva alfabetización pasa por la programación. *ReVisión*, *8*(2), 11–14.

Llorens Largo, F., García-Peñalvo, F. J., Molero Prieto, X., & Vendrell Vidal, E. (2017). La enseñanza de la informática, la programación y el pensamiento computacional en los estudios preuniversitarios. *Education in the Knowledge Society*, *18*(2), 7–17. doi:10.14201/eks2017182717

López Meneses, E., Vázquez-Cano, E., & Román Graván, P. (2015). Análisis e implicaciones del impacto del movimiento MOOC en la comunidad científica: JCR y Scopus (2010-13). *Comunicar*, *44*, 73–80. doi:10.3916/C44-2015-08

Lowenthal, P. R., & Wilson, B. G. (2010). Labels do matter! A critique of AECT’s redefinition of the field. *TechTrends*, *51*(1), 38–46. doi:10.1007/s11528-009-0362-y

Lye, S. Y., & Koh, J. H. L. (2014). Review on teaching and learning of computational thinking through programming: What is next for K-12? *Computers in Human Behavior*, *41*, 51–61. doi:10.1016/j.chb.2014.09.012

Martínez Abad, F., Rodríguez Conde, M. J., & García-Peñalvo, F. J. (2014). Evaluación del impacto del término “MOOC” vs “eLearning” en la literatura científica y de divulgación. *Profesorado. Revista de currículum y formación del profesorado, 18*(1), 185-201.

Martínez-Abad, F., Torrijos-Fincias, P., & Rodríguez-Conde, M. J. (2016). The eAssessment of key competences and their relationship with academic performance. *Journal of Information Technology Research*, *9*(4), 16–27. doi:10.4018/JITR.2016100102

Martínez-Núñez, M., Borrás-Gene, O., & Fidalgo-Blanco, Á. (2016). Virtual Learning Communities in Google Plus, implications and sustainability in MOOCs. *Journal of Information Technology Research*, *9*(3), 18–36. doi:10.4018/JITR.2016070102

Mens, T., Claes, M., Grosjean, P., & Serebrenik, A. (2014). Studying evolving software ecosystems based on ecological models. In T. Mens, A. Serebrenik, & A. Cleve (Eds.), *Evolving Software Systems* (pp. 297–326). Berlin: Springer. doi:10.1007/978-3-642-45398-4_10

Michavila, F., Martín-González, M., Martínez, J. M., García-Peñalvo, F. J., & Cruz-Benito, J. (2015). Analyzing the employability and employment factors of graduate students in Spain: The OEEU Information System. In G. R. Alves, & M. C. Felgueiras (Eds.), *Proceedings of the Third International Conference on Technological Ecosystems for Enhancing Multiculturality (TEEM'15)* (pp. 277-283). New York: ACM. doi:10.1145/2808580.2808622

Michavila, F., Martínez, J. M., Martín-González, M., García-Peñalvo, F. J., & Cruz-Benito, J. (2016). *Barómetro de Empleabilidad y Empleo de los Universitarios en España, 2015 (Primer informe de resultados)*. Madrid: Observatorio de Empleabilidad y Empleo Universitarios.

Minović, M., García-Peñalvo, F. J., & Kearney, N. A. (2016). Gamification in Engineering Education. *International Journal of Engineering Education*, *32*(1B), 308–309.

Moeller, S., Joseph, A., Lau, J., & Carbo, T. (2011). *Towards Media and Information Literacy Indicators*. Paris: UNESCO.

Molenda, M. (1997). Historical and philosophical foundations of instructional design: A North American view. In R. D. Tennyson, F. Schott, N. M. Seel, & S. Dijkstra (Eds.), *Instructional design. International perspectives: Theory, Research, and Models* (Vol. 1, pp. 41–54). Mahwah, NJ: Lawrence Erlbaum Associates.

Murteira Mendes, D. J., Pimenta Rodrigues, I., Rodríguez-Solano, C., & Fernandes Baeta, C. (2014). Enrichment/Population of Customized CPR (Computer-Based Patient Record) Ontology from Free-Text Reports for CSI (Computer Semantic Interoperability). *Journal of Information Technology Research*, *7*(1), 1–11. doi:10.4018/jitr.2014010101

National Research Council Committee on Information Technology Literacy. (1999). *Being Fluent with Information Technology*. Washington, DC: National Academy Press.

Oblinger, D. G., & Maruyama, M. K. (1996). *Distributed learning*. Boulder, CO: CAUSE. Retrieved from https://goo.gl/5Qw24y

Pinto-Llorente, A. M., Casillas-Martín, S., Cabezas-González, M., & García-Peñalvo, F. J. (2017). Building, coding and programming 3D models via a visual programming environment. *Quality & Quantity*. doi:10.1007/s11135-017-0509-4

Prensky, M. (2008). *Programming Is the New Literacy*. Retrieved from http://www.edutopia.org/literacy-computer-programming

Ramírez-Montoya, M. S., & García-Peñalvo, F. J. (2017). La integración efectiva del dispositivo móvil en la educación y en el aprendizaje. *Revista Iberoamericana de Educación a Distancia*, *20*(2), 29–47. doi:10.5944/ried.20.2.18884

Reimann, D., & Maday, C. (2017). Enseñanza y aprendizaje del modelado computacional en procesos creativos y contextos estéticos. *Education in the Knowledge Society*, *18*(3), 87–97. doi:10.14201/eks20171838797

Richey, R. C. (2008). Reflections on the 2008 AECT Definitions of the Field. *TechTrends*, *52*(1), 24–25. doi:10.1007/s11528-008-0108-2

Robinson, R., Molenda, M., & Rezabek, L. (2008). Facilitating Learning. In A. Januszewski & M. Molenda (Eds.), *Educational Technology: A Definition with Commentary* (pp. 15–48). New York: Routledge.

Rodríguez-de-Dios, I., & Igartua, J. J. (2016). Skills of Digital Literacy to Address the Risks of Interactive Communication. *Journal of Information Technology Research*, *9*(1), 54–64. doi:10.4018/JITR.2016010104

Román-González, M. (2014). Aprender a programar 'apps' como enriquecimiento curricular en alumnado de alta capacidad. *Bordón. Revista de Pedagogía*, *66*(4), 135–155. doi:10.13042/Bordon.2014.66401

Sánchez i Peris, F. J. (2015). Gamificación. *Education in the Knowledge Society*, *16*(2), 13–15. doi:10.14201/eks20151621315

Sánchez-Prieto, J. C., Olmos-Migueláñez, S., & García-Peñalvo, F. J. (2014). Understanding mobile learning: Devices, pedagogical implications and research lines. *Education in the Knowledge Society*, *15*(1), 20–42.

Selwyn, N. (2011). *Education and Technology: Key Issues and Debates*. London: Continuum International Publishing Group.

Seoane Pardo, A. M., & García-Peñalvo, F. J. (2014). Pedagogical Patterns and Online Teaching. In F. J. García-Peñalvo & A. M. Seoane Pardo (Eds.), *Online Tutor 2.0: Methodologies and Case Studies for Successful Learning* (pp. 298–316). Hershey, PA: IGI Global. doi:10.4018/978-1-4666-5832-5.ch015

Spector, J. M. (2015). *Foundations of educational technology: Integrative approaches and interdisciplinary perspectives*. New York: Routledge.

Steil, J. J., Röthling, F., Haschke, R., & Ritter, H. (2004). Situated robot learning for multi-modal instruction and imitation of grasping. *Robotics and Autonomous Systems*, *47*(2), 129–141. doi:10.1016/j.robot.2004.03.007

Suppes, P., & Morningstar, M. (1969). Computer-Assisted Instruction. *Science*, *166*(3903), 343–350. doi:10.1126/science.166.3903.343 PMID:5812034

TACCLE 3 Consortium. (2017). *TACCLE 3: Coding Erasmus + Project website*. Retrieved from http://www.taccle3.eu/

Vázquez-Ingelmo, A., Cruz-Benito, J., & García-Peñalvo, F. J. (2017). Improving the OEEU's data-driven technological ecosystem's interoperability with GraphQL. In J. M. Dodero, M. S. Ibarra Sáiz, & I. Ruiz Rube (Eds.), *Fifth International Conference on Technological Ecosystems for Enhancing Multiculturality (TEEM'17)*. New York, NY: ACM.

Vee, A. (2013). Understanding Computer Programming as a Literacy. *LiCS. Literacy in Composition Studies*, *1*(2), 42–64. doi:10.21623/1.1.2.4

Vila, A. R. (2016). Latin American and Caribbean literature transposed into digital: Corpus, Ecosystem, Canon. *Journal of Information Technology Research*, *9*(1), 34–53. doi:10.4018/JITR.2016010103

Walther, J. B. (1996). Computer-Mediated Communication: Impersonal, Interpersonal, and Hyperpersonal Interaction. *Communication Research*, *23*(1), 3–43. doi:10.1177/009365096023001001

Williams, T. C., & Zahed, H. (1996). Computer-based training versus traditional lecture: Effect on learning and retention. *Journal of Business and Psychology*, *11*(2), 297–310. doi:10.1007/BF02193865

Wilson, S., Liber, O., Johnson, M., Beauvoir, P., Sharples, P., & Milligan, C. (2007). Personal Learning Environments: Challenging the dominant design of educational systems *Journal of e-Learning and Knowledge Society, 3*(3), 27-38.

Wing, J. M. (2006). Computational Thinking. *Communications of the ACM*, *49*(3), 33–35. doi:10.1145/1118178.1118215

Yang, S. J. H. (2006). Context Aware Ubiquitous Learning Environments for Peer-to-Peer Collaborative Learning. *Journal of Educational Technology & Society*, *9*(1), 188–201.

Yaokumah, W., Brown, S., & Dawson, A. A. (2016). Towards Modelling the Impact of Security Policy on Compliance. *Journal of Information Technology Research*, *9*(2), 1–16. doi:10.4018/JITR.2016040101

Yu, E., & Deng, S. (2011). Understanding Software Ecosystems: A Strategic Modeling Approach. In S. Jansen, J. Bosch, P. Campbell, & F. Ahmed (Eds.), *IWSECO-2011 Software Ecosystems 2011. Proceedings of the Third International Workshop on Software Ecosystems* (pp. 65-76). Aachen, Germany: CEUR Workshop Proceedings.

Section 1

Information, Media, and Coding Literacy

Chapter 1
Assessment of Information Literacy and Its Relationship With Learning Outcomes

Fernando Martínez-Abad
University of Salamanca, Spain

Patricia Torrijos-Fincias
University of Salamanca, Spain

Adriana Gamazo
University of Salamanca, Spain

María José Rodríguez Conde
University of Salamanca, Spain

ABSTRACT

The global integration of competence-based education and training systems and the search for a generalized common framework for the incorporation of key competences in the curriculums of national education systems have generated a growing need for information literacy as a way of advancing to the awaited knowledge society. Large-scale assessments of student performance present criterion variables such as language, mathematics, or science, but it is noticeable how these assessments leave aside contents from other key competences such as information literacy. This chapter shows a theoretical approach to the subject and an example of an empirical study that aims to shed some light to the topic of information literacy by analysing the relationship between the level of information literacy shown by a student and their academic performance in subjects such as language and mathematics. The results suggest that it is possible to develop an instrument for the assessment of the complex information literacy competence, and which is also easy to administer in the classroom.

DOI: 10.4018/978-1-5225-4944-4.ch001

INTRODUCTION

Our current society is shaped by the use of Information and Communication Technologies (Castells, 1999), and citizens strive to move from an Information Society to a Knowledge Society (Area Moreira, 2001; Cassity & Ang, 2006; Cebrián Herreros, 2009; García_Peñalvo, 2014; Valimaa & Hoffman, 2008). Within this framework, new educational needs arise in all sectors of society, and these needs are mainly related to the adaptation and utilization of this new environment rich in information (Diehm & Lupton, 2012; Grizzle, Wilson, & UNESCO, 2011; Pinto Molina, Cordón, & Díaz, 2010; Price, Becker, Clark, & Collins, 2011), aiming to put up barriers to the so-called information overload or infoglut (Jarson & Taub-Pervizpour, 2015; Levitin, 2014; Zelder, 2009).

These emerging skills and knowledge related to the handling and treatment of information, especially in the digital field, are initially referred to within the field of information science (specifically the area of library science) as 'Information Literacy' (ALA/ACRL, 2000; Andretta, 2007; Bruce, 1997; Bundy, 1998). In the first few years of the new millennium, in which the development of the concept thrived, information literacy is understood as the set of key competences that every citizen must acquire in order to properly handle themselves within the information society, and it is emphasized that the experts in charge of its transmission and development should be librarians and information professionals. However, the limitations and determinants of the concept of literacy in the field of education, along with the significance that digital competences and information literacy have achieved in a few years, lead to a reconceptualization of the term, which starts to be known in the field of Educational Sciences as 'Information Skills' (Area Moreira, 2010; Bielba Calvo, Martinez-Abad, & Herrera García, 2014; García-Peñalvo, 2017;Martínez Abad, Olmos Migueláñez, & Rodríguez Conde, 2015; Rodríguez Conde, Olmos Migueláñez, & Martínez Abad, 2013). Thus, the name information skills provides the state of the art with a more educational, specific and definite approach to the term, overcoming the difficulties and biases related to the term 'literacy' (Cope & Kalantzis, 2009; Freire, 1981; Freire & Macedo, 1987; Lankshear & knobel, 2008; Wagner, 1998).

The global integration of information systems based on competences (Delors, 1996; Halász & Michel, 2011) and the generalization of a common framework on key competences in the curricula of the education systems of different countries (Official Journal of the European Union, 2006), have facilitated the study and development of information skills as a proper key (and transversal) competence which has been embedded within the principles of basic education. However, if we dig deeper into the need for the development of this competence, we encounter the general problem that teachers, who are experts in developing educational programs based on objectives and contents rather than competences, are not trained in the key competences themselves. Moreover, they do not always have the appropriate knowledge and skills to develop teaching-learning processes based on competences, as well as to assess the level student achievement regarding these key competences. This is the reason behind the important efforts that are being made on an institutional level to clearly define the indicators, learning outcomes and assessment criteria related to the key competences, and specifically, information skills (Ferrari, 2013; Grizzle et al., 2011; UNESCO, 2015).

Under the assumptions made above, it becomes apparent that there is a need to develop and implement tools to facilitate the work of teachers regarding the objective assessment of key competences, specifically information skills, and, ultimately, whether high levels of information skills achievement among basic education students are related to better academic performances, which is the object of study of the present chapter.

INFORMATION LITERACY SKILLS AND GLOBAL TECHNOLOGY

The introduction of the term "information literacy" in the scientific context can be understood through the classical definition provided by the American Librarian Association (ALA), which defines the term as a person's ability to "Recognize when information is needed and have the ability to locate, evaluate and use effectively the needed information" (1989, p. 1).

This initial definition guided the development of subsequent studies that contributed to the research field with multiple and innovative theoretical approaches. These approaches, in their turn, serve as a complement to the definition, while also adding their own contributions to the concept of information skills. Some notable examples of this are the studies that addressed the situation of information skills in relation to similar competences or skills, such as computer skills, digital skills, media skills, etc. (Andretta, 2007; Bawden, 2001; Byrne, 2005; Catts & Lau, 2009; O'Farrill, 2008). Some institutions also undertook intensive like-minded works which aimed to establish the standards and rules of information skills (ALA/ACRL, 2000; Bundy & ANZIIL, 2004; CAUL, 2001).

This exhaustive work towards the characterization and description of the term finally led to the development of a more global and appropriate definition, which was produced within the field of Educational Sciences. This definition can be summarized as follows:

Information skills are the set of knowledge, skills, dispositions and behaviors that enable individuals to recognize when they need information, where to locate it, how to assess their suitability and how to use them according to the problem at hand (CRUE-TIC & REBIUN, 2009, p. 5).

One of the main features of our current society is the easy access to all kinds of information in a quick and permanent way, which makes it necessary to provide the population with a minimum of technology-related competences. However, the easy nature of our access to information also entails a higher difficulty for the access to actual knowledge and in-depth subjects (Cerezo, 2008). Thus, citizens find themselves in a cognitively complex information environment which is demanding (Lankshear & Knobel, 2008) and where the construction of knowledge calls for skills which are quite different from those required in societies previous to the information revolution.

This perspective allows us to identify in a clear manner the dimensions that can be established within the complex construct entitled "information literacy skills": information searching, information assessment, information processing and information communication and dissemination. Such dimensions, which we could call sub-skills, are defined in the following terms:

1. **Information Searching:** It refers to the skill that must be developed by people in order to search for the information that they require according to their specific needs. This skill is exercised through the use of different search engines or databases, the access to different information formats, the distinction of desirable from non-desirable information, etc.
2. **Information Assessment:** A sub-skill referred to people's ability to evaluate the signs of quality (suitability, exhaustiveness, relevance, timeliness, sufficiency, etc.) of the information that has been located and selected in the previous phase. This skill entails that people should be able to decide whether the information is enough for their information needs or they should make a new search.

3. **Information Processing:** Once someone has been able to select and evaluate the necessary information, it is time for the treatment and management of said information. The terms "treatment and management" refer to the construction of knowledge based on the information. For this reason, a person must be able to organize, order and prepare the information in order to process it in an appropriate way.
4. **Information Communication and Dissemination:** Even though the process could end in the previous phase, from the point of view of information skills it is necessary for people to know how to communicate the information in different ways and through different media, in the appropriate language and with the most suitable communication system according to the people receiving the message. All this process must also respect the intellectual property laws and the ethical rules common to the professional or personal task being performed.

The use of media and information and communication technologies as didactic resources is a necessary tool to favour the acquisition of information literacy skills, which are deemed key in order to tackle the challenges presented by the society of the 21st century and to ensure a proper, comprehensive and quality education (Gutierrez & Tiner, 2012).

FACTORS ASSOCIATED TO STUDENT LEARNING OUTCOMES

The study of the factors related to learning outcomes and school effectiveness has a long history within the field of educational research. A notable milestone in this line of research is the publication of the Coleman report (Coleman, 1966), whose conclusions about the scarce impact of school intervention on the performance of students when taking into account socio-economic variables caused the development of plenty of studies aiming to delve into this innovative field of research. Later on, studies such as the one conducted by Weber (1971) contributed to the field by adding process variables like school climate or leadership to and object of study that up until that moment was solely focused on contextual factors.

Learning outcomes, understood within the framework of academic performance and school effectiveness, are identified as a phenomenon with multiple causes, involving a high number of intervening variables that are equally related among themselves (Lee & Shute, 2010). Thus, identifying and isolating these variables to promote improvements in the academic performance is a complex matter, and there are multiple strategies for their analysis. In the scientific literature, we can identify diverse studies that address the analysis of the factors related to academic performance, among which we can highlight personal, family, school and social factors (González Barberá, Caso Niebla, Díaz López, & López Ortega, 2012) which are the most commonly analyzed ones, and which have also proved to have a significant positive relationship with academic performance.

Regarding personal factors, we find that they relate to characteristics that are inherent to the student, such as contextual matters like gender, cognitive skills or age, as well as issues that are more related to the topic of competences, such as social skills, emotional competence, study strategies, etc. (Nath, 2012; Yıldırım, 2012).

Family factors seem to bear an important weight on academic performance as well, and we can identify essential factors such as family socio-economic and socio-cultural levels, parental level of education and job, family structure, parent support with school issues, or the parent's own involvement in their children's school community (Cervini, Dari, & Quiroz, 2014; De Witte & Kortelainen, 2013).

School factors have also proved to have relevant effects on student performance, although the evidence presented thus far seems to suggest that their weight on performance is lower than that of the aforementioned personal and family factors. Among school factors, we can underline the existence of a systematized teacher assessment, school and teaching practices, school curriculum, school climate and learning environment, or student-teacher ratio as factors that have presented a positive relationship with performance in come studies (De Witte & Kortelainen, 2013; Lee & Shute, 2010).

Finally, there are some factors with a less clear relationship with academic performance, but which are mentioned by some studies, such as social factors like the integration of the school in the community or area in which the students live (De Witte & Kortelainen, 2013; Nath, 2012).

There is not much scientific evidence on key competences other than the ones related to the 2 key instrumental areas (language and mathematics) as factors related to academic performance. A notable exception would be the studies derived from large-scale evaluations, specifically the PISA reports, which include in each edition a key competence apart from the two areas mentioned.

Large-scale assessment data as a source for scientific studies has gained a growing notoriety in the educational research community over the past few decades (Gamazo, Olmos-Migueláñez & Martínez-Abad, 2016; García Peñalvo 2017), due, to some extent, to the fact that datasets from these assessments (OCDE, 2017) are readily available to the general public, and there is an explicit permission to freely use them for research purposes. Apart from the performance of students in different competences, large-scale assessments also provide a wide range of contextual variables, regarding both the students and the schools, which enable the development of in-depth studies on the factors that exert influence on student performance, or on the effectiveness of the participating schools. However, in the Spanish context some authors claim that the validity and reliability of these instruments varies greatly, from the high level of technical guarantees presented by the performance tests, and the lacking design of the context questionnaires (De la Orden y Jornet, 2012).

There are several large-scale student performance assessments administered at an international level which can serve as a basis to carry out studies on factors related to student performance. Some examples are the Progress in International Reading Literacy Study (PIRLS), the Trends in International Mathematics and Science Study (TIMSS), both conducted by the International Association for the Evaluation of Educational Achievement (IEA), or the Programme for International Student Assessment (PISA), developed and analysed by the Organisation for Economic Co-operation and Development (OECD). All these international standardized tests measure the level of student competences, while at the same time collecting information about their personal, family, school and social backgrounds.

The open access to the results of the abovementioned tests greatly facilitates educational researchers the undertaking of the study of the impact of cultural, economic, social, educational and personal factors in student learning outcomes, and it also facilitates the analysis of the effectiveness of schools by studying to what extent they match the level of performance expected of them according to the contextual factors they present (the more a school exceeds this level with their actual performance, the more effective it is proving to be, and vice versa) (Murillo, 2003). This is reflected in the current definition of an effective school, which is a school that "achieves a comprehensive development of each and every student, higher than it would be expected by taking into account their previous performance and the social, economic and cultural situation of their families" (Murillo, 2005, p.25).

EXAMPLE OF AN EMPIRICAL STUDY ON THE RELATIONSHIP BETWEEN INFORMATION LITERACY SKILLS AND STUDENT LEARNING OUYTCOMES

Once the topics of information literacy and factors related to the effectiveness of schools have been introduced, this chapter goes on to present an empirical study on the relationships between information literacy skills and student learning outcomes. With the experience of the Educational Assessment and Guidance Research Group from the University Institute of Educational Sciences of the University of Salamanca (Spain) as a background (García-Valcarcel Muñoz-Repiso y Mena´-Marcos, 2016; Martínez-Abad, Torrijos Fincias y Rodríguez Conde, 2016; Hernández-Ramos, Martínez-Abad, García-Peñalvo, Herrera-García y Rodríguez-Conde, 2014), we present a strategy for the assessment of information literacy skills through a validated instrument administered entirely in an online form. The data from this questionnaire is then used to study the relationship between the level of information literacy skills demonstrated by the student and their academic performance in key instrumental subjects, such as language and mathematics.

Research Design

Given that the main interest of the study is of an exploratory and relational nature - the research design is a non-experimental, ex-post-facto design of a transversal nature (Kerlinger & Lee, 1999) - we do not assume an exhaustive control of the variables and the groups participating in the study. Instead, we study the facts and phenomena directly in their natural environment, where all kinds of intervening factors are involved.

The population of interest for this study was composed of Secondary Education students between 14 and 16 years of age from Spain, and the sample was obtained with a non-probability sampling by conglomerates in two stages. We considered all schools teaching secondary education in Spain as conglomerates, and we selected a total of 17 schools through incidental sampling. Within each school, students were selected according to sample accessibility criteria. The final sample was composed of a total of 285 students.

Regarding the variables, we identified the level of information literacy skills demonstrated by the students as the explaining variable of the study. This variable, in its turn, can be materialized in the skills that the subjects attain in the search, assessment, processing and communication of information. On the other hand, the criterion variable is defined in this study as the academic performance of students in the subjects of Spanish language and literature and mathematics, specifically the grade they obtained the last time they were formally assessed.

Building the Instrument for the Assessment of Information Literacy Skills

The instrument employed for the assessment of information literacy is based on the international standards for information literacy skills (ALA/ACRL, 2000; Bundy & ANZIIL, 2004; CAUL, 2001), from which we developed a diagram, seen in Table 1, which relates each sub-skill of information skills to the corresponding learning outcomes (up to a total of 14). Likewise, each learning outcome is assigned key indicators, which add up to a total of 42 (Bielva Calvo, Martínez Abad, Herrera García y Rodríguez Conde, 2015).

Table 1. Dimensions and learning outcomes of information literacy skills

Dimensions	Learning Outcomes
Information searching	• Knows and has a good command of information sources to access the information needed. (4 indicators) • Plans the search (builds and puts into practice effectively designed search strategies). (4 indicators) • Selects information (obtains information through analog and digital resources using diverse methods). (2 indicators) • Knows how to extract, register and manage the information and its sources. (3 indicators)
Information assessment	• Applies and evaluates (values the usefulness and relevance of the information obtained). (3 indicators) • Knows evaluation criteria (formulates and applies initial criteria to evaluate the information and its sources). (5 indicators)
Information processing	• Reviews the performance process, managing the gathered or generated information, and organizes the information. (2 indicators) • Synthesizes the main ideas to build new concepts. (3 indicators) • Compares new knowledge with the old one to determine the added value, the contradictions or other characteristics inherent to the information (determines whether the new knowledge has an impact on the personal value system and takes the proper measures to reconcile the differences (3 indicators) • Creates new knowledge. (2 indicators)
Information communication and dissemination	• Understands and applies the ethical, legal and social issues that surround information and information technologies (3 indicators) • Follows laws, regulations and institutional policies, as well as etiquette rules related to the access and use of information resources. (3 indicators) • Knows and has a good command of support tools for information communication (slide presentations, videos, images, graphs, etc.). (2 indicators) • Is able to effectively communicate the product or performance to others. (3 indicators)

Once the learning outcomes and indicators of the instrument were defined, the item formulation process was undertaken. The instrument initially includes 17 items integrated by dichotomous questions, multiple choice questions and ordinal scales which result in 47 dichotomous variables (0=error; 1=success). Each item was assigned to one of the dimensions and to one or more indicators and learning outcomes.

The content validation process, which was carried out by expert judges, consisted of the assessment, on a scale of 1 to 9, of three key aspects of the questionnaire, which could inform of the overall quality of the instrument and also point to some aspects that could be improved:

- The suitability of the items for the level of the educational stage the instrument is aimed at.
- The clarity of expression of the items.
- The relevance of the association between the items and the learning outcomes.

We deemed as acceptable all mean values above 7 in every aspect valued by the judges. The information gathered from the verification process led to the deep modification of 6 items, changes in the contents of 7 items, the full substitution of 3 items and the addition of a new one. Therefore, finally we obtained an instrument composed of 18 items divided into 61 dichotomous variables[1].

After the validation process by expert judges, we carried out the statistical validation of the instrument (Bielba Calvo, Martínez Abad, & Rodríguez Conde, In press), which consisted of the application of Rasch one-parameter models based on Item Response Theory models. This process led to the revision of the formulation of some response options that compose the items of the questionnaire, because we obtained acceptable fit values, both at a global scale level and at item level.

Once the final instrument was established, we contacted the schools to obtain the sample, to which we applied the full instrument, along with a questionnaire with socio-demographic and identification questions that included items about the grades obtained in the last global assessment for the subjects of Spanish Language and Literature and Mathematics (Figure 1).

Finally, regarding the data analysis techniques used, they were descriptive and correlational, consisting on the calculation of the mean as a central tendency statistic, the standard deviation as a dispersion statistic, the skewness and kurtosis indexes to study the shape of the distributions, and Pearson's correlation coefficient to study the relationship between pairs of variables. The data analysis process was carried out using the statistical package SPSS V.20 (License from the University of Salamanca).

Figure 1. Socio-demographical and identification questions from the assessment instrument[2]

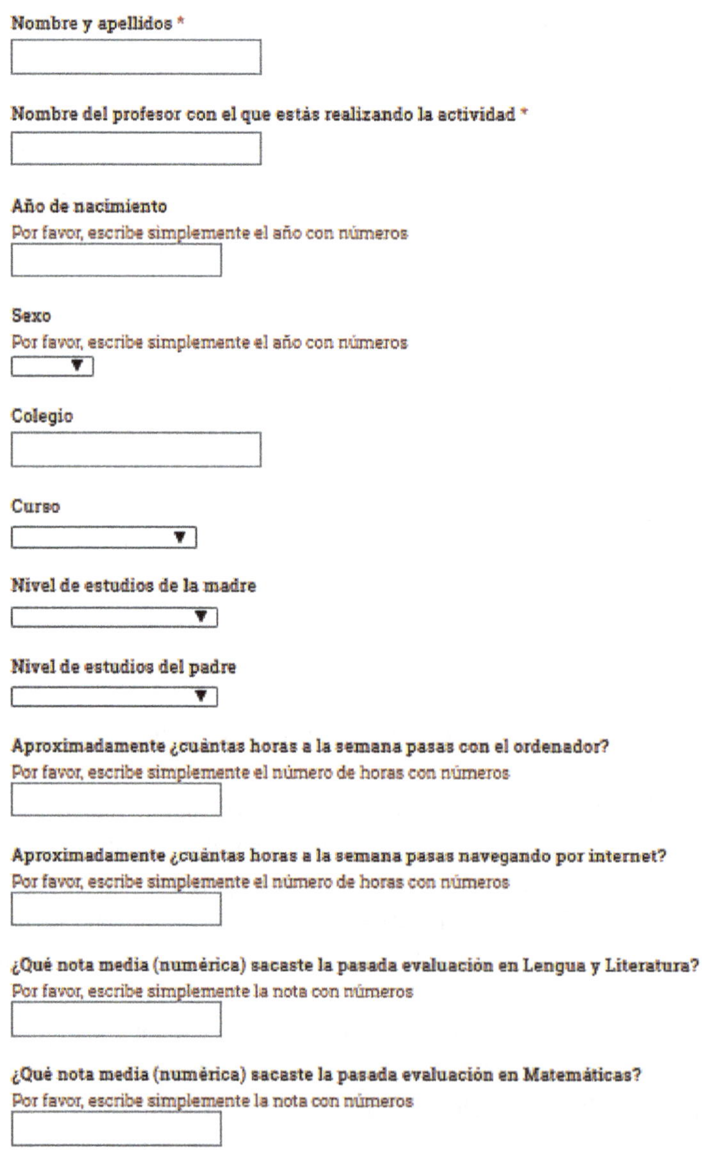

Nombre y apellidos *

Nombre del profesor con el que estás realizando la actividad *

Año de nacimiento
Por favor, escribe simplemente el año con números

Sexo
Por favor, escribe simplemente el año con números

Colegio

Curso

Nivel de estudios de la madre

Nivel de estudios del padre

Aproximadamente ¿cuántas horas a la semana pasas con el ordenador?
Por favor, escribe simplemente el número de horas con números

Aproximadamente ¿cuántas horas a la semana pasas navegando por internet?
Por favor, escribe simplemente el número de horas con números

¿Qué nota media (numérica) sacaste la pasada evaluación en Lengua y Literatura?
Por favor, escribe simplemente la nota con números

¿Qué nota media (numérica) sacaste la pasada evaluación en Matemáticas?
Por favor, escribe simplemente la nota con números

Some Results From the Empirical Study

The questionnaire was completed by a total of 285 students, of which 156 were men and 129 were women, which represents 54,7% and 45,3% of the sample respectively (Figure 2). These percentages of males and females are very similar and show the representativeness of the sample, at least according to this variable.

On the other hand, Table 2 shows the distribution of the sample according to the highest level of formal education attained by both the mother and father of the students. The distribution of the level of studies also appears to be coherent with Spanish demographic data provided by the National Statistics Institute[3] (*Instituto Nacional de Estadística*, INE).

Combining this information in a stacked column chart, shown in Figure 3, we observe how the distribution of the sample by the combined level of education of the father and mother also presents logical values at a demographic level, since it displays a clear endogamy regarding the level of studies of the parents.

When analysing the answers given by the participants about variables such as the number of hours a week that they spend using the computer or connected to the internet, or regarding their academic performance in the subjects of Language and Literature or Mathematics, we find that the students claim to spend more than 7 hours a week in front of a computer and almost 12 hours connected to the internet through some kind of electronic device (Table 3).

Figure 2. Distribution of the sample according to the variable "gender"

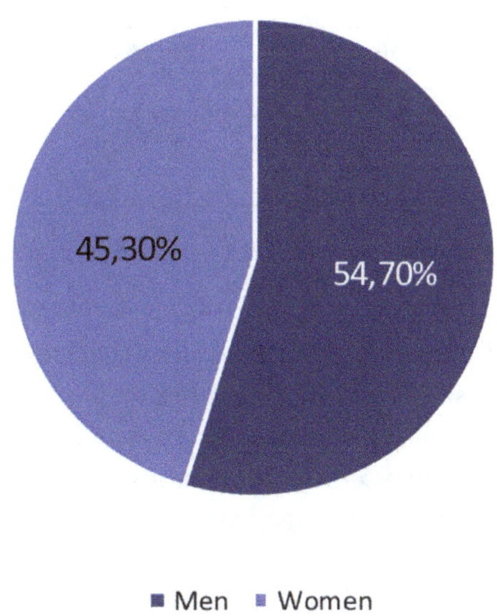

Table 2. Distribution of the sample according to parental levels of formal education attainment

	Mother	Father
No formal education	2.57%	4.07%
Primary Education	21.32%	24.81%
Secondary Education	44.85%	51.48%
Higher Education	31.25%	19.63%

Figure 3. Distribution of the sample according to the combined level of education of the mother and father

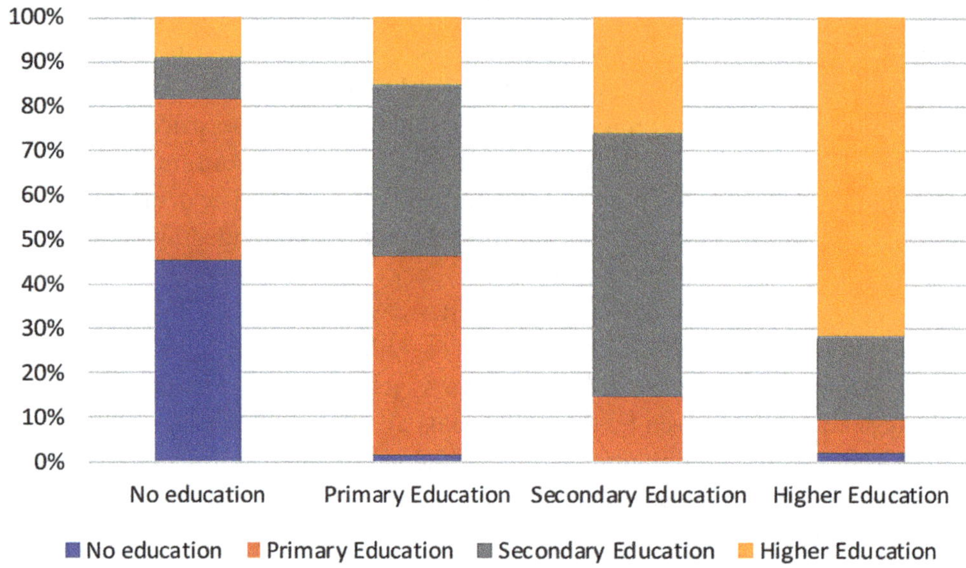

Table 3. Some descriptive statistics about the use of information and communication technologies and grade averages for the two subjects under study

	Mean	Standard Deviation	Skewness	Kurtosis
Hours computer	7.05	9.39	3.53	17.23
Hours Internet	11.78	17.47	4.29	23.59
Grade Language and Literature	5.77	1.86	0.05	-0.46
Grade Mathematics	5.54	1.93	0.05	-0.44

The high standard deviation of these two variables shows that the dispersion is quite high, and there is a great variability regarding the use of computers and Internet among the students. On the other hand, regarding the academic performance of the students, we find an average grade of 5.77 points out of 10 in Spanish Language and Literature and 5.54 in Mathematics, with more standard deviations which are more normalized than in the case of internet usage variables.

An in-depth study of the distribution of average scores reported by the students for the subjects of Language and Mathematics (Figures 4 and 5 respectively) show that they come close to a normal distribution in both variables, with a slight positive skewness and platykurtic kurtosis.

Finally, Table 4 shows the results of the study of the mean scores, dispersions and shape of the distributions in the information skills scale, both for the total scale and for the 4 dimensions. In this case we can observe how the distributions generally have a moderate variability and a more evident tendency to negative skewness and platykurtic kurtosis, which reach significant levels in some of the analysed dimensions.

In order to better understand the results shown in Table 4, it is worth noting the fact that the scoring scales do not range from 1 to 10, as it was the case for the grades in Language and Mathematics. The scales for these variables are the following:

Figure 4. Distribution of the sample according to the grade obtained in the subject of Spanish Language and Literature

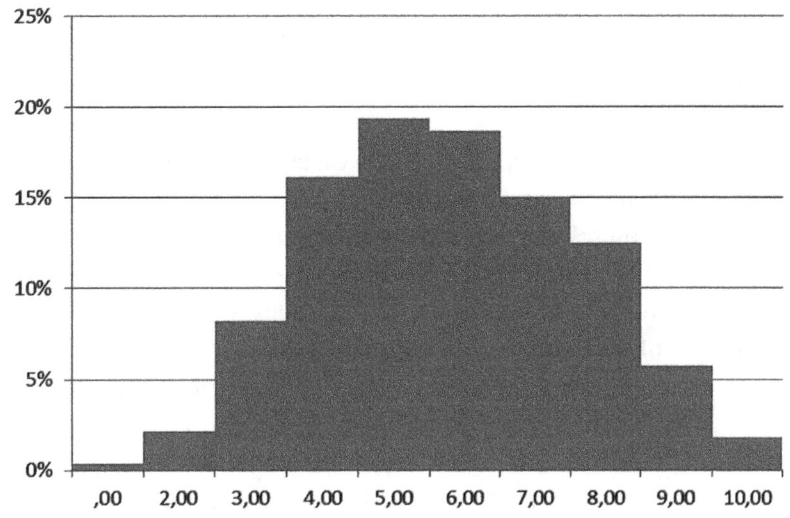

Figure 5. Distribution of the sample according to the grade obtained in the subject of Mathematics

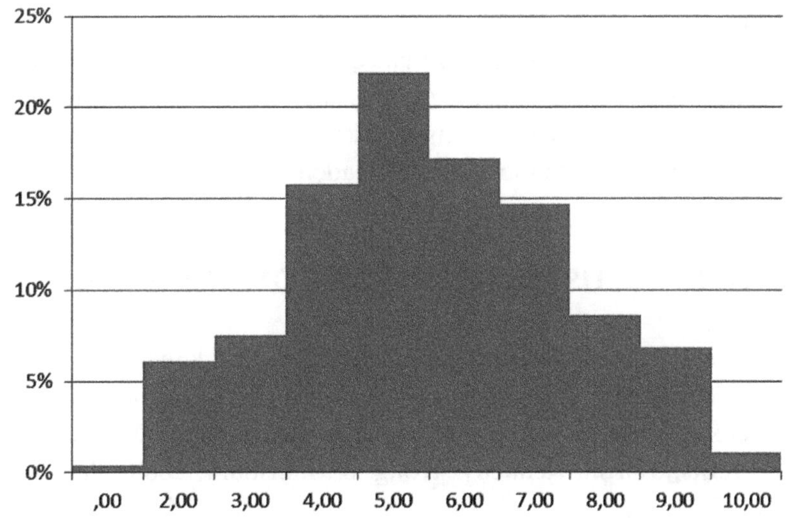

Table 4. Descriptive statistics on the level of information literacy skills demonstrated by students

	Mean	Standard Deviation	Skewness	Kurtosis
Information searching	11.51	2.85	0.24	-0.19
Information assessment	8.47	2.78	-0.10	-0.99
Information processing	8.56	2.33	-0.49	-0.38
Information communication	8.44	2.38	-0.59	-0.34
Total Information skills	**36.97**	**7.50**	**0.06**	**-0.67**

- Information searching (0-21 points)
- Information assessment (0-14 points)
- Information processing (0-13 points)
- Information communication (0-13 points)
- Total information skills (0-61 points)

Thus, the interpretation of the mean values in these variables shows that the participants were able to successfully answer more than half of the questions posed in the different items of each of the dimensions, especially in the dimension related to information searching.

Table 5 shows the results of the calculation of Pearson's correlation coefficient among the variables related to the level of information skills and the subjects' grades in Language and Literature and Mathematics and the hours of use of the computer and the Internet.

It is noteworthy that, while the hours the students spend with a computer have a positive and significant correlation with information skills in the dimension of information searching and at a global level, the correlation between the hours of Internet use and these variables is negative, reaching significant levels for the dimensions of information processing and communication, and also the global information skills score. These differences can be explained by the fact that, currently, the subjects are connected to the Internet through any electronic device, and these devices (mainly smartphones) do not perform the same activities that personal computers traditionally do, which have to do more with the information skills and abilities being evaluated in this study.

The relationship between the academic performance in Language and Mathematics and the level of information skills is clear and positive, reaching highly significant values in most of the dimensions. Therefore, it seems that the level of the subjects in a key competence such as information skills can indeed be considered a factor related to the subjects' academic performance.

DISCUSSION AND CONCLUSION DERIVED FROM THE STUDY

When dealing with the assessment of student academic performance, large-scale assessments and studies consider reading, mathematics and, in some cases, sciences, as the criterion variables, leaving aside the contents related to other key competences such as informational literacy (De Witte & Kortelainen, 2013; Lee & Shute, 2010; Mullis, Martin, Kennedy, Trong, & Sainsbury, 2009; Nath, 2012; OCDE, 2012;

Table 5. Results for Pearson's correlation coefficient

	Searching		Assessment		Processing		Communication		Total	
	r	p.	r	p.	r	p.	r	p.	r	p.
Hours computer	**0,17**	**0,00**	0,11	0,06	0,09	0,13	0,05	0,44	**0,15**	**0,01**
Hours Internet	-0,01	0,86	-0,07	0,22	**-0,18**	**0,00**	**-0,13**	**0,03**	**-0,13**	**0,03**
Grade Language	0,08	0,20	**0,30**	**0,00**	**0,25**	**0,00**	**0,24**	**0,00**	**0,30**	**0,00**
Grade Math	**0,16**	**0,01**	**0,16**	**0,01**	**0,25**	**0,00**	0,07	0,24	**0,22**	**0,00**

Yıldırım, 2012). In this sense, the present research was proposed with the aim of providing evidence about the relationship between the levels of information literacy skills and the academic achievement of students in key instrumental subjects, such as language or mathematics.

In the path from the information society to the long-desired knowledge society, the development of the citizens' information skills cannot be denied as a basic need (Ferrari, 2013. García-Peñalvo, 2017). Basic education teacher need to be properly trained in the development and assessment of a whole set of competences and skills necessary to provide a comprehensive, integral and quality education, while also addressing the challenges posed to education in the society of the 21st century. However, the study and the generalization of key competences in the education curricula of most of the developed countries has not been accompanied, in many cases, by efforts to provide teachers with the necessary knowledge to enable their actual implementation, especially in the case of information literacy skills and other competences that are regarded as transversal. This issue becomes even more pressing if we take into consideration the pressure put on teachers regarding the large amount of contents that core subjects have on the curriculum, which pushes these competences of a transversal nature, like information literacy, into the background in most cases. Given this pressure, teachers can sometimes feel that the most important thing is the contents of each one of the subjects, and that it is neither necessary nor operational to formally tackle the development of key competences in the classroom.

The work presented here revolves around this line of research, and it has been motivated by the need to provide support to secondary education teachers for the assessment of key competences, specifically of information skills, by presenting evidence about the existence of a significant (s.l. 0.05) relationship existent between the level of skills attained in the different aspects related to the treatment of information and digital skills, and the academic achievement or performance of students, understood in the classic sense (performance in language and mathematics). Therefore, its results endorse a current trend in scientific research, related to facilitating the education and assessment of key competences, particularly of information skills (Bielba Calvo et al., 2014; Martínez Abad et al., 2015; Rodríguez Conde et al., 2013).

Thus, the results obtained show, once again, how it is possible to develop an easy-to-use instrument to assess a competence as complex as information skills in the classroom, and they also clearly illustrate the close relationship between high levels of information skills and high levels of academic performance. Therefore, we propose to add to the group of factors related to academic performance those factors that have to do with the skills of a subject to search for information in virtual environments, to evaluate its quality and relevance, to process the desired information and to communicate in an efficient and ethical way.

Despite the strengths displayed by the results obtained in this study, it is worth mentioning some of its weaknesses, related to the restriction of the sample size, which had a moderate reach, and mainly the nature of the research design, which was non-experimental, and it does not allow us to establish a cause-effect relationship between information skills and formal academic performance. In fact, this design simply allows us to establish a relationship of joint tendency between pairs of variables.

Across Europe, we are currently witnessing a widespread effort to define and assess digital literacy skills, and also to propose educational intervention measures in order to foster the development of this competence among European citizens (Ferrari, 2013). In Spain, the National Institute for Educational Technologies and Teacher Training (*Instituto Nacional de Tecnologías Educativas y de Formación del Profesorado*, INTEF, 2017) has published a steering document that establishes the lines of intervention on the assessment and training of digital competences for the compulsory education teaching body. This working report elaborated by the education Administration, which was developed from 2012 to 2017, conveys the transcendence of this topic in the field of educational intervention, and even more in the field of educational research.

In sum, we can say that through the discussion posed in the theoretical framework, and the example of an empirical study presented, the present research opens up a wide spectrum of future studies in diverse fields, ranging from the development of research works to verify whether this positive relationship between performance and information skills occurs in the same way with other key competences, to the future need to develop studies that implement experimental designs that allow us to attribute the possible causality of the level of information skills and other key competences on the students' academic performance. Likewise, we present a specific experience in the field of key competence assessment, which can be useful for the development of future studies within the line of the design and implementation of valid and reliable evaluation instruments.

REFERENCES

ALA/ACRL. (2000). *Information literacy competency standards for higher education*. Chicago, IL: ACRL.

American Library Association. (1989). *American Library Association Presidential Committee on Information Literacy: final report*. Chicago: American Library Association.

Andretta, S. (2007). Phenomenography: A conceptual framework for information literacy education. *Aslib Proceedings*, *59*(2), 152–168. doi:10.1108/00012530710736663

Area Moreira, M. (2001). *Educar en la sociedad de la información*. Bilbao: Desclée De Brouwer.

Area Moreira, M. (2010). ¿Por qué formar en competencias informacionales y digitales en la educación superior? *Revista de Universidad y Sociedad del Conocimiento*, *7*(2), 2–5.

Bawden, D. (2001). Information and digital literacies: A review of concepts. *The Journal of Documentation*, *57*(2), 218–259. doi:10.1108/EUM0000000007083

Bielba Calvo, M., Martínez-Abad, F., & Herrera García, M. E. (2014). Selection of key standards to create an instrument for information literacy assessment in Compulsory Secondary Education. In *Proceedings TEEM' 14. Technological Ecosystems for Enhancing Multiculturality* (pp. 475-482). Salamanca: ACM.

Bielba Calvo, M., Martínez Abad, F., Herrera García, M. E., & Rodríguez Conde, M. J. (2015). Diseño de un instrumento de evaluación de competencias informacionales en educación secundaria obligatoria a través de la selección de indicadores clave. *Education in the Knowledge Society*, *16*(3), 124–143. doi:10.14201/eks2015163124143

Bielba Calvo, M., Martínez Abad, F., & Rodríguez Conde, M. J. (En prensa). Validación psicométrica de un instrumento de evaluación de competencias informacionales en la educación secundaria. *Bordón. Revista de pedagogía*.

Bruce, C. (1997). *The seven faces of information literacy*. Adelaide: Auslib Press.

Bundy, A. (1998). Information literacy: the key competency for the 21sr century. In *Annual Conference of the International Association of Technological University Libraries* (p. 16). Pretoria, South Africa: Academic Press.

Bundy, A., & Anziil. (2004). *Australian and New Zealand information literacy framework principles, standards and practice*. Australian and New Zealand Institute for Information Literacy. Retrieved from http://www.anziil.org/resources/Info%20lit%202nd%20edition.pdf

Byrne, A. (2005). La alfabetización informacional desde una perspectiva global: Eldesastre agudiza nuestras mentes. *Anales de Documentación*, *8*, 7–20.

Cassity, E., & Ang, I. (2006). Industry partnerships and the «Knowledge Society»: The australian experience. *Minerva*, *44*(1), 47–63. doi:10.1007/s11024-005-5412-9

Castells, M. (1999). *The information age: economy, society and culture*. Oxford, UK: Blackwell.

Catts, R., & Lau, J. (2009). *Hacia unos indicadores de alfabetización informacional*. Madrid: Ministerio de Cultura.

CAUL. (2001). *Information literacy standards*. Canberra: University of South Australia.

Cebrián Herreros, M. (2009). *La sociedad de la informacion y del conocimiento en los paises nordicos*. Barcelona: Gedisa.

Cerezo, J. M. (2008). Hacia un nuevo paradigma: la era de la información fragmentada. *Revista Telos*, (76). Retrieved from http://sociedadinformacion.fundacion.telefonica.com/telos/articulocuaderno.a sp@idarticulo=6&rev=76.htm

Cervini, R., Dari, N., & Quiroz, S. (2014). Estructura familiar y rendimiento académico en países de América Latina: Los datos del Segundo Estudio Regional Comparativo y Explicativo. *rmie. Revista Mexicana de Investigación Educativa*, *19*(61), 569–597.

Coleman, J. S. (1966). *Equality of educational opportunity* (Vol. 2). Washington, DC: US Department of Health, Education, and Welfare, Office of Education.

Cope, B., & Kalantzis, M. (2009). Multiliteracies: New literacies, new learning. *Pedagogies*, *4*(3), 164–195. doi:10.1080/15544800903076044

CRUE-TIC & REBIUN. (2009). *Competencias informáticas e informacionales en los estudios de grado*. Retrieved from http://crue-tic.uji.es/index.php?option=com_remository&Itemid=28&func=startdown&id=226

De la Orden, A., & Jornet, J. M. (2012). La utilidad de las evaluaciones de sistemas educativos: El valor de la consideración del contexto. *Bordón*, *64*(2), 69–88.

De Witte, K., & Kortelainen, M. (2013). What explains the performance of students in a heterogeneous environment? Conditional efficiency estimation with continuous and discrete environmental variables. *Applied Economics*, *45*(17), 2401–2412. doi:10.1080/00036846.2012.665602

Delors, J. (1996). *Learning, the treasure within: report to UNESCO of the International Commission on Education for the Twenty-first Century*. Paris: UNESCO.

Diehm, R.-A., & Lupton, M. (2012). Approaches to learning information literacy: A phenomenographic study. *Journal of Academic Librarianship, 38*(4), 217–225. doi:10.1016/j.acalib.2012.05.003

Ferrari, A. (2013). *DIGCOMP: a framework for developing and understanding digital competence in Europe* (Y. Punie & B. N. Brečko, Eds.). Luxembourg: Publications Office of the European Union.

Freire, P. (1981). *Literacy and revolution*. New York: Continuum.

Freire, P., & Macedo, D. P. (1987). *Literacy: reading the word & the world*. South Hadley, MA: Bergin & Garvey Publishers.

Gamazo, A., Olmos-Migueláñez, S., & Martínez-Abad, F. (2016, November). Multilevel models for the assessment of school effectiveness using PISA scores. In *Proceedings of the Fourth International Conference on Technological Ecosystems for Enhancing Multiculturality* (pp. 1161-1166). ACM. doi:10.1145/3012430.3012663

García-Peñalvo, F. J. (2014). Managing the Knowledge Society Construction. *International Journal of Knowledge Management, 10*(4), iv–vii.

García-Peñalvo, F. J. (2015). Information Technology Research. *Journal of Information Technology Research, 8*(1), iv–v.

García-Peñalvo, F. J. (2016). Digital Humanities Data Processing. *Journal of Information Technology Research, 9*(1), v–viii.

García-Peñalvo, F. J. (2017). Multidisciplinary and interdisciplinary approaches in information technology research. *Journal of Information Technology Research, 10*(1), vi–viii.

García-Valcárcel Muñoz-Repiso, A., & Mena Marcos, J. J. (2016). Information Technology As a Way To Support Collaborative Learning: What In-service Teachers Think, Know and Do. *Journal of Information Technology Research, 9*(1), 1–17. doi:10.4018/JITR.2016010101

González Barberá, C., Caso Niebla, J., Díaz López, K., & López Ortega, M. (2012). Rendimiento académico y factores asociados: aportaciones de algunas evaluaciones a gran escala. *Bordón. Revista de pedagogía, 64*(2), 51-68.

Grizzle, A., Wilson, C., & UNESCO. (2011). *Media and information literacy curriculum for teachers*. Paris: UNESCO. Retrieved from http://www.unesco.org/new/fileadmin/MULTIMEDIA/HQ/CI/CI/pdf/publications/Media%20and%20information%20literacy%20curriculum%20for%20teachers.pdf

Gutierrez, A., & Tiner, K. (2012). Educación para los medios, alfabetizaciónmediática y competencia digital. *Comunicar, 38*(9), 31–39.

Halász, G., & Michel, A. (2011). Key Competences in Europe: Interpretation, policy formulation and implementation. *European Journal of Education, 46*(3), 289–306. doi:10.1111/j.1465-3435.2011.01491.x

Hernández-Ramos, J. P., Martínez-Abad, F., García-Peñalvo, F. J., Herrera García, M. E., & Rodríguez-Conde, M. J. (2014). Teachers' attitude regarding the use of ICT. A factor reliability and validity study. *Computers in Human Behavior*, *31*, 509–516. doi:10.1016/j.chb.2013.04.039

Instituto Nacional de Tecnologías Educativas y de Formación del Profesorado (INTEF). (2017). *Marco Común de Competencia Digital Docente*. Author.

Jarson, J., & Taub-Pervizpour, L. (2015). Infoglut or Information Literacy? Using Blended Learning for Student Research in «New Information Technologies». *Blended Learning in the Liberal Arts Conference*. Retrieved from http://repository.brynmawr.edu/blended_learning/2015/2015/3

Kerlinger, F., & Lee, H. (1999). Foundations of behavioral research (4th ed.). Wadsworth Publishing.

Lankshear, C., & Knobel, A. (2008). *Nuevos alfabetismos su práctica cotidiana y el aprendizaje en el aula*. Madrid: Morata.

Lee, J., & Shute, V. J. (2010). Personal and Social-Contextual Factors in K-12 Academic Performance: An Integrative Perspective on Student Learning. *Educational Psychologist*, *45*(3), 185–202. doi:10.1080/00461520.2010.493471

Levitin, D. J. (2014). *The Organized Mind: Thinking Straight in the Age of Information Overload*. Penguin.

Martínez Abad, F., Olmos Migueláñez, S., & Rodríguez Conde, M. J. (2015). Evaluación de un programa de formación en competencias informacionales para el futuro profesorado de E.S.O. *Review of Education*, (370): 45–70.

Martínez-Abad, F., Torrijos-Fincias, P., & Rodríguez-Conde, M. J. (2016). The eAssesment of key competences and their relationship performance. *Journal of Information Technology Research*, *9*(4), 16–27. doi:10.4018/JITR.2016100102

Mullis, I. V., Martin, M. O., Kennedy, A. M., Trong, K. L., & Sainsbury, M. (2009). *PIRLS 2011 Assessment Framework*. International Association for the Evaluation of Educational Achievement. Retrieved from http://www.iea.nl

Murillo, F. J. (2003). El movimiento de investigación de Eficacia Escolar. In *La investigación sobre Eficacia Escolar en Iberoamérica. Revisión internacional del estado del arte*. Bogotá: Convenio Andrés Bello.

Murillo, F. J. (2005). *La Investigación sobre Eficacia Escolar*. Barcelona: Octaedro.

Nath, S. R. (2012). Factors Influencing Primary Students' Learning Achievement in Bangladesh. *Research in Education*, *88*(1), 50–63. doi:10.7227/RIE.88.1.5

O'Farrill, R. (2008). Information Literacy and knowledge management: Preparations for an arranged marriage. *Libri*, *58*(3), 155–171. doi:10.1515/libr.2008.017

OECD. (2012). *PISA 2009 technical report*. Paris: OECD Publishing. Retrieved from http://proxy.library.carleton.ca/login?url=http://dx.doi.org/10.1787/9789264167872-en

OECD. (2017). *PISA: Programme for International Student Assessment, OECD Education Statistics (database)*. DOI: 10.1787/data-00365-en

Official Journal of the European Union. (2006). *Key competences for Lifelong Learning. European Reference Framework*. Brussels: European Parliament and of the Council. Retrieved from http://eur-lex.europa.eu/LexUriServ/site/es/oj/2006/l_394/l_ 39420061230es00100018.pdf

Pinto Molina, M., Cordón, J. A., & Díaz, R. G. (2010). Thirty years of information literacy (1977-2007): A terminological, conceptual and statistical analysis. *Journal of Librarianship and Information Science*, 17.

Price, R., Becker, K., Clark, L., & Collins, S. (2011). Embedding information literacy in a first-year business undergraduate course. *Studies in Higher Education*, *36*(6), 705–718. doi:10.1080/03075071003725350

Rodríguez Conde, M. J., Olmos Migueláñez, S., & Martínez Abad, F. (2013). Evaluación de competencias informacionales en educación secundaria: Un modelo causal. *Cultura y Educación*, *25*(3), 361–373. doi:10.1174/113564013807749687

UNESCO. (2015). *Keystones to foster inclusive knowledge societies: Access to information and knowledge, freedom of expression, privacy, and ethics on a global internet*. París: UNESCO. Retrieved from http://apo.org.au/node/55441

Valimaa, J., & Hoffman, D. (2008). Knowledge society discourse and higher education. *Higher Education: The International Journal of Higher Education and Educational Planning*, *56*(3), 265–285. doi:10.1007/s10734-008-9123-7

Wagner, D. A. (1998). *Alfabetización: construir el futuro*. Paris: UNESCO.

Weber, G. (1971). *Inner-city children can be taught to read: four successful schools*. Washington, DC: Council for Basic Education.

Yıldırım, S. (2012). Teacher Support, Motivation, Learning Strategy Use, and Achievement: A Multilevel Mediation Model. *Journal of Experimental Education*, *80*(2), 150–172. doi:10.1080/00220973.2011.596855

Zelder, N. (2009). Infoglut: Its the disease of the new millennium. How do we treat it? *IEEE Spectrum*, *46*(10), 30-32, 52.

ENDNOTES

1 The final assessment instrument used for the study can be consulted at: http://goo.gl/forms/e1ZmF24Igr.

2 Name and surname; Name of the teacher you are doing this activity with; Year of birth; Sex; School; Grade; Mother's level of formal education; Father's level of formal education; Approximately, how many hours a week do you spend surfing the web?; What is the average grade that you obtained last term in the subject of Language and Literature?; What is the average grade that you obtained last term in the subject of Mathematics?

3 More information on the level of formal education attained by the adult population in Spain can be consulted in the following link: http://www.ine.es/ss/Satellite?c=INESeccion_C&p=1254735110672&pagename=ProductosYServicios%2FPYSLayout&cid=1259925481659&L=0.

Chapter 2
Developing a Research Method to Analyze Visual Literacy Based on Cross-Cultural Characteristics

Felicidad García-Sánchez
University of Salamanca, Spain

José Gómez-Isla
University of Salamanca, Spain

Roberto Therón
University of Salamanca, Spain

Juan Cruz-Benito
University of Salamanca, Spain

José Carlos Sánchez-Prieto
University of Salamanca, Spain

ABSTRACT

This chapter presents a new approach of a quantitative analysis used to research the understanding of visual literacy issues. The objective of the research is to find common patterns, opinions, and behaviors between different people regarding the use of visual communication and people's state of visual literacy, while also considering the possible cultural differences related. To explain visual literacy and its implications, the theoretical background about the visual literacy research field is presented first. Then, also within the section on background, the chapter presents the main concepts related to culture, and how it and visual literacy can be analyzed together to enable cross-cultural analysis. To conduct these cross-cultural analyses, this chapter proposes a new kind of quantitative questionnaire-based instrument that includes a section to measure the cultural characteristics of the individual and their level of literacy. This instrument proposal is the main result, since the research field of visual literacy lacks this kind of quantitative approach.

DOI: 10.4018/978-1-5225-4944-4.ch002

1. INTRODUCTION

The concept of "visual literacy" is not a brand-new term. Visual literacy as a term and a field of study was introduced in 1969 by the authors Fransecky & Debes (1972). They define, within their role steering the International Visual Literacy Association, visual literacy in the following way:

"a group of vision competencies a human being can develop by seeing and at the same time having and integrating other sensory experiences. The development of these competencies is fundamental to normal human learning. When developed, they enable a visually literate person to discriminate and interpret the visual actions, objects, and/or symbols, natural or man-made, that are [encountered] in [the] environment. Through the creative use of these competencies, [we are] able to communicate with others. Through the appreciative use of these competencies, [we are] able to comprehend and enjoy the masterworks of visual communications" (Fransecky & Debes, 1972; International Visual Literacy Association, 2003).

The fact that the individuals of the new generations have greater access to information does not imply that such access is relevant or significant. In order to achieve this, it is necessary to have an intentional intervention of the different educational agents in the different fields, who must ensure that this intervention is not done from an optics contrary to their culture (Hernández Serrano & González Sánchez, 2011).

Prensky (2001) discusses how digital "natives", who grow and develop around new technologies, think differently than "digital emigrants" (those who come from a handwriting-based culture). The author argues that, while the minds of the migrants are linear, those of the natives can be considered hypertextual, which means that they can read in a discontinuous, global and interconnected way. The natives are capable of processing up to three different information screens simultaneously, and have multidimensional spatial-visual abilities for the processing of images and three-dimensional representations, mental maps or interactive figures, responding quickly to both expected and unexpected stimuli with very surprising selective attention dimensions. All of these notes on different kind of consumptions and productions of images, and different levels of visual literacy gathered in literature, support the main idea of this paper and the goals and research work presented in the following sections.

There are many approaches and theories in the literature that highlight the importance of using visual literacy and related aspects and competencies in education, communication, etc. Despite of that, the authors of this paper consider that this field of research lacks (at least partly) proper research methods and quantitative/qualitative approaches. This is the aim of this paper, to contribute to the visual literacy knowledge field by proposing a quantitative instrument that could help researchers to measure people's understanding level and state of visual literacy, allowing them to compare this state among different people. The main result of this paper is the proposal of a questionnaire that: measures different aspects related to understanding and visual literacy and measures some cultural aspects of people. These cultural aspects in the questionnaire can be used as a lever to establish different categories of visual literacy depending on the people that respond, or to investigate about cultural factors and their relationship to visual communication or other visual aspects (that could help to support many of these theoretical approaches detected in the literature).

The paper is structured as follows: the second section presents the background that supports this research work. The third section presents the different hypotheses and goals (both specific and general ones) for this research and future research based on the experiment proposed in this paper. The fourth section outlines the methodology used to build the research. The fifth section shows the results achieved. Finally, the last section (sixth) finalizes the paper with some conclusions and future work.

2. BACKGROUND

When talking about cross-culture differences, cross-cultural analysis, etc., the first issue should be to define the term *culture.*

According to the work of (Rizo García & Romeu Aldaya, 2006), culture is a term that has been investigated in many moments throughout history, and which has varied according to the interpretation and social changes occurred over time. In this work, these authors present approaches for elementary definitions, which, despite their conceptual ambiguity, manage to define and situate the term in a more understandable way. This kind of understandable definitions are what the authors of this paper use as the basis for their research.

In this sense, authors Rizo García and Romeu Aldaya (2006) handle three definitions of culture:

1. In a first definition, they talk about the relationships that arise between people with different "organizing principles, different places in which individuals or groups are positioned and recognized" (González, 1987). This author, González, through his definition, brings culture into the field of identity.
2. In the second definition, they include a more political vision of the term: relating culture with Cultural Studies that also allow us to find differences in the identities of the subjects.
3. To finish the definitions, they add the definition from (Comaroff & Comaroff, 1992), which presents culture as a structure in which the social order is reproduced and transformed.

Other authors, like Geert Hofstede (Geert Hofstede, 2017) present culture as:

Simply said, culture is how you were raised. It developed while you grew up. With a computer metaphor, culture is the software of our minds. We need shared software in order to communicate. So, culture is about what we share with those around us.

All of these definitions presented could help the reader to understand what culture is. In this case, the authors prefer this last one by Hofstede, since it provides a more accurate and suitable example for the research presented in this paper.

Once the concept of culture is defined, it is time to introduce its implications for the human life and environment.

Regarding culture and its acquisition, given the importance of the cognitive (mental) and communicative (relational) character of culture, it could be stated that acquiring culture depends to a great extent on what we can learn (Bodley, 1994) through social relations, and to the communication thanks to what we (humans) share, either generated by us or not. Regarding the concept of culture generated or not by people, the authors bear in mind the two dimensions of Geertz (Geertz, 1987). These two dimensions are tradition - what is given to us - and innovation - what is built or generated by us-. The tools that people use to acquire that culture paradoxically serve to people to bind to their peers, just because they share the culture between them (García Castaño & Pulido Moyano, 1992). When the attempt is made to find the differences of interpretation that arise using the same message in two different cultures with the same possibilities of literacy, and which coexist in the same time and often in the same place, as is the case in this research work, it is possible to include the concept of culture proposed by Freud (1970, 2012). Freud recognizes culture as a "unitary regulation of life in common". This, together with the concepts of ego (the individual) and superego (the collective), introduces an important item in defining culture, such as the relationship of the individual with society: a relationship that also arises through the difference of ideas, traditions and customs in a social class, over time, etc.

Regarding the measurement of the cultural aspects of people to establish cross-cultural differences, it is possible to understand cross-cultural as "the differences of interpretation that arise from the same message in two literate cultures." Cross-cultural differences explain how people could have characteristic ideas, traditions and customs, that are not shared either among social classes or across time (Spanish Royal Academy of Language, 2017). According to the literature, cross-cultural differences can be measured using the following dimensions (a reproduction of the definitions that author provided in 2011) proposed by (Geert Hofstede, 2011):

1. **Power Distance Index (PDI):** The power distance index is defined as "the extent to which the less powerful members of organizations and institutions (like family) accept and expect that power is distributed unequally." In this dimension, inequality in power is perceived by the followers, or the lower level. A higher degree of the index indicates that hierarchy is clearly established and executed in society, without doubt or reason. A lower degree of the index signifies that people question authority and attempt to distribute power.
2. **Individualism vs. Collectivism (IDV):** This index explores the "degree to which people in a society are integrated into groups." Individualistic societies have loose ties that often only relate an individual to his/her immediate family. They emphasize the "I" versus the "we." Its counterpart, collectivism, describes a society in which tightly-integrated relationships tie extended families and others into in-groups. These in-groups are laced with undoubted loyalty, and they support each other when a conflict with another in-group arises.
3. **Uncertainty Avoidance Index (UAI):** The uncertainty avoidance index is defined as "a society's tolerance for ambiguity," in which people embrace or avert an event containing something unexpected, unknown, or far from the status quo. Societies that score a high degree in this index opt for stiff codes of behavior, guidelines, laws, and generally rely on absolute Truth, or the belief that one single Truth dictates everything and people know what it is. A lower degree in this index shows more acceptance of differing thoughts/ideas. Society tends to impose fewer regulations, people are more used to ambiguity, and the environment is more free-flowing.
4. **Masculinity vs. Femininity (MAS):** In this dimension, masculinity is defined as "a preference in society for achievement, heroism, assertiveness and material rewards for success". Its counterpart represents "a preference for cooperation, modesty, caring for the weak and quality of life." Women in the respective societies tend to display different values. In feminine societies, they share modest and caring views equally with men. In more masculine societies, women are more emphatic and competitive, but notably less emphatic than the men. In other words, they still recognize a gap between male and female values. This dimension is frequently viewed as taboo in highly masculine societies.
5. **Long-Term Orientation vs. Short-Term Orientation (LTO):** This dimension associates the connection of the past with the present and future actions/challenges. A lower degree of this index (short-term) indicates that traditions are honored and kept, while steadfastness is valued. Societies with a high degree in this index (long-term) views adaptation and circumstantial, pragmatic problem-solving as a necessity. A poor country that is short-term oriented usually has little to no economic development, while long-term oriented countries continue to develop up to a point.

6. **Indulgence vs. Restraint (IND):** This dimension is essentially a measure of happiness; whether or not simple joys are fulfilled. Indulgence is defined as "a society that allows relatively free gratification of basic and natural human desires related to enjoying life and having fun." Its counterpart is defined as "a society that controls gratification of needs and regulates it by means of strict social norms." Indulgent societies believe themselves to be in control of their own life and emotions; restrained societies believe other factors dictate their life and emotions.

On the other hand, literacy is the teaching of reading and writing a language to a person. "In 1968 the IVLA (International Visual Literacy Association) was founded when a group of people interested in various aspects of visuals in education and communication met to plan a conference on visual literacy" (García-Sánchez, Therón, & Isla, 2014). This moment (1968) marked the first time that the term visual literacy appeared in the theoretical panorama, but the real beginning of the reflection on this topic can be traces back to the appearance of photography at the end of the 19th century. Just as it happened with the appearance of the printing press in the 15th century (written language begins to be accessible to the public, new readers begin to grow and so does the number of people can read and write), people that "prosume" -a term later coined by (McLuhan & Nevitt, 1972)- began to appear. These actors (subjects) begin to produce images thanks to the development of increasingly accessible technologies, and finally the photograph passes from the silver frame to the digital frame (Gómez Cruz, 2012), finally coming to be that "everybody today with a mobile phone with a built-in camera automatically becomes a photographer" (Manovich, 2015).

The communicational changes produced new languages and codes that influenced the beginning of Knowledge Society Information (Expósito Ortiz, 2013). Thanks to the growth generated by the information society and the technologies, currently almost everyone handles images in their day-to-day lives (Manovich, 2015). It is possible to find images included in the media, social networks or our own phones.

Since Apple introduced the iPhone in 2007, which included a 2-megapixel camera, mobile devices have become increasingly intelligent (thus being called smartphones). Thanks to them, people have been including in their habits of communication and consumption some tasks that were not previously a part of their routine, such as the generation of images or videos.

It is possible to talk about the construction of a visual language (which, like verbal language, will be considered at some point an evolutionary leap) and to affirm that it has been developing since the beginning of human history (Dondis, 1974), but it becomes democratic the moment that anyone can generate codes, leaving aside their graphomotor skills or their sociocultural or socioeconomic level.

Therefore, if there is a visual language, it is also assumed that communication exists.

Intentionally, someone elaborates and diffuses a message to make it possible that another person, depending on their level of visual literacy, is able to receive and interpret it (Felbo, 2017; Hug, 2013).

In order for this communication to take place, already known concepts come into play, thanks to verbal communication: "transmitter – message – receiver" (Watzlawick & Watzlawick, 1976). In the case of visual communication, according to Dondis (1974), when generating a visual message we use the basic elements of visual communication, but also shared perceptual mechanisms. With the latter, the author refers to the existence of visual perception, which is a mechanical operation of the human body that unites us, but which can sometimes separate us because of the message decoding process. Message decoding depends on some kind of extension in the interpretation of what we observe, which includes subjectivity. The subjectivity of someone who recounts something includes infinite formulas to tell something. The fact of recounting is what shows that there is a visual language, but at this point some questions might be posed: are there differences and similarities in the ways of recounting things related to the cultural differences of the communicator? Can we all read visual messages in the same way? Who is visually literate? What is the quality regarding visual literacy? Are there cultural differences in this quality? In order to answer these questions, the authors of this article propose the generation of a quantitative instrument (and some kind of basis for a future qualitative instrument) to measure visual literacy, visual communication understanding and other related issues. This measurement will occur using opinions from people and using concepts like self-perceived understanding of visual communication.

3. HYPOTHESES AND GOALS

This section presents the different research hypotheses proposed by the authors and the different goals pursued in this research (general goals, and specific ones). It is worth to note that all of these hypotheses and goals are not only proposed for this paper, but also as the desired goals to be accomplished by using the results of this research and the instrument proposed.

3.1 Research Hypothesis

The main hypothesis is that it is possible to propose a method to research if there are cross-cultural differences between different people regarding visual literacy and the understanding in visual communication.

Understanding cross-cultural as previously explained in the background section, the specific hypotheses to conduct a study based on this proposal would be:

H1: There are differences in people's visual literacy depending on gender.

H2: There are differences in people's visual literacy depending on their age or generation (Baby Boom, Z, X, Y).

H3: There are differences in visual literacy depending on the cultural provenience or ethnic aspects of the people. As authors like (Jones & Healing, 2010; Jones, Ramanau, Cross, & Healing, 2010) present, there are some cultural changes in the information usage and practice in newer generations like the "Google generation".

3.2 Goals

3.2.1 General Goals

- To design a measurement instrument to assess an individual's level of visual literacy, with specific attention to the cross-cultural aspects.
- To define the proper analysis required for this kind of instrument (goal for future analysis based on this proposal). The main analysis techniques that could be applied are based on structural equation models (SEM), such as those used by the authors on previous research works (Sánchez Prieto, Olmos Migueláñez, & García-Peñalvo, 2017; Sánchez-Prieto, Olmos-Migueláñez, & García-Peñalvo, 2017).
- To validate the instrument (goal for future analysis based on this proposal).

3.2.2 Specific Goals

- To establish a quantitative method that allows researchers to find differences regarding the consumption, production and reading of images in a cross-cultural way (goal for this paper).
- Test if, apart from the cultural differences, there are differences regarding the individuals' age or other personal details (goal for future research based on this paper).

4. METHODOLOGY

In order to measure the different variables presented in the survey, the authors have designed an instrument divided in three main sections: the first section asks for the identification data of the person that is filling out the questionnaire (age, gender, tendencies using the Internet, apps that use images or audiovisual contents as part of their business core, etc.). The second section tries to gather information about different aspects related to visual literacy (use of images, visual representations, *visual thinking*, etc.). The third and last section of the survey tries to get information about the cultural aspects of the participant, in this case using the cultural dimensions and questions presented in the methods section.

Specifically, the first part is composed by six questions that ask about gender, age, country of origin, the educational level of the individual, or how many hours the respondent uses the Internet for, or which apps they use (the apps proposed in the questionnaire are selected because they include image or video as part of their main features or functionalities and could serve for visual communication).

The second and third parts are made up of a set of items formulated in a five-point Likert-type scale. The range of values varies from 1 to 5 (1= Totally disagree; 2= Disagree; 3= neither agree nor disagree; 4= Agree; 5= Totally agree). Regarding the questions that compose the questionnaire, it should be outlined that the questions in general will not be mandatory.

The second part is composed by 62 items grouped in seven different constructs related to visual literacy and elements of visual communication. As previously explained, not many examples of how to measure visual literacy understanding and state can be found in the literature, so authors have defined their own instrument. The items and constructs proposed are based on the work by (Dondis, 1974). The authors proposed several practical exercises and experiments to evaluate people's visual literacy level but based on a qualitative approach. For this paper, the authors have adapted some parts of the exercises proposed, but using a quantitative approach suitable to be reproduced in future research.

In the third part of the questionnaire there are 30 items that belong to the six main constructs about cultural differences proposed by (Geert Hofstede, 2011). These items are taken directly or adapted from the work of Hofstede and many other authors like (Obeidat et al., 2016; Sánchez-Prieto, Huang, Teo, Olmos-Migueláñez, & García-Peñalvo, 2017; Sánchez-Prieto, Olmos-Migueláñez, & García-Peñalvo, 2016; Tarhini, 2016; Tarhini, Elyas, Akour, & Al-Salti, 2016; Tarhini, Hone, & Liu, 2015; Tarhini, Teo, & Tarhini, 2016).

Considering the three parts, the questionnaire is composed by 88 questions. Table 1 summarizes all the constructs and items that appear in the questionnaire (excluding the identification variables).

Regarding other methodological aspects of the experiment to be run using the questionnaire proposed, the population of the experiments should be controlled. In depth, to find cross-cultural differences using the questionnaire proposed, the experiments should be developed using populations with similar features (age for example) in different cultural contexts and countries. For that reasons, the authors propose to

Table 1. Distribution of the number of items per construct

Construct	Number of Items
Character and content (CC)	5
Composition (COM)	9
Basic elements of visual communication (EVC)	9
The anatomy of the visual message (AVM)	10
Contrast Dynamics (CD)	9
Visual techniques: communication strategies (VTCS)	5
Synthesis of visual style (SVS)	5
Individualism vs. collectivism (IDV)	5
Power distance index (PDI)	4
Uncertainty avoidance index (UAI)	5
Indulgence vs. restraint (IND)	4
Long-term orientation vs. short-term orientation (LTO)	5
Masculinity vs. femininity (MAS)	7

run the same tests in different cities, countries or different schools, universities, etc. to similar groups of people (students, old people, young people, etc.).

As a final remark, regarding the methodology and its implications, it should be outlined that, according to several studies (Baartman & Ruijs, 2011), the level of self-perceived competence stated by a person is similar to the actual level of competence demonstrated by that person. Assuming that, in this research the people that will respond the proposed questionnaire are self-evaluating their understanding related to visual literacy issues, it could be stated that this self-evaluation is similar to their real competence level.

5. RESULTS

The main result of this research is the instrument proposed for the assessment of the level of visual literacy in people of different ages and cultures.

So, taking into account the considerations presented in the previous sections, the questionnaire proposed is displayed across different tables (Tables 2, 3 and 4).

6. CONCLUSION AND FUTURE STEPS

This paper presents a new approach of a quantitative analysis to be used in the research about the understanding of visual literacy and visual communication issues. The goal of this new approach is to find common patterns, opinions and behaviors between different people regarding the usage of visual communication and people's level of visual literacy, while also considering possible cultural differences related to it. the theoretical background about the visual literacy research field is presented in the first place. Also, regarding the background of this research, the paper presents the main concepts related to

Table 2. Personal information requested in the first part of the questionnaire

Information Required: Gender	
Text of the question	Gender
Possible responses (chose only one)	Male
	Female
	Prefer not to answer
Information Required: Age	
Text of the question	Age
Possible responses	Open answer
Information Required: Nationality	
Text of the question	Country
Possible responses	Open answer
Information Required: Educational Level	
Text of the question	What is your educational level?
Possible responses (chose only one)	Less than compulsory secondary education
	Compulsory secondary education
	Higher education (vocational)
	Higher education (undergraduate)
	Higher education (graduate)
Information Required: Consumption of the Internet per Day (in Hours)	
Text of the question	How many hours a day do you use the Internet (it includes messaging apps, social networks, email, etc.)?
Possible responses (chose only one)	None
	Less than 1 hour
	1-2 hours
	2-4 hours
	4-6 hours
	More than 6 hours
Information Required: Kind of Apps Used by the Respondent of the Questionnaire	
Text of the question	Do you use the following applications (mark all that proceed)?
Possible responses (chose all that apply)	Audiovisual content apps based on subscription (Netflix, HBO, Amazon Prime Video, etc.)
	Blogs and forums (Tumblr, Baidu Tieba, etc.)
	Dating apps (Tinder, Badoo, eHarmony, Happn, Match, Dating Direct, etc.)
	Messaging apps (Whatsapp, WeChat, Facebook Messenger, QQ / Qzone, Telegram, Line, Signal, Google Hangouts, etc.)
	Online photo libraries (Google Photos, Apple Photos, iOS Photos, iCloud, etc.)
	Professional social networks (Linkedin, Slideshare, Xing, etc.)
	Social networks (Facebook, Twitter, Sina Weibo, etc.)
	Social networks for designers and artists (Behance, DeviantArt, Dribbble, Domestika, etc.)
	Video apps and social networks (Youtube, Vimeo, etc.)
	Videomeeting apps (Skype, Google Hangouts, Apple Facetime, Viber, Google Hangouts, Google Duo, Zoom, etc.)
	Videostreaming apps (Youtube Live, Twitch, Facebook Live, Periscope, etc.)
	Visual social networks (Snapchat, Pinterest, Instagram, etc.)
	Any other similar to some of the previous ones (type the name) ________________

Table 3. Questions about visual literacy and visual aspects in the second part of the questionnaire

Construct: Character and Content	
Definition: About the ability to decode a visual message (Dondis, 1974)	
CC1	When I have to represent something, I first visualize its image in my mind
CC2	When I see an emoji (emoticon) I am able to understand what the emitter of that message wants to tell me
CC3	When I see a GIF animation I am able to understand what the emitter of that message wants to tell me
CC4	When I see an Internet meme (image) I am able to understand what the emitter of that message wants to tell me
CC5	I can find something in a space, thanks to the previous spatial memory I have of it
Construct: Composition	
Definition: About the relation (ordering) of lines, colors, contours, directions, textures, scales, movements and dimensions that thanks to the tone, the light, give meaning to the visual message. Knowing how to use these perceptual resources "educates our compositional ability and allows us to use syntactic criteria for those who begin to learn visual literacy." To arrange these elements in a plane is known as compose (Dondis, 1974).	
COM1	When I am watching an image, I am able to perceive the inclination of the motif, and sometimes I recognize myself inclining the head to see it totally horizontal
COM2	When I look at an image I need to find the main visual motif supported by (or aligned with) the frame (axis) of the image, this makes it easier for me to find interesting what I see
COM3	When I get a paper sheet, the first thing I look at is the center of the paper, then the bottom and then I set out to see/read
COM4	At some moment, I have looked an image and I do not know why I have focused something in the first place
COM5	When I see an image or a visual representation, I watch it from left to right
COM6	When you take a photo with your mobile phone or camera, you use the "grid" feature of the camera to organize what you are seeing
COM7	When I see similar figures in an image, I tend to group them in my mind and consider them as a set.
COM8	When I see a monochromatic image (black and white, etc.) I feel some kind disturbed if there is something different (a colored object in a black and white image, etc.)
COM9	I tend to find images or shapes in the clouds
Construct: Basic Elements of Visual Communication	
Definition: The basic elements of visual communication are: point, line, contour, direction, tone, color, texture, dimension, scale and movement. By joining them we produce the visual information that forms our visual message (Dondis, 1974).	
EVC1	When I look at a landscape I recognize in it what is far and what is near and I am aware of its size difference
EVC2	I have occasionally discovered when approaching an image on any screen that is composed of small dots
EVC3	I have ever seen an image of the Eiffel Tower and in it I have been able to see lines that make up a triangular outline
EVC4	When I look at a photograph I am able to perceive whether it is taken at night, during the day or in an interior because I see a difference of clarity or darkness
EVC5	I am able to recognize that an element has a shadow and see where it is headed
EVC6	I am aware that I can only take photos when there is light
EVC7	When I see a color photograph and the same in black and white, I find differences in the information received
EVC8	When I see a photo of a cactus I find myself thinking about the sensation of the skewers pricking me
EVC9	Sometimes I have seen something in photos that I later saw in reality and I was surprised by its size
Construct: The Anatomy of the Visual Message	
Definition: People express and receive visual messages, people prosume, and people have different levels regarding the attainment of information: representationally, abstractly and symbolically. All can exist in the same image. Sight is the only thing needed to understand visually, as with speech, people do not need to be cultured to use language (Dondis, 1974).	
AVM1	I am able to transmit a message visually without using language
AVM2	When I see an image, I can find messages that make me feel
AVM3	I can see the sun, a drawing of the sun and a painting of the sun and understand that the three are the same concept
AVM4	I am able to find the toilets in the airport of any country, whatever it is the written language
AVM5	I understand that not all the symbols used by my peers, can be understood by other people

continued on following page

Table 3. Continued

AVM6	I think that "a picture is worth a thousand words"
AVM7	Based on my experience, when I see a photograph of snow I know what kind of element is, its composition, its different states and other details (possible temperature, etc.)
AVM8	All the profile photos I have in social networks present the best-looking version of me
AVM9	I take more than 10 photos per week
AVM10	I find elegance in some images thanks to the content they use
Construct: Contrast Dynamics	
Definition: To be able to express visually people have to take into account that a visual message is composed of two factors: meaning and message. In order to construct the message people must know techniques that support the meaning of the message we want to convey. Contrast is one of the most important techniques for controlling the message. It is the process of visual articulation, a force to create a coherent whole. Contrast intensifies meaning and simplifies communication. Just as light (tone) allows people to see and is one of the basic elements of visual communication, contrast is also important. Without contrast people would not be able to distinguish the characteristics of what surrounds us, if we saw through a veil of gray light, no motive would have characteristics already seen as texture, contours... (Dondis, 1974)	
CD1	I am able to understand a contour although there is a missing part of the line that draws it
CD2	Sometimes, when retouching my photos (even automatically or simply including a filter) I like them more than at the beginning
CD3	I can recognize that a series of elements harmonize in a space
CD4	On occasion, having seen the same thing as other people, it means a different thing to me than it does for the others
CD5	If I see a balance that is loaded on both sides, and the load on one side is bigger than the other, I say that side weighs more
CD6	When I have to work on a document, I like to have the sheet organized to be able to see the content better
CD7	The images where there are lights and shadows draw my attention because of their mystery
CD8	Sometimes I find the proportions between several elements of some image exaggerated
CD9	In some of the images I have consumed in my life, I have recognized opposite concepts, such as war and peace, faced
Construct: Visual Techniques: Communication Strategies	
Definition: Content and form are basic and irreducible components of the expression media (music, dance, prose, art, photo...). Content is the message, which is expressed directly or indirectly. The form, the final result of the message, lies in the interaction of pairs of opposites, of polarities: first in forces of content (message and meaning) and form (design, medium and ordering); and second in the reciprocal effect of the articulator (designer, artist, craftsman) and receiver (audience). In all cases, the first and second go together. The message and method for expressing it depend greatly on the understanding and ability to use visual techniques: the visual composition tools. Visual techniques are not exactly choices to construct or analyze everything we see. Some of them may be: equilibrium-instability, symmetry-asymmetry, regularity-irregularity, simplicity-complexity, unity-fragmentation, economy-profusion, reticence-exaggeration, predictability-spontaneity, activity-passivity, subtlety-boldness, variation, realism-distortion, flat-deep, singularity-juxtaposition, sequence-randomness, acuity-diffusivity, continuity-episodicity (Dondis, 1974).	
VTCS1	When I set out to shoot a photograph, I reflect to try to convey a message through the resulting image
VTCS2	When I take a photograph, I am aware that someone is going to see it and interpret it
VTCS3	When I take a photograph, I act by intuition
VTCS4	When I take a photograph, I try to find positions that make the image somewhat different from the others
VTCS5	Sometimes I find in the advertising pictures characteristics that make me want to consume something specifically
Construct: Synthesis of Visual Style	
Definition: The style is the visual synthesis of all elements, techniques, syntax, instigation, expression and basic purpose. The style becomes a category of visual expression, which depends on culture. Therefore, it is an aspect that implies a consensus accepted by those who share it. The style influences the artistic expression and represents individuals that fall within a common characteristics (Dondis, 1974).	
SVS1	I am able to find within some animated films similarities in their style characteristics
SVS2	In some of the photographs I see on Instagram (or other social network based on images), I find repeated elements and that makes me think that some people copy others
SVS3	When I see a photograph, and like it, I try to repeat it somehow
SVS4	I am able to recognize a company thanks to its advertising without having to see the brand
SVS5	When expressing myself visually I find points in common with my peers

Table 4. Cultural information and dimensions asked in the third part of the questionnaire

Construct: Individualism vs. Collectivism (IDV)	
Definition: The extent to which individuals are integrated into groups (Tarhini, 2016)	
IDV1	Individuals should sacrifice self-interest for the group interest that they belong to.
IDV2	Individuals should stick with the group even with difficulties.
IDV3	Group interest/welfare is more important than individual interest
IDV4	Group success is more important than individual success.
IDV5	Being accepted as a member of a group is more important than having autonomy and independence.
Construct: Power Distance Index (PDI)	
Definition: The extent to which individuals expect and accept differences in power between different people. Questions partially adapted from adapted from (Tarhini, 2016)	
PDI1	People should make the most decisions by consulting/discussing with administrators/superiors
PDI2	People should have social interactions with administrators/superiors.
PDI3	Administrators/superiors should use authority and power when dealing with people
PDI4	People should agree with administrators/superiors' decisions.
Construct: Uncertainty Avoidance Index (UAI)	
Definition: The extent to which ambiguities and uncertainties are tolerated. Questions partially adapted from adapted from (Tarhini, 2016)	
UAI1	Specific rules or regulations are important to me.
UAI2	Detailed requirements are important to me.
UAI3	Detailed instructions are important to me.
UAI4	Standardized operating procedures are helpful so that I can follow suit.
UAI5	The best is to closely follow requirements, instructions and procedures.
Construct: Indulgence vs. Restraint (IND)	
Definition: The extent to which people try to control their desires and impulses, based on the way they were raised (Hofstede, 2011)	
IND1	Overall I consider myself to be a very happy person.
IND2	I have complete free choice over my life.
IND3	Leisure time is a very important part in my life.
IND4	Wellbeing is very important to me.
Construct: Long-Term Orientation vs. Short-Term Orientation (LTO)	
Definition: The extent related to the choice of focus for people's efforts: the future or the present and past (Hofstede, 2011). Questions partially adapted from (Obeidat et al., 2016)	
LTO1	Respect for tradition is important to me
LTO2	The thrift is important to me
LTO3	What is good and evil depends upon the circumstances
LTO4	The success and failure in life depends on luck
LTO5	The family life must be driven by shared tasks
Construct: Masculinity vs. Femininity (MAS)	
Definition: The extent of division of emotional roles between women and men. Adapted from (Hofstede, 2011)	
MAS1	In my environment, the women work
MAS2	In my environment, the men are kind and gentle
MAS3	In my environment, men know how to balance family and work
MAS4	In my environment, women know how to balance family and work
MAS5	When someone in my environment needs help, because their conditions put him/her in a worse position than mine, I understand and worry about him
MAS6	I think there should be more women commanding companies or governments
MAS7	Girls cry, boys don't; boys should fight back, girls shouldn't fight

culture and how culture and visual literacy can be related or analyzed together to enable cross-cultural comparisons and analysis. To conduct these cross-cultural analyses, this paper proposes a new kind of quantitative questionnaire-based instrument that includes a section that measures the cultural characteristics of the individual (the participant) and their level of literacy regarding visual communication. This instrument proposal is the main result of this research, since the research field of visual literacy lacks this kind of quantitative approaches.

As future work based in this research, the authors plan:

- To validate this instrument by asking experts.
- To propose new goals and hypotheses based on the feedback provided by the experts.
- To validate the resulting instrument in a statistical way.
- To use the instrument to run an experiment in several countries and cultural contexts to study visual literacy and its cross-cultural issues.
- To propose other instruments and experiments based on qualitative approaches to evaluate how people express themselves using visual communication and visual literacy competences. Also, these quantitative approaches would serve to find out what are the keys and determinants that produce differences in the level of visual literacy, depending on the variables identified in the research hypotheses presented in this paper.

ACKNOWLEDGMENT

The authors Felicidad García-Sánchez and José Carlos Sánchez-Prieto would like to thank the University of Salamanca and the Santander Bank for funding their pre-doctoral research contract. Also, they want to acknowledge the support received by the PhD in Knowledge Society of the University of Salamanca. The author Juan Cruz-Benito would like to thank the European Social Fund and the Consejería de Educación of the Junta de Castilla y León (Spain) for funding his pre-doctoral fellow contract.

REFERENCES

Baartman, L., & Ruijs, L. (2011). Comparing students' perceived and actual competence in higher vocational education. *Assessment & Evaluation in Higher Education*, *36*(4), 385–398. doi:10.1080/02602938.2011.553274

Bodley, J. H. (1994). An anthropological perspective. *Cultural Anthropology: Tribes, States, and the Global System, 3*.

Comaroff, J., & Comaroff, J. (1992). *Of totemism and Etnicity*. Ethnography and the Historical Imagination.

Dondis, D. A. (1974). *A primer of visual literacy*. Mit Press.

Expósito Ortiz, S. (2013). *La importancia de la alfabetización visual en educación primaria*. Academic Press.

Felbo, B. (2017). *What can we learn from emojis?* Retrieved from https://www.media.mit.edu/posts/what-can-we-learn-from-emojis/

Fransecky, R. B., & Debes, J. L. (1972). *Visual Literacy: A Way to Learn--A Way to Teach.* Academic Press.

Freud, S. (1970). *El malestar en la cultura.* Ediciones AKAL.

Freud, S. (2012). *El porvenir de una ilusión (Serie Great Ideas 16).* Taurus.

García Castaño, F. J., & Pulido Moyano, R. A. (1992). *Educación multicultural y antropología de la educación.* Academic Press.

García-Sánchez, F., Therón, R., & Isla, J. G. (2014). Semiotic and technological analysis of photography: a visual literacy study in the educative area. *Proceedings of the Second International Conference on Technological Ecosystems for Enhancing Multiculturality.* doi:10.1145/2669711.2669978

Geertz, C. (1987). El desarrollo de la cultura y la evolución de la mente. *La interpretación de las culturas.*

Gómez Cruz, E. (2012). *De la cultura Kodak a la imagen en red: una etnografía sobre fotografía digital.* Academic Press.

González, J. A. (1987). Los frentes culturales. Culturas, mapas, poderes y lucha por las definiciones legítimas de los sentidos sociales de la vida. *Estudios sobre las Culturas Contemporaneas, 1*(3), 5–44.

Hernández Serrano, M. J., & González Sánchez, M. (2011). *La generación Google. Evolución en las predisposiciones y comportamientos informativos de los jóvenes.* Academic Press.

Hofstede, G. (2011). Dimensionalizing cultures: The Hofstede model in context. *Online Readings in Psychology and Culture, 2*(1), 8. doi:10.9707/2307-0919.1014

Hofstede, G. (2017). *Why is culture so important?* Retrieved from http://www.geerthofstede.nl/

Hug, T. (2013). Competencia mediática y alfabetización visual. Hacia consideraciones más allá de las alfabetizaciones. *Razón y Palabra, 18*(83).

International Visual Literacy Association. (2003). *What is visual literacy?* Retrieved from https://web.archive.org/web/20031204235851/http://ivla.org:80/org_what_vis_lit.htm

Jones, C., & Healing, G. (2010). Net generation students: Agency and choice and the new technologies. *Journal of Computer Assisted Learning, 26*(5), 344–356. doi:10.1111/j.1365-2729.2010.00370.x

Jones, C., Ramanau, R., Cross, S., & Healing, G. (2010). Net generation or Digital Natives: Is there a distinct new generation entering university? *Computers & Education, 54*(3), 722–732. doi:10.1016/j.compedu.2009.09.022

Manovich, L. (2015). Data Science and Digital Art History. *International Journal for Digital Art History,* (1).

McLuhan, M., & Nevitt, B. (1972). Take today; the executive as dropout. Don Mills.

Obeidat, B., Al-Sarayrah, S., Tarhini, A., Al-Dmour, R. H., Al-Salti, Z., & Sweis, R. (2016). Cultural influence on strategic human resource management practices: A Jordanian case study. *International Business Research*, *9*(10), 94. doi:10.5539/ibr.v9n10p94

Prensky, M. (2001). Digital Natives, Digital Immigrants Part 1. *On the Horizon*, *9*(5), 1–6. doi:10.1108/10748120110424816

Rizo García, M., & Romeu Aldaya, V. (2006). Hacia una propuesta teórica para el análisis de las fronteras simbólicas en situaciones de comunicación intercultural. *Estudios sobre las Culturas Contemporaneas*, *12*(24).

Sánchez-Prieto, J. C., Huang, F., Teo, T., Olmos-Migueláñez, S., & García-Peñalvo, F. J. (2017). *Versión en español de una herramienta para el análisis de la influencia de los factores socioculturales en la aceptación de TIC de los docentes universitarios.* doi:10.5281/zenodo.582401

Sánchez-Prieto, J. C., Olmos-Migueláñez, S., & García-Peñalvo, F. J. (2016). Informal tools in formal contexts: Development of a model to assess the acceptance of mobile technologies among teachers. *Computers in Human Behavior*, *55*, 519–528. doi:10.1016/j.chb.2015.07.002

Sánchez Prieto, J. C., Olmos Migueláñez, S., & García-Peñalvo, F. J. (2017). ¿Utilizarán los futuros docentes las tecnologías móviles? Validación de una propuesta de modelo TAM extendido. *RED. Revista de Educación a Distancia*(52).

Sánchez-Prieto, J. C., Olmos-Migueláñez, S., & García-Peñalvo, F. J. (2017). MLearning and pre-service teachers: An assessment of the behavioral intention using an expanded TAM model. *Computers in Human Behavior*, *72*, 644–654. doi:10.1016/j.chb.2016.09.061

Tarhini, A. (2016). *The Effects of Cultural dimensions and Demographic Characteristics on E-learning acceptance.* arXiv preprint arXiv:1607.01492

Tarhini, A., Elyas, T., Akour, M. A., & Al-Salti, Z. (2016). Technology, Demographic Characteristics and E-Learning Acceptance: A Conceptual Model Based on Extended Technology Acceptance Model. *Higher Education Studies*, *6*(3), 72–89. doi:10.5539/hes.v6n3p72

Tarhini, A., Hone, K., & Liu, X. (2015). A cross-cultural examination of the impact of social, organisational and individual factors on educational technology acceptance between British and Lebanese university students. *British Journal of Educational Technology*, *46*(4), 739–755. doi:10.1111/bjet.12169

Tarhini, A., Teo, T., & Tarhini, T. (2016). A cross-cultural validity of the E-learning Acceptance Measure (ElAM) in Lebanon and England: A confirmatory factor analysis. *Education and Information Technologies*, *21*(5), 1269–1282. doi:10.1007/s10639-015-9381-9

Watzlawick, P., & Watzlawick, P. (1976). *Teoría de la comunicación humana; interacciones, patologías y paradojas*. Barcelona: Herder.

Chapter 3
Latin American and Caribbean Literature Transposed Into Digital:
Corpus, Ecosystem, Canon, and *Cartonera* Publishing

Adrian R. Vila
Universidad de Buenos Aires, Argentina & Universidad de Salamanca, Spain

ABSTRACT

The aim of the present chapter is the study of the action performed by the publishing industry in the context of the transposition into digital format of printed books comprising a Latin American and Caribbean literary corpus. The designed corpus includes works and authors labelled as a Latin American segment of the Western Canon, in addition to those segments provided by feminist, queer, postcolonial, and/or decolonization critical theories. It is described/defined the digital ecosystem to which the corpus is transposed as well as some of the strategies implemented by the major e-book trade platforms and the main digital libraries to offer Latin American and Caribbean literature transposed into digital format are described/defined. Likewise, a reading extended to the influence of the postcolonial turn is proposed by considering cartonera publishing as a device of the postcolonial turn but, this time, widening the range of typographic forms as well as the postcolonial effect on the publishing field.

INTRODUCTION

Firstly, the research was focused on the transposition of printed Latin American and Caribbean literature into digital format. Secondly, it was established the terminological and conceptual clarification of definitions inherent to the literary and geo-cultural fields, such as Canon/Corpus; Latin/Iberian/ Latin America/ Latin America and the Caribbean; Latin American and Caribbean literature, among others, was established. Thirdly, a literary corpus comprising Latin American and Caribbean titles (LA&CLC) was selected to undertake the research. Once the LA&CLC was defined and circumscribed, and the

DOI: 10.4018/978-1-5225-4944-4.ch003

agent components and the features of the ecosystem to which the LA&CLC is transposed, as well as the digital platforms and libraries where the search will be conducted. Finally, a decision was made regarding how to actually carry out the field work and how to display the data obtained based on the elucidation of strategies implemented by the publishing industry for the transposition of the LA&CLC into digital format.

DESIGNING A LATIN AMERICAN AND CARIBBEAN LITERARY CORPUS (LA&CLC)

The LA&CLC arises from the survey of heterogeneous Latin American and Caribbean literary documentary sources (repertoires, essays, anthologies, articles, memoirs, catalogs, dictionaries) which differ from each other, as well as the perspectives and criteria for selecting authors and works produced by the sources themselves.

- The 18 authors and 36 works that Harold Bloom includes in the Latin American segment of his Western Canon were included as well as the 5 authors and 7 works he introduces in the Caribbean segment (West Indian), i.e. 43 titles and 23 authors.
- The *Diccionario de autores latinoamericanos* (Dictionary of Latin American Authors) by César Aira contributes substantially to the incorporation of authors, playwrights, poets, novelists, short story writers, scientists, historians, chroniclers, journalists and essayists. Authors emerged since 1985 are excluded. It furnishes 11,102 titles by 1,322 authors.
- The *Diccionario de la literatura latinoamericana* (Dictionary of Latin American Literature) by Susana Cella brings a list of authors and literary works which completes Aira's *Diccionario de autores latinoamericanos* since it includes authors born until 1940 or, if born after the 40's, those who were deceased in 1998. It provides 4,737 titles by 595 authors.
- The *Encyclopaedia of Latin American and Caribbean Literature 1900-2003*, edited by Daniel Balderston and Mike Gonzalez includes works produced between 1900 and 2003.
- *Saraus. Movimiento - Literatura - Periferia - São Paulo. Antología* (Movement- Literature-Periphery-São Paulo. Anthology) by Lucía Tennina brings authors and works by contemporary *Paulista* poets from the so called "marginalia". It provides 58 titles by 38 authors.
- *De la Tricolor a la Wiphala. Narrativa contemporánea de Bolivia,* (From Tricolor to Wiphala. Contemporary Narrative from Bolivia) by Sergio Di Nucci, Nicolás Recoaro and Alfredo Grieco y Bavio furnishes short stories by Bolivian authors in a selection that displays the emergence of a new independent Bolivian narrative. It supplies 67 titles by 14 authors.
- *McOndo (una antología de nueva literatura hispanoamericana)* (An Anthology of New Hispanic American Literature) edited by Alberto Fuguet and Alberto Gómez contributes to this LA&CLC stories by authors born from 1959 on as well as not-so-well-known authors until the mid-1990's, and whose selection lays aside all traces of magic realism for programmatic reasons. It brings 17 titles by 15 authors.
- *Manifiesto del Crack* (Crack Manifesto) supplies a list of 6 Mexican authors (Jorge Volpi, Pedro Ángel Palou, Eloy Urroz, Ignacio Padilla, Ricardo Chávez Castañeda and Vicente Herrasti), with a poetic program of Mexican narrative dislocated from time and space, and works produced after 1996 which display a split narrative stance for both Latin American magical realism and certain

'light' poetry (Ángeles Mastretta, Marcela Serrano, Laura Esquivel). It provides 22 titles by 6 authors.

- *1.000 millones: poesía en lengua española del siglo XXI* (1 Billion: XXI Century Poetry in Spanish) edited by Daniel Helder, Daiana Henderson and Bernardo Orge, provides a list of 33Hispanic-American authors born between 1980 and 1995, whose works were previously edited in digital format or print. It brings 95 titles by 32 Latin American authors (one is from Spain).
- *Bogotá 39* is the result of a selection by the Colombian authors Piedad Bonnet, Héctor Abad Faciolince and Óscar Collazos. It incorporates 133 titles by 33 authors.
- *La gran novela latinoamericana* (The Great Latin American Novel) by Carlos Fuentes (2011) provides a repertoire of authors and novels from the perspective of an author of the 'boom'. It brings 124 titles by 65 authors.
- A section including titles by Caribbean, Antillean and Guyanese authors, designed on the basis of the combination of the titles and authors comprising the different compilations, anthologies and selections: *West Indian Narrative: An Introductory Anthology*; the anthology *They Came in Ships: An Anthology of Indo-Guyanese Writing*; the selection *The Routledge Reader in Caribbean*; *The Companion to Anglophone Caribbean Literature*; the selection *The Oxford Book of Caribbean Verse* and *Jahaji Bhai: An Anthology of Indo-Caribbean Literature*, and the selection *Sugar Is All: Caribbean Short Stories and Poems from St. Kitts and Nevi*. It provides 2,597 titles by 559 authors.
- In addition, a specific segment for Colombian literature was designed to be represented in the LA&CLC. This segment provides 50 titles of Colombian literature and 8 authors (many of the works are anonymous or of collective authorship).

If Aira's *Diccionario* proposes authors and works in Spanish and Portuguese, the anthology *Los chongos de Roa Bastos* (Di Nucci, Recoaro, Grieco, & Bavio, 2014) supplies 33 titles by 9 authors, and the *Saraus* anthology brings works in Jopara, Guarani and different heteroglossia and hybrid forms blending Guarani, Spanish, and other indigenous languages and dialects; and additionally, other texts in Portuguese produced by the Paulista youth *perifatividade*. In turn, if Aira's *Diccionario* only offers authors and works in Spanish and Portuguese, the *Diccionario* by Susana Cella includes pre-Columbian texts written in non-Hispanic languages. The anthology *De la Tricolor a la Wiphala* incorporates texts produced in different varieties of Bolivian Spanish, except for the case of Alison Spedding (produced in Aymar). *Latin American Writers on Gay and Lesbian Themes: a Bio-Critical Sourcebook* by David Foster Wallace brings texts produced in Spanish and Portuguese although the author works with texts produced in Spanish and Portuguese and other works translated into English. The *Bogotá 39* selection includes authors producing works in Spanish, Portuguese and English.

While most of the sources under study do not highlight gender perspectives, *Latin American Women Writers: an Encyclopaedia* (André & Bueno, 2008) brings a record of narratives by Latin American writers framed by the theoretical approach of feminine literature. This Encyclopaedia brings 2,951 titles by 303 female authors. At the same time, the essay *Latin American Writers on Gay and Lesbian Themes* (Foster, 1991) offers 27 titles and 25 Latin American multi-gender authors in an attempt to furnish the LA&CLC with a perspective articulating the bond between homosexuality and society in Latin American literature.

McOndo and *De la Tricolor a la Wiphala* offer short stories; *Manifiesto del Crack* brings novels as well as Carlos Fuentes' *Gran novela latinoamericana*; Saraus supplies poetry as the anthology *1.000 millones*.

The *Catálogo de Editoriales Cartoneras Latinoamericanas de la Universidad de Wisconsin* (University of Wisconsin Digital Collections, n.d.) includes titles and authors from *cartonera publishers* which collaborate with the structure of an LA&CLC formed by editorial genres integrating printed material as cultural artefacts typical of Latin American literature. *Editoriales Cartoneras* are characterized by employing both handmade formats such as covers and sketches colored book by book, and by printing cardboard covers and mostly photocopied pages. It brings 469 titles by 358 authors.

Authors from the Caribbean, the Antilles and from Guyana were removed from the *Encyclopaedia of Latin American and Caribbean Literature* to be included in the LA&CLC section devoted to the Caribbean, Antilles and Guyana's authors. The remaining authors were added to an additional section. It brings 2,672 titles by 651 authors.

Languages of the works comprising the LA&CLC: different varieties of Latin American Spanish and Portuguese, French, English, Dutch, Guarani, Nahuatl, Aymara, Tomáraho, Ashlushlay, Ybytozo, Toba quom, Sanapaná, Maká, Axe-guayaki, Ayoreo Quiché, Quiché, Maya Yucateca, Chontal, Cakchiquel, Pokomchí, Mam, Zutuhil, Patois, Creole, Jorapa, Portuñol (a blend between Spanish and Portuguese) and heteroglossia between different languages.

Nationalities of the authors comprising the LA&CLC (43 different nationalities): 1. Argentina, 2. Bolivia, 3. Brazil, 4. Chile, 5. Colombia, 6. Costa Rica, 7. Cuba, 8. Ecuador, 9. El Salvador, 10. Guatemala, 11. Honduras, 12. Mexico, 13. Nicaragua, 14. Panama, 15. Paraguay, 16. Peru, 17. Puerto Rico, 18. Dominican Republic, 19. Uruguay, 20. Venezuela, 21. Trinidad and Tobago, 22. Virgin Islands, 23. St. Vincent and the Grenadines, 24. St. Lucia, 25. Martinique, 26. Jamaica, 27. Suriname, 28. Guyana, 29. Dominica, 30. Barbados, 31. Saint Maarten, 32. Haiti, 33. Bahamas, 34. Grenada, 35. Guadalupe, 36. Belize, 37. Montserrat, 38. Antigua and Barbuda, 39. Aruba, 40. St. Kitts and Nevis, 41. Spain, 42. Portugal, 43. U.S.A. The presence of authors from Spain, the United States and Portugal is due to two types of inclusion. On the one hand, they are authors (or '*conquistadores*') from the times of the Conquest; on the other, they are authors included by the authors from different sources but from the abovementioned nationalities.

WHAT ECOSYSTEM IS THE CORPUS TRANSPOSED INTO?

The ecosystem of the digital book is a sub-system embedded within the digital data general system which affects all kinds of electronic documents, their entrepreneurial and dynamic structures of creation, production, distribution and reception. All this falls within the general field of ICT (Information & Communication Technology sharing characteristics between the spheres. But the digital book (ebook) possesses a series of intellective, sociological and entrepreneurial features pointing to the existence of a kind of autonomous set hence we talk about ecosystem. The concept of 'field' derives from Pierre Bourdieu (1966: 2), with the methodological intention of inserting a certain relative autonomy (from economic, religious and political powers), based on analyzing a system ruled by its own regulations (inherent to Structuralism).

The impact of the globalization process has various classes and different speeds and scales which García Canclini (1999: 49) describes as 'a set of homogenization processes and, in turn, articulated division of the world.' The technological frontier into which the 'cloud' (cloud computing) has become: impacts on the productivity patterns of the whole international economic system at different levels. At

Figure 1. Composition of the LA&CLC
Produced by the author.

the manufacturing, finance and various subsidiary levels or classes which comprise the economy and the organization of work, the estates and the subjectivities. The digital ecosystem functions as a central articulator or agent that coordinates and powers entrepreneurial actions related to entertainment, browsers, commerce, email, Web, social networks, data transfer and storage services, applications, edition software. These actions and services, in turn, overlap and juxtapose to meet the users' needs and taste in terms of the consumption of technological goods and mobile applications, digital libraries, retail consumption and digital signals articulated by the computer. (Refer to Figure 2).

Figure 2. Digital ecosystem
Production based on Infuz Digital (2012).

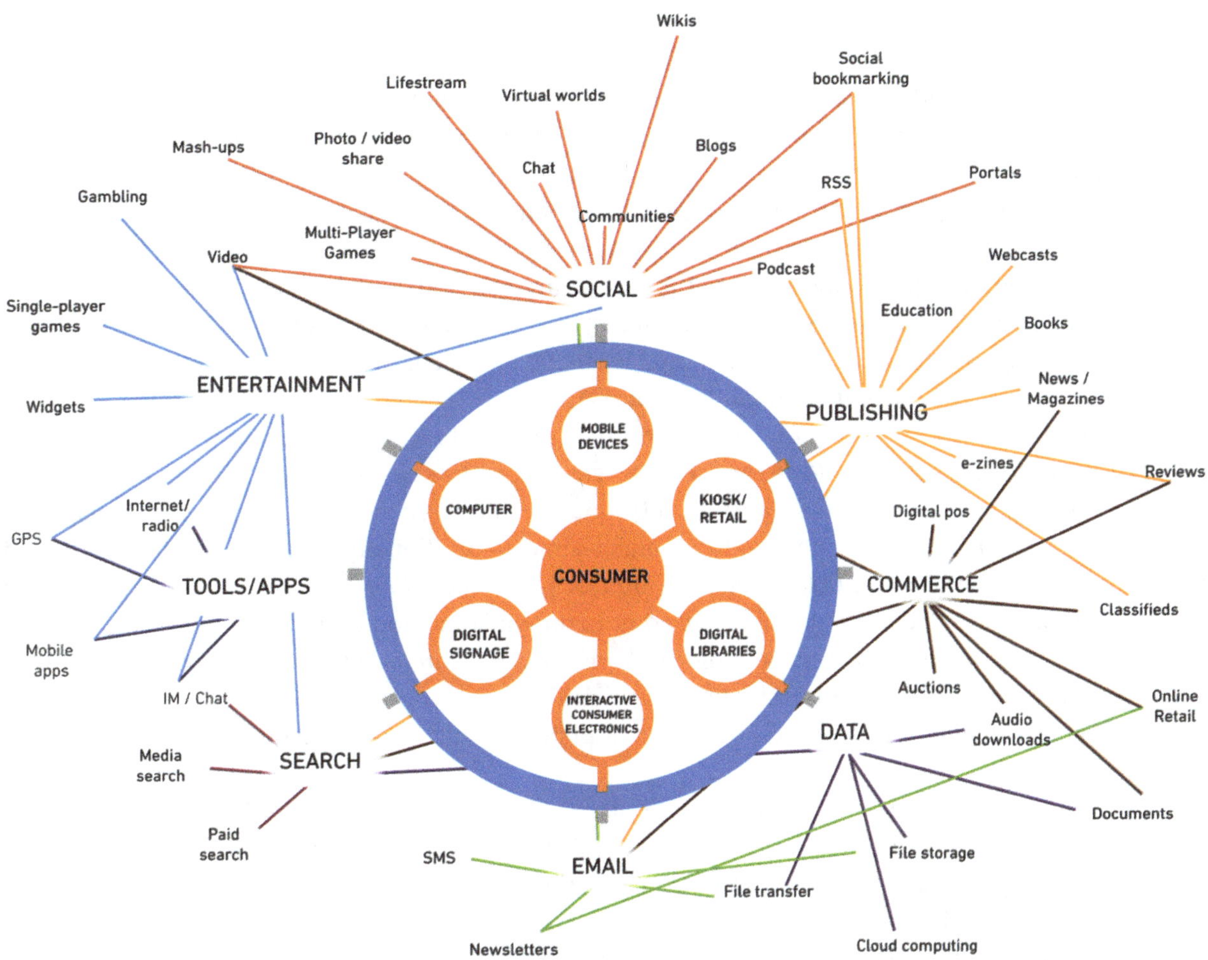

According to Ros García (1994: 53), an information system 'comprises multiple processes which are, at the same time, agents in other sub-systems of the organization and because the information system participates in any activity undertaken by that organization.' In our research, we emphasize that general systems of analogue and digital data hold overlapping areas where the component elements are shared. The Gutenberg galaxy belongs to the analogue data general system. The digital book ecosystem belongs to the general digital data system. Both sub-systems, Gutenberg galaxy and the digital book ecosystem, hold overlapping areas where their components are shared. At the same time, the systems, their sub-systems and their components are part of the overall field of communication information.

DEFINITION OF COMMERCIAL PLATFORMS AND LIBRARIES WHERE THE SEARCH WILL BE CONDUCTED

In order to disclose the transposition of the LA&CLC into digital format, a search of the titles in the major global platforms and in various libraries acknowledged in the field of linguistics and which excel in performance, architecture and digital availability was launched. The process of expansion of digital books is maintained worldwide, but the United States leads the market due to titles availability, a developed trading system, and a marked trend in the global business.

Hence, authors and titles were selected from the Amazon platform (representing 70% of the e-book market in the U.S.A.) and its competitors Barnes & Noble, Google Books, iBooks/Apple platforms.

The selection of Free-eBooks is due to the need to obtain samples of the state of free download digital platforms.

Likewise, Scribd is presented as an innovative marketing service system, file interchange and access to digital books at a flat fee for subscribers such as Nubico. *Livraria Cultura, Gandhi* and *Librería Santa Fe*, which are prominent platforms in the Latin American book trade.

At the library level, the New York Public Library and the Columbia University Libraries are part of the most prestigious group at the international level due to the quality of the resources they provide. Besides, they have in place multiple digital resources.

The *Biblioteca Nacional de España* is the most recognized library in the field of Spanish as well as Gallica; the *Bibliothèque Nationale de France* is the most internationally recognized library in the French language since it possesses approximately 900,000 digitalized books, newspapers, manuscripts, maps and drawings. *Europeana* is a digital library that contains almost 4 million digital objects including printed books coming from more than 1,000 European Union archives. The *Bibliothek des Lateinamerikanischen Institut* in Berlin was selected as it is specialized in Latin American topics and authors.

The *Biblioteca Nacional de Perú* and the *Biblioteca de São Paulo* are part of the most recognized Latin American libraries which offer digital collections.

Details of the leading criteria for the selection of platforms and libraries and their features:

- In early 2015, Amazon stated that it has 1,700,000 titles in digital format and a dominant position in the U.S. market (Milliot, Wischenbart *et al.*, 2013: 18). In late 2013, Amazon expanded its e-book catalog in Spanish to 48,000 titles (Wischenbart *et al.*, 2013: 38) and to 1,500 free classics in Spanish. The Kindle Unlimited flat free subscription service has a catalog comprising 700,000 titles, 25,000 of which are in Spanish. At the same time, it offers a free multi-platform application to access the catalog of e-books on sale. Kindle is available for iPad, iPhone, iPod touch, Android operating system devices, Windows Phone, BlackBerry, WebOS, offering versions for specific devices and more traditional operating systems. All versions can be downloaded from the same website. The application is compatible with several e-book formats, as well as newspapers, magazines and PDF files. The reader formats are Kindle. For the first Kindle generation, the accepted formats are Mobipocket (MOBI, PRC), plain text files (TXT),Topaz books (TPZ) and Amazon AZW format. For *Kindle 2*, the ability to read PDF and HTML files was added. Then *KindleTouch, KindleTouch 3G and Kindle Paperwhite*can display AZW, TXT, PDF, MOBI unprotected, and HTML, DOC, DOCX, JPEG, GIF, PNG and BMP files through conversion. *Touch*and*Touch 3G* can also process Audible Enhanced (AA, AAX) and MP3 files. Acquiring Goodreads scanning platform (25 million members) strengthens the influence of Amazon in the publishing field.
- Barnes & Noble has almost 3 million titles of digital books available with approximately 27% digital book consumers (Milliot, 2011). The company's businesses include the operation of about 700 specialty stores and more than 600 university libraries in the 50 states of the U.S.A., online retailing on its web BN.com and a market leadership in the digital industry market through Nook –its reading devices, tablets and digital contents brand. The Nook application is available on the

major platforms including Android, iOS and Windows. Besides, Nook HD –a 7' high resolution tablet– and Nook HD + (an ultra-light HD full tablet) was launched.

- iBookstore and iTunes. In June 2011, Apple registered sales up to 130 million in digital books via iTunes. In October 2012, sales rose to 400 million. Since 2013, iBooks has been available in more than 50 countries. In most of Latin America, iBookstore has been operating since mid-2012. In Brazil, it holds an estimated market portion of about 30% (Wischenbart*et al.*, 2014: 62). The iBooks application (available in 30 languages) is the resident in iPad tablets. By December 2012, iTunes Store became available in 119 countries with large markets such as India and Indonesia, Russia, Saudi Arabia and Turkey. However, the iTunes universe is not exactly the same in every market.
- Google Books serves almost 350,000 titles combining direct sales with a format similar to traditional wholesale distributors. More than 50,000 publishers hold distribution agreements with Google Books. Google has digitalized more than 20 million books.
- Kobo is a Canadian open standard e-book company operating in 190 countries, reaching 10 million users and counting on a catalog comprising more than 4 million titles in 68 languages as well as a remarkable e-reader. It has stores in Germany and France, and it is associated with Mondadori in Italy. In the United Kingdom it is associated to WHSmith, the publisher, distributor and creator of the ISBN (International Standard Book Number) codes, which were then internationally adopted as ISBN. In regard to social reading, Reading Life is the name of the system that collects statistics about reading habits in Kobo, counting number of the read pages, books started and finished, and time of day in which we read, among others. Users have the ability whether to share this information with the community. Kobo sells its own e-readers and tablets: Kobo Touch, Kobo Glo, Kobo Mini, Kobo Arc. The Reading devices can be bought in France, Italy, Germany, Austria, Portugal, United Kingdom and *online* stores such as Redcoon.es in Spain. In addition to their own devices, Kobo has free reading applications for other environments such as iOS, Android, BlackBerry and Windows.
- *Casa del Libro* offers a catalog with about 1,000,000 titles and one of the largest catalogs of e-books in Spanish: 135,000 titles with approximately 1 million registered users and 118 stores in Spain.
- The searches in Free-Ebooks were conducted as a way of establishing a state-of-the-art digital book free download platforms.
- *Gandhi* is the main bookshop for printed and digital books in Mexico, as well as a platform for self-publication; *Livraria Cultura* is Brazil's most important bookstore for printed and digital books. The latter, has a catalog including about 8 million titles in printed version, and a million titles in digital format. Perubookstore.com is Peru's main online bookstore; and *Librería Santa Fe* is the leading trade platform for digital books –5,000 books– in Argentina.

A research was also conducted in platforms which represent "new sales models": Nubico and Scribd. Nubico is a digital platform for sale by subscription which has 100,000 customers and 7,700 titles from 83 publishers, and offers flat rates. Scribd is an online application where documents and presentations can be uploaded to be shown online or to be downloaded. It has a catalog of 500,000 books that can be accessed at a "flat rate" and offers 150,000 new e-books from Harlequin publisher. Scribd allows exploring, reading, and sharing a million works.

The research was also conducted in the major libraries acknowledged from the linguistic point of view, and their high standards of performance, architecture and digital availability (Chowdhury y Foo, 2012; Klibanski, 2006; Tramullas Saz, 2002; Turiel, 2010).

- The New York Public Library and the Columbia University Library are part of the groups comprising the most prestigious libraries internationally due to their academic level and the quality of the resources they possess as well as recognition in the academic field.
- The *Biblioteca Nacional de España* is the most recognized library in the field of Spanish as well as Gallica; the *Bibliothèque Nationale de France* is the most internationally recognized library in the French language and *Europeana*, a European continental scale library.
- The *Biblioteca Nacional de Perú* and the *Biblioteca de São Paulo* were selected based on the criterion of incorporating the most recognized Latin American libraries offering digital collections.
- The *Bibliothek des Lateinamerikanischen Institut* in Berlin was selected as it is specialized in Latin American topics and authors.

In short, 9 digital libraries, and 14 digital trading platforms and streaming and downloading reading services were selected to perform a comparative study of the LA&CLC Latin American and Caribbean titles availability. This selection of platforms and libraries allows the analysis and comparison of the LA&CLC titles present in them to analyze the transposition from print into digital format.

SEARCH OF TITLES AND AUTHORS PRESENT IN THE CORPUS IN PRINTED VERSIONS AND DIGITAL FORMATS IN COMMERCIAL PLATFORMS AND DEFINED LIBRARIES, AND DATA COLLECTION

Actions

Searching for corpus titles and authors in printed version and digital format in trading platforms and selected libraries as well as data collection;

- Searches conducted in thirteen platforms: Amazon, Barnes & Noble, Google Books, iBookstore/iTunes, Kobo, Free-Ebooks, *Casa del Libro* (Spain), Gandhi (Mexico), *Librería Santa Fe* (Argentina), *Livraria Cultura* (Brazil), Perubookstore.com (Peru), Scribd, Nubico (Spain).
- Searches conducted in eight libraries: Columbia University Libraries, the *Bibliothek des Lateinamerikanischen Institut* in Berlin, *Biblioteca Nacional de España*, New York Public Library (U.S.A.), Gallica (*Bibliothèque Nationale de France*), Europeana (European Union Digital Library), *Biblioteca Nacional de Perú, Biblioteca de São Paulo* (Brazil).

The process of search and data collection in commercial platforms and libraries

1. A 'boot' was constructed to collect the search outcomes by title and by author present in the LA&CLC in the different platforms. The search mechanism varies according to the features portrayed by each platform and/or library.

2. The data collected comprise: author, title, print/digital format. This information was stored on a database for subsequent filtering.
3. Searching mechanism
    a. Information about author + title + source was uploaded to a database. Those titles included in more than one segment of the LA&CLC were only searched once.
    b. The authors included in the LA&CLC were searched in each platform and library.
    c. Outcomes were verified by the following criteria:
        i. Exact match: if the author found was exactly the same as that sought, it was considered valid.
        ii. Flexible comparison: if the found author name contained the same words as the sought one ignoring punctuation but in different word order, it was considered valid, for example, 'Borges, Jorge Luis' equals 'Jorge Luis Borges.'
        iii. Lax similarity: values found containing at least one of the searched words, regardless of word order, punctuation, special characters (considering characters as numbers, accented letters, etc.) will be considered valid only after manual verification on platforms and libraries. This lax similarity criterion was applied in cases when platforms and libraries did not incorporate characters of the authors and works languages of origin, for example, the case of accented characters in titles and author names as found on Amazon. From the search results, 30% of titles met the 3.a criterion (exact match); 20% met criterion 3.b (Flexible comparison) and the remaining 50% met the 3.c criterion (Lax similarity). In turn, after manual verification, 70% of the latter were considered valid (approximate percentage).
    d. Then the process was repeated but, this time, searching author + title in each platform – only the authors verified in 3.
    e. Verification was performed following the same method as in 3 but, this time, searching for a coincidence in author + title.
        i. From the results obtained in this stage, 10% were valid according to 3.a criterion, 30% in relation to criterion 3.b and the remaining 60% through criterion 3.c. Out of the latter, and after manual verification, only 20% of them were considered valid (approximate percentage).
    f. Based on the results validated in 5, a series of comparative tables were drawn up to display the results.
4. **Problems Encountered and Solutions Found:** Content issues:
    a. **Data Quality:** The metadata for each platform –on which the search was conducted– were heterogeneous, imprecise and inaccurate. They displayed spelling errors, incomplete words, and the formats vary per platform and library, for example; name-surname/name-surname orders. Solution: it was decided that two additional search criteria would be added: flexible comparison and lax similarity, described in 3.b and 3.c, respectively. Thus, the margin of error was considerably reduced.
    b. **Data Location:** Platforms and libraries not always locate the title with the author and/or the author with the title. There are platforms that do not even have the set of results on the same page. This error is similar to the problem in 1 (data quality). There are platforms that offer a result 'title: Jorge Luis Borges' and 'author: The Aleph'. This result is considered valid despite

the inversion. Solution: it was decided that the location relevant would not be considered on the condition that the searched value complied with any of the criterions in 3 for both fields (title or author). It was considered correct only if the other field complies with the same criterion for the other value.

c. **Data Encoding:** Not all platforms (especially libraries) have in place a good support for UTF-8, i.e., special characters (accents, Ò, ç, etc.) so they omit from the search those words containing any of these characters. For example: if you are searching for 'Antología poética' and you find 'Antologia poetica', the result is perfectly valid but the platforms' browsers will not return them since they have no support for these characters. This problem affects searches for Portuguese authors and works as they include special characters. Solution: in the platforms where the problem was detected, the search terms were altered by replacing conflicting characters by equivalents. Thus, outcomes were broadened. It was also necessary to implement the search criterion 3.c. (lax similarity) where this issue is ignored to include possible results coincidence.
d. **Type of Digital Format:** The digital format or electronic resource comprises a wide category, from e-books to software, videos, CDs and audio. Several platforms provide resources of various kinds which do not allow direct filtering. Solution: it was decided that only the results matching e-books explicitly, excluding any other type of resource that might appear, would be collected.
e. **Filtering Power:** Not every platform has an advanced browser. Several allow searching a single uniform token (a text where author and title must go together), therefore it is not possible to obtain highly accurate results. This makes it necessary to analyze more results added to ensure reliable outcomes. Solution: all data provided by the platforms were collected, and the boot was added a mechanism to identify and exclude those results not matching the basic patterns; as an example, the case in issue 4, where video format results were excluded.
f. **Volume of Relevant Results:** Multiple platforms, primarily the commercial ones, implement what is known as 'Infinite Pager.' It consists in displaying the first and second pages of relevant outcomes and, from then on, the rest of the catalog with thousands of results for a single search for which only the first fifty ones are relevant. Solution: only the results displayed in the first three pages were collected.

5. **Creating Tables:** All the tables include the following fields:
   a. **Platform/Library:** Website of platform/library to be analyzed;
   b. **Country of Origin of the Platform/Library:** Country of residence;
   c. **Author and Work:** The author is mentioned and the existence or absence of the author included in the LA&CLC work in digital format is indicated. The order of author searched is established by surname alphabetical order. The order of the titles of the works is that included in the sources used.
   d. **Literary Work Presentation:** Printed version/digital format.
   e. Number of platforms/libraries where the work can be found.
   f. **Results Out of the Total:** From the surveyed titles, the number obtained in the platform/library.
   g. **Source:** LA&CLC segment bringing work and author (there are 15 sources).

Every search was conducted in the original language of the works. It is also emphasized that all general tables contain the source or origin for every title and, in case of juxtaposition, the sources where the works come from.

A total of 61,551 searches were conducted only for authors and 207,570 searches matching author + book, all of them between December 2014 and March 2015.

In the built database, 2,262,364 search records were stored (each one related to a tuple –an ordered list or sequence of elements comprised by site + searched book+ author sought + title found+ print/ digital + author found).

We conducted 61,551 searches only for authors.

The searches conducted by author + book amounted to 207,570.

Overlapping works:

- Works included in one segment of the LA&CLC: 16,515.
- Works included in two segments of the LA&CLC: 3,095.
- Works included in three segments of the LA&CLC: 714.
- Works included in four segments of the LA&CLC: 70.
- Works included in five segments of the LA&CLC: 14.

Authors and works:

- 746 authors with one work in the LA&CLC
- 325 authors with two works in the LA&CLC
- 241 authors with three works in the LA&CLC
- 207 authors with four works in the LA&CLC
- 185 authors with five works in the LA&CLC
- 157 authors with six works in the LA&CLC
- 128 authors with seven works in the LA&CLC
- 95 authors with eight works in the LA&CLC
- 84 authors with nine works in the LA&CLC
- 82 authors with ten works in the LA&CLC
- 82 authors with eleven works in the LA&CLC
- 71 authors with twelve works in the LA&CLC
- 54 authors with thirteen works in the LA&CLC
- 62 authors with fourteen works in the LA&CLC
- 75 authors with fifteen works in the LA&CLC
- 36 authors with sixteen works in the LA&CLC
- 31 authors with seventeen works in the LA&CLC
- 22 authors with eighteen works in the LA&CLC
- 23 authors with nineteen works in the LA&CLC
- 31 authors with twenty works in the LA&CLC
- 23 authors with twenty-one works in the LA&CLC
- 25 authors with twenty-two works in the LA&CLC
- 22 authors with twenty-three works in the LA&CLC
- 12 authors with twenty-four works in the LA&CLC

- 17 authors with twenty-five works in the LA&CLC
- 9 authors with twenty-six works in the LA&CLC
- 9 authors with twenty-seven works in the LA&CLC
- 8 authors with twenty-eight works in the LA&CLC
- 12 authors with twenty-nine works in the LA&CLC
- 8 authors with thirty works in the LA&CLC
- 5 authors with thirty-one works in the LA&CLC
- 8 authors with thirty-two works in the LA&CLC
- 4 authors with thirty-three works in the LA&CLC
- 2 authors with thirty-four works in the LA&CLC
- 7 authors with thirty-five works in the LA&CLC
- 3 authors with thirty-six works in the LA&CLC
- 3 authors with thirty-seven works in the LA&CLC
- 2 authors with thirty-eight works in the LA&CLC
- 1 author with thirty-nine works in the LA&CLC
- 1 author with forty works in the LA&CLC
- 4 authors with forty-one works in the LA&CLC
- 3 authors with forty-three works in the LA&CLC
- 1 author with forty-four works in the LA&CLC
- 1 author with forty-six works in the LA&CLC
- 1 author with forty-seven works in the LA&CLC
- 1 author with forty-eight works in the LA&CLC
- 1 author with fifty works in the LA&CLC
- 1 author with fifty-eight works in the LA&CLC

In short, the ratio author/work in the searches conducted: 50% of authors have less than 5 works; 75% less than eleven works, and 25% more than ten works present in the LA&CLC.

General tables comparing the offer in print/digital as found in trading platforms/libraries by titles and authors included in the LA&CLC.

Amazon – Barnes & Noble – Google Books – iBookstore/iTunes – Kobo – Free-Ebooks – *Casa del Libro* – Gandhi – *Librería Santa Fe* – *Livraria Cultura* – Perúbookstore.com – Scribd – Nubico / Columbia University Libraries – *Bibliothek des Lateinamerikanischen Institut* in Berlin – *Biblioteca Nacional de España* – New York Public Library– Gallica (*Bibliothèque Nationale de France*) – Europeana (European Union Digital Library) – *Biblioteca Nacional de Perú* – *Biblioteca de São Paulo*

The database stores 2,262,364 result records (by tuples: website+ searched book+ sought author+ title found +print/digital + author found).

As for the outcomes produced by the search in digital format, it is highlighted that the platform offering more titles is the Exchange web Scribd, confirming that the user's participation causes an increase in accessibility: 1,718 out of 4,842 digitalized titles found (35.48% of the existing total). In libraries, the *Biblioteca Nacional de España* is where more titles were surveyed in digital version from the searches conducted in this research: 1,749, thus becoming the site providing the largest amount of digitalized titles as included in the LA&CLC. Refer to Table 1.

Table 1. Comparison of outcomes for titles in printed/digital format and authors in the LA&CLC

Platform/ Library	Authors in printed	Authors only in printed	% Authors only in printed	Titles in printed	Titles only in printed	% Titles in printed	Authors in digital	Authors only in digital	% Authors only in digital	Titles in digital	Titles only in digital	% Titles only in digital
Amazon	1728	35	2.03 %	7144	485	6.79 %	432	31	7.18 %	1101	119	10.81 %
Barnes & Noble	358	2	0.56 %	795	23	2.89 %	250	6	2.4 %	600	48	8 %
Biblioteca de São Paulo	78	0	0 %	218	2	0.92 %	0	0	0 %	0	0	0 %
Biblioteca I. Lat. Berlín	1479	24	1.62 %	4865	198	4.07 %	175	27	15.43 %	319	85	26.65 %
Biblioteca D. Unión Europea	0	0	0 %	0	0	0 %	281	31	11.03 %	582	0	0 %
Biblioteca Nacional de España	727	0	0 %	2473	40	1.62 %	522	172	32.95 %	1749	1137	65.01 %
Biblioteca Nacional de Francia	1186	9	0.76 %	3685	89	2.42 %	20	3	15 %	23	9	39.13 %
Biblioteca Nacional de Perú	913	4	0.44 %	2855	208	7.29 %	9	2	22.22 %	10	0	0 %
Casa del Libro	729	1	0.14 %	2211	30	1.36 %	198	0	0 %	512	5	0.98 %
Columbia University Libraries	1257	4	0.32 %	4663	88	1.89 %	275	0	0 %	572	164	28.67 %
Free E-Books	0	0	0 %	0	0	0 %	45	0	0 %	88	7	7.95 %
Gandhi	231	0	0 %	666	4	0.6 %	240	3	1.25 %	617	10	1.62 %
Google Books	2230	173	7.76 %	11140	1877	16.85 %	92	12	13.04 %	143	37	25.87 %
iBook Store	0	0	0 %	0	0	0 %	173	1	0.58 %	368	10	2.72 %
Kobo	0	0	0 %	0	0	0 %	123	5	4.07 %	262	26	9.92 %
Librería Santa Fe	548	0	0 %	1890	69	3.65 %	150	2	1.33 %	336	20	5.95 %
Livraria Cultura	534	0	0 %	1785	103	5.77 %	337	21	6.23 %	1035	185	17.87 %
Nubico	0	0	0 %	0	0	0 %	182	0	0 %	496	1	0.2 %
Perú Book Store	36	0	0 %	76	3	3.95 %	45	1	2.22 %	154	2	1.3 %
Public Library New York	1297	4	0.31 %	5760	171	2.97 %	8	0	0 %	21	0	0 %
Scribd	0	0	0 %	0	0	0 %	715	209	29.23 %	1718	864	50.29 %

Produced by the author.

THE MAIN ANALYSIS: THE TRANSPOSITION TO DIGITAL FORMAT OF LATIN AMERICAN AND CARIBBEAN CARTONERA PUBLISHERS (CP)

The phenomenon of CPs began with Eloísa Cartonera in Argentina in 2001. It works as a publishing cooperative producing hand-made books. Eloísa Cartonera was triggered by the Argentine political crisis and the 2001 economic crash followed by the interruption of post-neoliberal policies implemented in the early 2000s when unemployment forced thousands of persons to rake through the garbage cans in city streets. Many of these survivors engaged in classifying the garbage for recycling and this is why they were called *"cartoneros"* (cardboard collectors). The cultural industry was affected by that social and economic crisis: book sales plummeted, bookstores and publishing houses were closed, authors were unable to publish their works. Thus, the writer Washington Cucurto and visual artists Javier Barilaro and Fernanda Laguna started manufacturing artistic books using waste cardboard. As a way of supporting the unemployed, they changed the original goal of creating artistic books for creating literature that would be within everyone's reach selling books at affordable prices and paying the *cartoneros* a better price than the recycling centers. The cooperative Eloísa Cartonera was created after some *cartoneros* were incorporated to the production team. Little by little the publishing house started attracting Latin American authors, who donated their works for the purpose of building a catalog formed by short tales, isolated short stories, poem collections, vanguard texts and children's literature.). At present, their collective catalog includes Bloomian authors, such as Reinaldo Arenas, acclaimed writers of the so-called Generation of '72 (Nicholson and McClennen, 2013), such as Tomás Eloy Martínez, Ricardo Piglia, Diamela Eltit, Luisa Valenzuela, and César Aira, and celebrated writers of different literature styles, such as Rodolfo Walsh, Haroldo Campos, and José Emilio Pacheco, among others. Ksenija Bilbija (2010) mentions especially the Bolivian writers Adolfo Cárdenas and Christian Jiménez, the Colombian writers Andrés Caicedo, Andrés Nieva and Adriana Martínez, the Argentine writers Ricardo Piña and Viridiana Pistorio, the Paraguayan writer Maggie Torres, and the Mexican writer Antonio Ramos Revilla.

Consequently, CPs became cultural distribution platforms remaking traditional publishing forms in hard copy:

- The cardboard is obtained, on the one hand, from some of the millions of workers collecting paper and cardboard from trash bags and cans in Latin American cities and, on the other hand, it is directly sought by the members of the publishing house.
- The published titles are donated by the authors.
- Some of these publishing houses publish unknown or new authors from the suburbs thus allowing authors with no access to mainstream publishing houses to be published.
- With hand-crafted art and design (covers are hand-colored with watercolor, tempera or acrylic paint, or using other techniques such as collage, pasting and cutting of pictures, stencil, stamping, etc.).
- They lack any traditional distribution channels of their own –most of the titles are distributed by the authors themselves, sold at their stores or in fairs (even though a few of the titles are distributed using traditional distribution companies).
- Most CPs do not obtain ISBNs for the works they publish (although some of them do: Ultramarina and Meninas [Spain] and Animita [Chile]).
- Since binding tasks are present constraints, they usually publish short stories, novels, and essays, as well as poetry. However, certain CPs publish books in several volumes.

Michel de Certeau described the forms of cultural appropriation by popular sectors, their ways of "consumption" as characterized by "its ruses, its fragmentation (the result of the circumstances), its poaching, its clandestine nature, its tireless but quiet activity, in short by its quasi-invisibility, since it shows itself not in its own products (where would it place them?) but in an art of using those imposed on it" (de Certeau, 1984: 49) in the light of an "expansionist" cultural industry (de Certeau, 2010: XLIII). Although CPs start from a hand-crafted production level, the resulting books may be directly sold to readers or they may also be included in the traditional publishing industry distribution and sale channels. At the production level, Latin American CPs offer a semi-clandestine production reversing the notion of serialized and standard industrialization.

As mentioned by Cordón García and García-Figuerola, the economy of publishing "suffers from the definitive implementation of a market governed by GILT (globalization, internationalization, localization and translation) processes" (Cordón García and García-Figuerola, 2012: 12). However, at this time, we need to mention that the material forms whereby we access Latin American literature entail wider and more diversified styles, typography and ways of conceiving writing, which are broader and more diverse than before the inrush of the digital phenomenon.

In Mathematics and Computer Sciences a graph is a set of objects joined by links, which allow representing binary relations between the elements of that set. These relations also have an attribute called "weight", which determines the value of such relations. The so-called "degree" of an object is the number of relations it has with the objects of the highest degree being those related to the largest number of different objects while the "weighted degree" is the aggregate amount of weights of all the relations of an object being the object with the highest weighted degree the one with the strongest relations. From a practical standpoint, graphs allow studying interrelations between objects interacting with one another. They also allow representing the structure of a network where the vertices (nodes) are joined by lines (they may be called edges or loops based on whether they are directed or bidirectional) (Alonso Berrocal et al, 2011:87). Considering the relations and their related weights, it is mathematically possible to group the objects in the abovementioned communities, that is to say, subsets of objects which are more related among themselves than with the rest of the objects. In the particular case of this study, the set is the corpus of CP titles, the objects in the graph are each of the titles and the link between two titles shows that both them appear in the same platform in the same support. Such link shall be assigned a 1 weight and if two titles appear in more than one platform in the same support, the weight of the relation would increase proportionately. Likewise, the graphs obtained by using software considering the abovementioned database enable reaching the conclusion that the titles of the first *cartonera* publishing house –Eloísa Cartonera– operate as a "*cartonero* subcanon": they are located at the center in the distribution in platforms and libraries in both types of support. The central position in the graphs of these titles seems to be consistent with the concept of "consolidated visibility" proposed by Cordón García for highly recognized works "because they are works with a high symbolic value", among other things (Cordón García, 2004: 17). Lotman stated that "in the history of literature there are cases in which the perception of certain works by the readers was determined by the reputation of the edition in which they were published" (Lotman, 1979: 80). The weight of the tradition generated inside the movement by Eloísa Cartonera, the phenomenon giving rise to its publishing experience and the quality of the published titles could be the bases for such recognition.

The graphs related to the searches in printed and digital support have been included in Figures 3 and 4, Graphs 1 and 2, respectively. The titles in the same color belong to the same community.

Figure 3. Graph 1: Titles and authors with the highest degree of titles printed by cartoneras publishing houses; degree with loops and labels
Produced by the author.

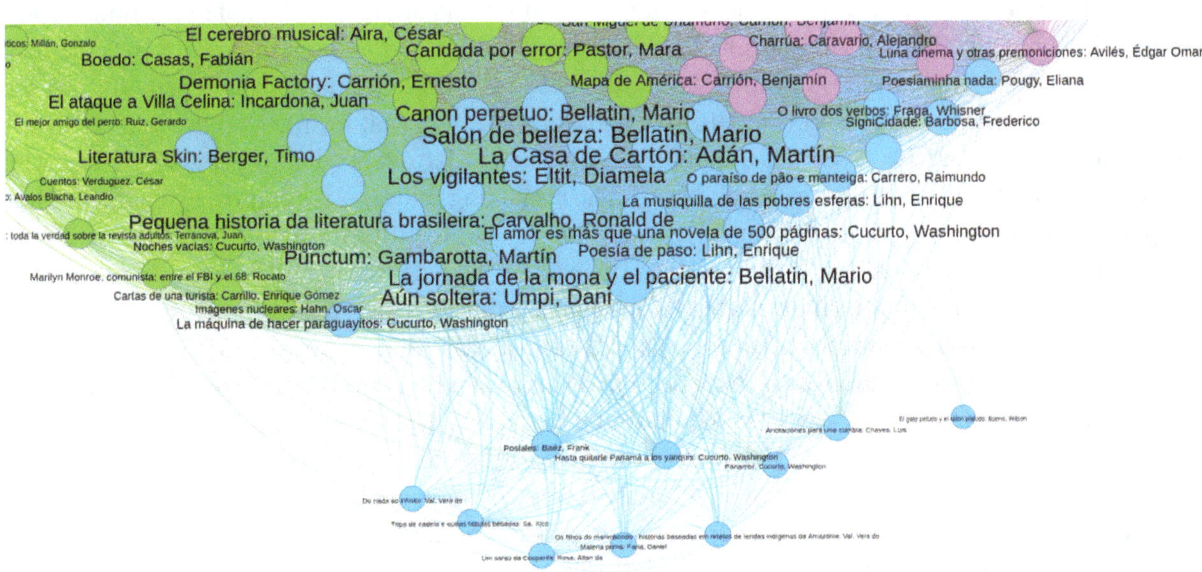

Figure 4. Graph 2: Cartoneras in digital format; degree with loops and labels
Produced by the author.

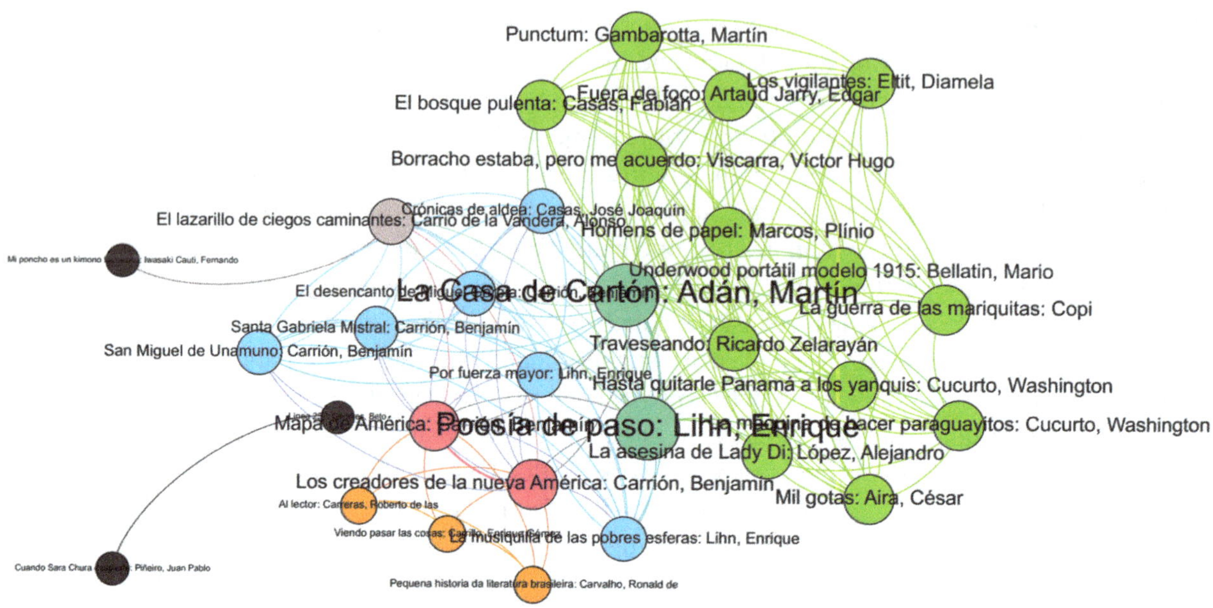

In the case of the graphs of Latin American and Caribbean titles in the *Cartonera* Publishers Database of the University of Wisconsin (found in printed form in 21 libraries and platforms), since the number of related titles (resulting from having found more titles available in printed form), the differences in weighted degree are subtler. The titles are found with their writers in three large communities (green, pale blue and pink nodes). In a high-definition resolution computer screen (see Figures 3 and 4, Graphs 1 and 2), the titles *Salón de belleza* and *La jornada de la mona y el paciente*, both of them by Mario Bellatín; *La casa de cartón*, by Adán Martín; *Punctum*, by Martín Gambarotta; *Aún soltera*, by Dani Umpi; and *Los vigilantes*, by Diamela Eltit (the six titles and authors in the pale blue nodes community) show a higher degree with weight (in the graph, the titles and authors in larger font). We may thus infer from the graph of CP in printed form that these two works rarely appear separately and the they are usually distributed as a set (in platforms and libraries, in digital form). And that out of the six titles with the highest grade, five titles are published by Eloísa Cartonera (the titles by Mario Bellatín, Dani Umpi, Adan Martín and Martín Gambarotta). The remaining title (*Los vigilantes*, by Diamela Eltit) was published by Sarita Cartonera from Peru. With a lower degree, further apart from the center, the titles in the pale blue community: *Um saráu de Cooperifa*, by Alan de Rosa and *El gato peludo y el gato platudo*, by Wilson Bueno. It is inferred that these titles are mostly separated from the rest of the titles and that the platforms and libraries offer them separated randomly in hard copy.

In the case of the graph of the Latin American and Caribbean titles in the Cartonera Publishers Database of the University of Wisconsin (found in digital form in 21 libraries and platforms), the titles *La casa de cartón*, by Adán Martín, and *Poesía de paso*, by Enrique Lihn, have the highest weighted degree as they belong to the same community which is only formed by them; thus, these are the most representative titles in the set. We may thus infer from the graph that these two works rarely appear separately and the they are usually distributed as a set (in platforms and libraries, in digital form). These two titles were published by Eloísa Cartonera. Two titles by Benjamín Carrión are next in weighted degree (the two nodes in pink): *Los creadores de la nueva América* and *Mapa de América* (both of which are in the public domain).

SOME CONCLUSIONS

For the analysis of the transposition of Latin American and Caribbean literature from print to digital version, a corpus (LA&CLC) was designed to display the diversity of these literatures (Figure 1). Ana Pizarro (1987: 190-191) states that as part of the aesthetic-cultural formalization literature, the notion of literature might be expanded 'to include in its scope speeches such as oratory, sermons, and historical, political and philosophical discourses, etc.' If the idea of Lotman-Uspenski (1979: 77) is to talk of 'culture as a mechanism that creates a set of texts, and to talk about texts as a realization of culture', it may be implied that culture works as a generative and structuring mechanism which bases its borders on its own self-consciousness and self-organization. Thus, the LA&CLC is proposed as the result of forms of self-conception of Latin American and Caribbean literature. It arises then both, from the survey of heterogeneous Latin American and Caribbean repertoires and literary sources far apart from each other in time span and in the incorporation of the traditional canonical version of Latin American and Caribbean literature as well as an alternative canon added up to repertoires displaying the various language forms and dialects spoken in the region. This LA&CLC initially included 20,409 titles, out of which 14,835 were found in printed version (almost 75% of the corpus under search) after conducting the searches using the methodology described in the section *Search of titles and authors present in the corpus in printed and digital format in commercial platforms and defines libraries, and data collection* in the present paper. The detailed results of the 50 titles searched are available in the Table 2.

But, in turn, out of the total of books found in printed version, 50.62% has been transposed into digital format: 4,842 books (almost 24% or the initial corpus). 10,190 titles (almost 50% of the books found in printed or digital format) were only found in printed version. The most highlighted results found after the search for authors and titles of the *LA&CLC,* display titles whose copyright is in the public domain and are leading the transposition into digital format of the *LA&CLC.* When searching for books in printed version, the first 50 works found whose copyright is of public domain; do not have much weight in the composition of the offered élite (Table 2). However, from the first 50 titles found in digital format (those that appear 14, 13, 12 & 11 times in the same amount of platforms and libraries), 41 authors whose works are in the public domain (Table 3). This confirms one of the transposition criteria implemented by the publishing industry: works with no quoted copyright are mostly transposed. In addition to what may be addressed as 'a problem inherent to the Latin American and Caribbean literary field', the action of the publishing industry when transposing printed into digital formats can also set its strategic action criteria that would enter into conflict with the autonomy of the academic field when setting titles and authors. Publisher intervention forms linked to marketing disciplines framed by the international book market concentration processes, as well as the trading platforms' own strategies (especially Amazon, Barnes & Noble, Google Books, Kobo and Apple/iBooks) make up a series of actions that unsettle the digital catalog building strategies based on the canon aesthetic authority (Bloom, 1995: 47).

All this, together with strategies for acquiring digital books prompted by the user (Patron Driven Acquisition-PDA), the involvement of users in works as a window through which poetic and aesthetic criteria would participate in the building of the general digital catalog.

Table 2. The LA&CLC50 titles most common in printed version

Author	Title	Source	Times
Carpentier, Alejo	*Concierto Barroco*	2,13,14	14
Carpentier, Alejo	*Los pasos perdidos*	1,2,3,13,14	14
Vargas Llosa, Mario	*Los cachorros*	2,3,14	14
Allende, Isabel	*La casa de los espíritus*	3	13
Arlt, Roberto	*El juguete rabioso*	2	13
Azuela, Mariano	*Los de abajo*	2,3,14	13
Benedetti, Mario	*Inventario*	2,14	13
Bolaño, Roberto	*Los detectives salvajes*	6	13
Borges, Jorge Luis	*El Aleph*	1,2,3,14	13
Cambaceres, Eugenio	*Sin rumbo*	2,14	13
Carpentier, Alejo	*El reino de este mundo*	1,2,3,13,14	13
Neruda, Pablo	*Veinte poemas de amor y una canción desesperada*	1,2,14	13
Roncagliolo, Santiago	*Abril rojo*	3,8	13
Rulfo, Juan	*Pedro Páramo*	2,3,14	13
Alegría, Ciro	*El mundo es ancho y ajeno*	2,14	12
Altamirano, Ignacio Manuel	*Clemencia*	2	12
Altamirano, Ignacio Manuel	*El Zarco*	2,14	12
Amado, Jorge	*Gabriela, cravo e canela*	2,14	12
Amado, Jorge	*Sudor*	2	12
Arguedas, Alcides	*Raza de bronce*	2,14	12
Arguedas, José María	*Los ríos profundos*	2,14	12
Arguedas, José María	*Yanar fiesta*	2	12
Arlt, Roberto	*Aguafuertes porteñas*	2	12
Arlt, Roberto	*Los lanzallamas*	2	12
Arlt, Roberto	*Los siete locos*	2	12
Asturias, Miguel Ángel	*El señor Presidente*	2,14	12
Asturias, Miguel Ángel	*Hombres de maíz*	1,2,14	12
Asturias, Miguel Ángel	*Leyendas de Guatemala*	2,14	12
Benedetti, Mario	*La borra del café*	14	12
Benedetti, Mario	*La muerte y otras sorpresas*	2,14	12
Benedetti, Mario	*La tregua*	2,14	12
Benedetti, Mario	*Pedro y el capitán*	14	12
Benedetti, Mario	*Primavera con una esquina rota*	2	12
Bioy Casares, Adolfo	*Historias de amor*	2	12
Borges, Jorge Luis	*Ficciones*	1,2,3,14	12
Borges, Jorge Luis	*Inquisiciones*	2,14	12
Borges, Jorge Luis	*Obras completas*	2,14	12
Cambaceres, Eugenio	*En la sangre*	2,14	12
Cané, Miguel	*Juvenilia*	2,14	12
Carpentier, Alejo	*El acoso*	2,13,14	12
Carpentier, Alejo	*El recurso del método*	1,2,3,13,14	12
Castellanos, Rosario	*Oficio de tinieblas*	2,14	12
Conti, Haroldo	*Sudeste*	2,14	12
Cortázar, Julio	*Bestiario*	2,3	12
Cortázar, Julio	*Las armas secretas*	2,14	12
Cortázar, Julio	*Octaedro*	2,14	12
Donoso, José	*El Mocho*	2,14	12
Filloy, Juan	*Op Oloop*	2,14	12
Neruda, Pablo	*Canto general*	1,2,14	12
Neruda, Pablo	*Residencia en la tierra*	1,2,14	12

Produced by the author.

Out of the total 4,842 titles found as a result of the search for digital formats, Scribd provides 1,718 titles, i.e. 35.48% titles existing in digital format. From these titles, 864 are only provided by Scribd. It offers 715 authors, 52.84% of the 1,353 authors surveyed in digital format, out of which 209 are brought only by the mentioned interchange website. In sum, Scribd provides 15.44% of the authors found in the

Table 3. The LA&CLC50 titles most common in digital version

Author	Title	Source		
Cieza de León, Pedro	*Crónica del Perú*	2	14	09
Mármol, José	*Amalia*	2	14	10
Alberdi, Juan Bautista	*Bases y puntos de partida para la organización política...*	2,14	13	09
Alberdi, Juan Bautista	*El crimen de la guerra*	2,14	13	11
Altamirano, Ignacio Manuel	*Clemencia*	2	13	12
Benedetti, Mario	*La tregua*	2,14	13	12
Darío, Rubén	*Prosas profanas*	2,3	13	09
Lillo, Baldomero	*Sub Terra*	2,14	13	08
Villaverde, Cirilo	*Cecilia Valdés*	2,14	13	12
Acevedo Díaz, Eduardo	*Brenda*	14	12	04
Barros Arana, Diego	*Historia general de Chile*	2,14	12	08
Blest Gana, Alberto	*Martín Rivas*	14	12	10
Borges, Jorge Luis	*El Aleph*	1,2,3,14	12	13
Cruz, Sor Juana Inés de la	*Obras*	2	12	11
Darío, Rubén	*Cantos de vida y de esperanza*	2,3,14	12	10
Ercilla y Zúñiga, Alonso de	*La araucana*	2	12	09
Garcilaso de la Vega	*Comentarios reales*	2	12	10
Ingenieros, José	*El hombre mediocre*	2,14	12	10
Lillo, Baldomero	*Sub sole*	2	12	08
López de Gómara, Francisco	*Historia general de las Indias*	2	12	10
Rodó, José Enrique	*Ariel*	2,14	12	10
Aguinis, Marcos	*La gesta del marrano*	14	11	08
Asturias, Miguel Ángel	*El señor Presidente*	2,14	11	12
Borges, Jorge Luis	*Ficciones*	1,2,3,14	11	12
Borges, Jorge Luis	*Historia universal de la infamia*	2,3,14	11	11
Cané, Miguel	*Juvenilia*	2,14	11	12
Cieza de León, Pedro	*Señorío de los Incas*	2	11	08
Darío, Rubén	*Canto errante*	2	11	06
Darío, Rubén	*Tierras solares*	2,14	11	07
Díaz del Castillo, Bernal	*Historia verdadera de la conquista de la Nueva España*	2	11	11
García Márquez, Gabriel	*La hojarasca*	2,14	11	08
García Márquez, Gabriel	*La Mala Hora*	2,14	11	08
Güiraldes, Ricardo	*Don Segundo Sombra*	2,14	11	08
Las Casas, Bartolomé de	*Historia de las Indias*	2	11	06
Lugones, Leopoldo	*Las fuerzas extrañas*	2,14	11	09
Montenegro, Carlos	*Hombres sin mujer*	2,13	11	05
Neruda, Pablo	*Canto general*	1,2,14	11	12
Nervo, Amado	*Los jardines interiores*	2,14	11	06
Nervo, Amado	*Místicas*	2	11	07
Palma, Ricardo	*Tradiciones*	2	11	12
Palma, Ricardo	*Tradiciones peruanas*	2,14	11	12
Rabasa, Emilio	*La bola*	2	11	09
Altamirano, Ignacio Manuel	*El Zarco*	2,14	10	12
Altamirano, Ignacio Manuel	*La Navidad en las montañas*	2,14	10	11
Arlt, Roberto	*El juguete rabioso*	2	10	13
Borges, Jorge Luis	*El Hacedor*	1,2,14	10	10
Cané, Miguel	*El viaje*	2	10	11
Carriego, Evaristo	*Misas herejes*	2,14	10	10
Cervantes de Salazar, Francisco	*Crónica de Nueva España*	2	10	08
Cortázar, Julio	*Final del juego*	2	10	09

Produced by the author.

search of digital formats (excluding the titles overlapped with other platforms, represented in high-grade searches). At this point, the user intervention in Scribd works as another window through which poetic, aesthetic and artistic criteria about Latin American and Caribbean literature enter the construction of the general digital catalog. Hence, if the publishing industry develops strategies for the propagation of criteria related to taste (Bourdieu, 1988) by building an offer of Latin American and Caribbean literature in digital format, the scheme complements the criteria related to taste by the participation of users in the exchange website Scribd contributing titles and widening the offer. At this stage, we extend the offer of the LA&CLC in digital format of the titles designed by the reasoning of Pierre Bourdieu (1988: 228) about the relationship between the production field and taste:

The field of production –which could obviously not operate without having the existing taste, propensities more or less intense to consume more or less strictly defined goods– permits taste to realize by offering, at all times, the universe of cultural goods as a possible stylistic system where the stylistic feature system constituent of a lifestyle can be selected (Bourdieu, 1988: 22)

If according to Bourdieu (1988: 228), 'the universe of goods offered by each of the producing fields tend to limit the universe of forms of experience (aesthetics, ethics, politics, etc.) which are objectively possible at a moment in time', we should note that in developing such a limitation, those resulting propensities constitute a cutback in the offer that can be projected into the construction of the Latin American and Caribbean general digital catalog built by the transposition of the printed works into digital format. This 'possible stylistics' partially erects the trading platforms offer and the library offer within the framework of strategies adopted by the agents involved in both sectors. The digital format purchasing strategies based on the acquisition model prompted by the user (Patron Driven Acquisition-PDA) (Cordón García, Gómez Díaz, Alonso Arévalo and Alonso Berrocal, 2014; Esposito et al., 2012; Gammon & O'Neill, 2011; Vicente García, 2013; Walters, 2012) allow users to participate directly in the building of a collection and constitutes an area of negotiation and competition among users and decision-maker librarians. Most aggregators provide access to both the e-book selected titles and the titles selected by the library according to profile and interest. If the publishing industry shapes its model for transposition into digital format as a combination of strategies based on commercial vectors linked to market demands as well as on proceedings derived from criteria linked to taste, the construction of a digital library offer through the model of book acquisition prompted by the user (Patron Driven Acquisition-PDA) is a window through which taste, poetic, aesthetic and artistic criteria of literature enter the forms of construction of the general library catalog.

Then, as already discussed (Vila, 2016: 44) it follows that the outcome derived from the search in the Latin American and Caribbean literature general offer is constructed by combining the following strategies (Refer to Figure 5):

1. Strategies for transposing into digital format by the publishing industry + strategies related to copyright + marketing strategies related to cost + strategies derived from poetic and aesthetic criteria.
2. The action of the user/reader by adding titles attached to 'taste' (Scridb in platforms, PDA in libraries).
3. Titles self-published by authors (through *KDP*, *iBooks Author*, *Smashwords, LULU,* etc.).
4. It is worth noting that (in the results of the graphs obtained by CPs both in printed and digital form), the titles of Eloísa Cartonera work as a "*cartonero* subcanon" occupying the main space in the distribution in platforms and libraries in both forms.

Figure 5. From the printed LA&CLC to the digital corpus
Produced by the author.

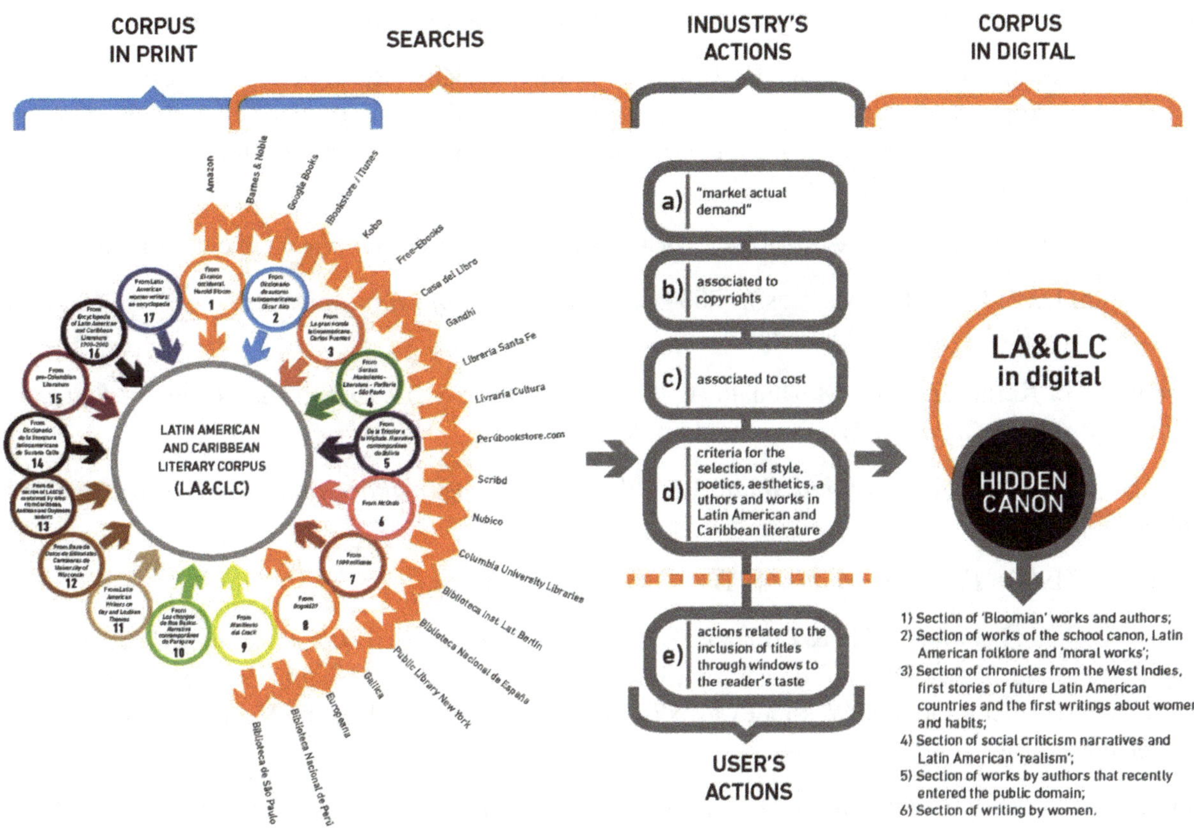

That is, the publishing industry would appear juxtaposing commercial strategies: those related to sales statistics of titles in printed version considered 'actual market demand' (Cordón García, 2004: 150) to run the business risk of transposing into digital format; those strategies associated to the need to recover financial returns given the investments made in terms of purchasing copyrights and the cost of marketing strategies related to titles and authors. Likewise, the overall strategic action seems to add up procedures derived from prevailing criteria for the selection of style, poetics, aesthetics, authors and works in Latin American and Caribbean literature.

The use of new technologies and especially the digitization of a portion of CP catalogs through open and/or copyleft license access forms added to the free availability of the titles digitized by the users of the Scribd network also result in a double strategy: the incorporation by the reader of esthetic and poetic ("liking") criteria upon building a general digital catalog and, on the other hand, the maintenance of the resistance ritual of CPs also in digital format by promoting their own titles with open access.

The publishing realm is thus expanded to more material forms from the publishing formulas of Latin American CPs to the Kindle AZW3 format. The inrush of the *cartonera* support –with its incipient entry into the digital ecosystem– is broader in the ways of considering writing and literature, which also relate to material, printed, "typographic" forms.

REFERENCES

Aira, C. (2001). *Diccionario de autores latinoamericanos*. Buenos Aires: Emecé & Ada Korn.

Alonso Berrocal, J. L., Gómez Díaz, R., García Figuerola, C., Zazo Rodríguez, A., & Cordón García, J. A. (2011). Propuesta de estudio del campo semántico de los libros electrónicos en Twitter. Zaragoza: *Scire: representación y organización del conocimiento, 17*(2), 87-97. Retrieved from http://ibersid. eu/ ojs/index.php/scire/article/view/3939/3710

André, M., & Bueno, E. (2008). *Latin American women writers: an encyclopaedia*. New York: Routledge.

Balderston, D., & Gonzalez, M. (Eds.). (2004). *Encyclopaedia of Latin American and Caribbean Literature 1900-2003*. London: Routledge.

Benjamin, J., Kallicharan, L., McDonald, I., & Searwar, L. (Eds.). (1998). *They Came in Ships: An Anthology of Indo-Guyanese Writing*. London: Peepal Tree P.

Birbalsingh, F. (Ed.). (1988). *Jahaji Bhai: An Anthology of Indo-Caribbean Literature*. Toronto: TSAR Publications.

Bloom, H. (1995). *El canon occidental*. Barcelona: Anagrama.

Bonett, P., Abad Faciolince, H., & Collazos, O. (n.d.). *Bogotá39. 39 escritores menores de 39*. Bogotá: Hay Festival. Secretaría Distrital de Cultura, Recreación y Deporte de Bogotá. Retrieved from http://www.hayfestival.com/bogota39/es-index.aspx?skinid=7&localesetting=es-ES&currencysetting=EUR&resetfilters=true

Bourdieu, P. (1966). *Campo de poder, campo intelectual. Itinerario de un concepto*. Buenos Aires: Montressor.

Brown, S., & McWatt, M. (Eds.). (2004). *The Oxford Book of Caribbean Verse*. Oxford, UK: OUP.

Cella, S. (1998). *Diccionario de literatura latinoamericana*. Buenos Aires: El Ateneo.

Chowdhury, G., & Foo, S. (2012). *Digital libraries and information access: Research perspectives*. London: Facet Publishing.

Cordón García, J. A. (2004). La visibilidad en los circuitos de la creación: literatura y traducción. In C. Gonzalo García & V. García Yebra (Eds.), *Manual de documentación y terminología para la traducción especializada* (pp. 127–169). Madrid: Arco/Libros.

Cordón García, J. A., & García-Figuerola, C. (2012). Introducción. Aventuras, inventos y mixtificaciones del libro electrónico. In Libros electrónicos y contenidos digitales en la sociedad del conocimiento. Mercado, servicios y derechos. Madrid: Pirámide.

Cordón García, J. A., Gómez Díaz, R., Alonso Arévalo, J., & Alonso Berrocal, J. L. (2014). *El ecosistema del libro electrónico universitario*. Salamanca: Ediciones Universidad de Salamanca.

de Certeau, M. (1996). *La invención de lo cotidiano. Tomos I y II*. Universidad Iberoamericana / ITESO / Centro de Estudios Mexicanos y Centroamericanos.

Di Nucci, S., Recoaro, N., Grieco, & Bavio, A. (2014). *De la Tricolor a la Wiphala. Narrativa contemporánea de Bolivia*. Buenos Aires: Santiago Arcos Editor.

Di Nucci, S., Recoaro, N. and Grieco y Bavio, A. (2014). *Los chongos de Roa Bastos. Narrativa contemporánea de Paraguay*. Buenos Aires: Santiago Arcos Editor.

Donnell, A., & Bucknor, M. (Eds.). (2011). *The Companion to Anglophone Caribbean Literature*. Routledge.

Donnell, A., & Lawson Welsh, S. (Eds.). (1996). *The Routledge Reader in Caribbean Literature*. London: Routledge. doi:10.4324/9780203431054

Esposito, J., Walker, K., & Ehling, T. (2012). *PDA and the university press. A report prepared for The Andrew W. Mellon Foundation*. Baltimore, MD: The Johns Hopkins University Press/Cornell University Library. Retrieved from https://jscholarship.library.jhu.edu/bitstream/handle/1774.2/36210/PDA%20and%20the%20University%20Press%20FINAL%202012-10-15.pdf

Foster, D. W. (1991). *Latin American Writers on Gay and Lesbian Themes: A Bio-Critical Sourcebook*. Austin, TX: University of Texas Press.

Fuentes, C. (2011). *La gran novela latinoamericana*. Madrid: Alfaguara.

Fuguet, A., & Gómez, A. (Eds.). (1996). *McOndo (una antología de nueva literatura hispanoamericana)*. Barcelona: Gijalbo-Mondadori.

Gammon, J., & O'Neill. (2011). *Ohio LINK-OCLC Collection and Circulation Analysis Project 2011*. Dublin, OH: OCLC Research, September 2011. Retrieved from http://www.oclc.org/content/dam/research/publications/library/2011/2011-06.pdf

García, R. (1994). *Documentación general (sistemas, redes, centros).Guía del alumno*. Madrid: Síntesis.

García Canclini, N. (1999). *La globalización imaginada*. Buenos Aires: Paidós.

Helder, D., Henderson, D., & Orge, B. (2014). *1.000 millones: poesía en lengua española del siglo XXI*. Rosario: Editorial Municipal de Rosario.

Klibanski, M. (2006). *Grandes bibliotecas digitales y virtuales del mundo*. Retrieved from http://portal.educ.ar/debates/educacionytic/super-sitios/grandes-bibliotecas-digitales-y-virtuales-del-mundo.php

Lotman, I., & Uspenski, B. (1979). Sobre el mecanismo semiótico de la cultura. In *Lotman y Escuela de Tartu: Semiótica de la cultura* (pp. 67–92). Madrid: Cátedra.

Milliot, J. (2011). The Latest E-Book Buying Trends. *Publisher's Weekly*. Retrieved: April 5, 2015, from http://www.publishersweekly.com/pw/by-topic/digital/content-and-e-books/article/48418-the-latest-e-book-buying-trends.html

Nicholson, B., & McClennen, S. (Eds.). (2013). The Generation of '72. Raleigh, NC: Editorial Contracorriente.

Palou, P. A., Urroz, E., Padilla, I., Chávez-Castañeda, R., & Volpi, J. (2002). Manifiesto Crack I, II, III, IV, V. México: Lateral, 70.

Pizarro, A. (1987). *Hacia una historia de la literatura latinoamericana*. México: El Colegio de México/ Universidad Simón Bolívar.

Ramchand, K. (Ed.). (1966). *West Indian Narrative: An Introductory Anthology*. London: Nelson.

Tennina, L. (Ed.). (2014). Antología Saraus. Movimiento - Literatura - Periferia - São Paulo. Buenos Aires: Tinta Limón.

Tramullas Saz, J. (2002). Propuestas de concepto y definición de la biblioteca digital. In *e-LIS, 2002*. Retrieved: April 4, 2015, from http://eprints.rclis.org/15118/1/04_2002.pdf

Tsang, J. (2013). Doble personalidad. *Entrevista a Alberto Fuguet.Buen Salvaje, 5*, 20-23. Retrieved: April 2, 2015, from http://buensalvaje.com/tag/mcondo/

Turiel, J. (2010). La manía de las listas: 100 bibliotecas digitales y 'recolectores' de libros electrónicos. *bid - textos universitaris de biblioteconomia i documentació, 24*, 1-10. Retrieved: April 4, 2015, from en: http://bid.ub.edu/24/turiel2.htm

University of Wisconsin Digital Collections. (n.d.). *The Art Collection. Base de Datos de Editoriales Cartoneras*. Madison, WI: University of Wisconsin. Retrieved: May 5, 2015 from http://digicoll.library.wisc.edu/WebZ/SearchOrBrowse?sessionid=01-53965-684142241

Vicente García, R. (2013). *Los libros electrónicos en las colecciones de las bibliotecas públicas*. Barcelona: Revista bid, textos universitarios de biblioteconomia documentació nº 30, June 2013. Retrieved: May 10, 2015, from http://bid.ub.edu/es/30/devicente.htm#sthash.UPe1oIq8.dpuf

Vila, A. R. (2016). Latin American and Caribbean Literature Transposed into Digital. Journal of Information Technology Research, 9(1). doi:10.4018/JITR.2016010103

Walters, W. H. (2012). Patron-Driven Acquisition and the Educational Mission of the Academic Library. *University Libraries, Library Resources & Technical Services Vol. 56*. Piscataway, NJ: Rutgers. Retrieved: May 7, 2015, fromhttp://journals.ala.org/lrts/article/view/5528/6796

Walwyn Michael, A. (Ed.). (2001). *Sugar Is All: Caribbean Short Stories and Poems from St. Kitts and Nevis*. Morris Publisher.

Wischenbart, R., Carrenho, C., Licher, V., Kovac, M. & Vinhuta, M. (2013). *Global eBook: Current Conditions & Future Projections*. London: O'Reilly.

Chapter 4
Teaching Formation to Develop Computational Thinking

Klinge Orlando Villalba Condori
Universidad Nacional San Agustín, Peru & Universidad Católica de Santa María, Peru

ABSTRACT

In order to develop computational thinking, it is essential to modify the initial training of teachers, that is to say, to restructure the curricula of the faculties or professional schools of education in any field or context it is going to be possible to identify areas of basic or general training and areas of specialized training. As well as reading or writing, computer thinking is currently essential because its application in any professional context is necessary.

INTRODUCTION

Computational thought is considered as a competence of the 21st century, which has several developmental possibilities from the perspective of initial and in-service teacher training. In this chapter, we present the conceptual meanings of computational thought from the perspective of pedagogical optics, establishing the link between the levels of mathematical thought (sensory, concrete and graphic) with the phases of analysis and abstraction, modeling and organization, code, technological system, programming and permanent reflection, then.

Teacher training must contribute to the formation of capable citizens for developing critical and rigorous thought processes, capable of "learning to learn", an expression that connotes a radical transformation about the way which teachers and students do their daily lives in the classroom, and which requires the search for new forms of access to knowledge (Snalune, 2015).

In all the specialties of teacher training (initial, primary or some specific of the secondary level) we are going to find the differentiation in terms of the areas of general or basic training and the areas about the specialized training, the percentages of the subjects destined to the areas of training or specialized areas vary finding recurrence differences between the area of basic training and the specialized training, finding recurrently greater percentage in the specialized area.

DOI: 10.4018/978-1-5225-4944-4.ch004

To promote an awareness and attitude so that future teachers harmonize their relations with nature, within a framework of respect for cultural diversity, social and ethnic diversity, with a responsibility sense for the future generations needs (Román-González, 2015), (Zapata, 2015). This is what the basic training of future teachers pursues, a very important aspect of basic training has to do with pedagogical practice, understanding it as a systematic, inclusive, interactive and flexible process, if this interaction is successful the achievement of the learning purpose is assured (Figure 1).

Contemporary training models use the TPACK (Technological Pedagogical Content Knowledge) model as a reference, in which the teacher seeks to show the disciplinary, technological and pedagogical mastery of the subject in a joint way.

Pedagogical practice is part of what is called the Pedagogical Domain and is developed within the general areas, here it is important to mention the Technological Domain complements the pedagogical domain, since it guarantees the effective incorporation of ICT in the teaching and learning process (Rojas López, García-Peñalvo, 2016).

The Disciplinary Domain consists of the teaching specialization corresponding to initial teacher raining (Román-González, Pérez-González & Jiménez-Fernández, 2017). That is to say, specialties such as pre-school, primary or some specific secondary level.

Adapt the training model to the new skills of the 21st century. Among them, the computational thinking is a fundamental skill for everyone and not just for computer scientists (Romero, Usart & Ott, 2015). (Lye, & Koh, 2014). In addition to reading, writing and arithmetic, we must add computational thinking to the analytical capacity of each child (Llorens-Largo, 2015). It involves problem solving, system design and understanding of human behaviour, based on the fundamental concepts of computer science. Computational thinking includes a range of mental tools that reflect the computing field breadth. It is influencing research in both the sciences and the humanities. From the different meanings we dimensioned the concept of Computational Thought in:

Figure 1. Fields of teaching practice

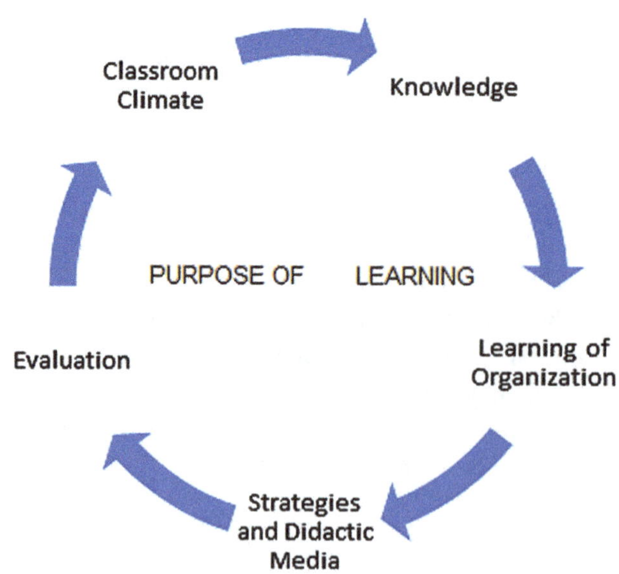

- Knowledge of the computing fundamental concepts,
- Formal use of this knowledge and
- Provision in use for problem solving
- Undertaking to solve a problematic situation.

This dimensioning could be linked in some way to the levels of mathematical thinking proposed by Piaget (1980) that should be part of the teacher's general training.

Even Computational Thought has been understood in some cases as progressive sequences to learn how to program (Li, Liu, Shu, & Ieee, 2015), that is to say, to elaborate an algorithm and pass it on to a programming language, in many cases this activity is the most outstanding evidence on Computational Thought, however Computational Thought has a greater scope, it is closely linked to the solution of problems, it implies: observing the situation, planning the solution of the observed situation, executing.

In short, this process in educational contexts in which it has already been incorporated into the school curriculum, in the case of Argentina, Estonia, to name but a few, the development of Computational Thought is made explicit, there are movements that involve a series of actions that are linked to the development of Computational Thought such as the TACCLE 3 Project (García-Peñalvo, 2016).

Next we will detail aspects to be taken into account for the basic or general training of teachers closely linked to Computer Thought, the development of mathematical logical thinking opens the doors not only to numbers, operations or reasoning, it is better to say that it constitutes a very wide bridge that allows communication with the surrounding environment, the environment in which the human being is in constant interaction. Therefore, mathematical logical thinking must be perceived in school as the basis for the integral formation of the human being (Grgurina, N., Barendsen, Zwaneveld, Van Veen & Stoker, 2014). Based on this inference, the school must prepare and ensure the development of essential competencies.

To consider that each individual learns in a different way the importance of fostering an environment of investigation, inventions and manipulation that promotes the development of the child's abilities (Gross, Kim, Schlosser, Lluch, Mohtadi, Schneider, & Ieee, 2014).

Stages of the Development of Mathematical Thinking

For Computational Thought, stages such as the following have not yet been defined: on (Gonzales, 1998) the Piaget levels of intellectual operations:

1. **Motor Sensory Stage:** The discovery of new situations (13 to 18 months) The circular reactions. Throughout these first two years of the child's life, an intellectual revolution takes place, characterized by four processes that construct the practical categories of:
    a. Object.
    b. Space.
    c. Causality.
    d. Time.

The acquisition of these four categories, such as language, is closely linked to their psychomotor evolution.

2. **Preoperative Stage (2-7 Years):** There is significant differentiation meaning, but without logical restrictions.
3. **Logical-Concrete Stage (7-12 Years):** Operations on the level of representations, structures the present according to the past without deformations, dislocations or contradictions of the preoperative child. The symbol comes to play an important role in addition to language, this occurs between the ages of 2-4 years.

At the second level, which covers approximately 4-6 years of age, the child develops the ability to symbolize reality, building more complex thoughts and images through language and other signifiers. However, certain limitations in the child's thinking are presented, such as: egocentrism, focus, realism, animism, artificiality, precaution, irreversibility, transductive reasoning.

Sub-stage of concrete operational thinking. From approximately 7-11 years old. At this level the child achieves reversibility of thought and can solve problems if the object is present. The ability to serialize, classify, mentally order sets is developed. Progress is being made in the process of socialization as relationships become more complex.

The children who have started the stage of specific operations have not changed. An object may appear larger, longer or heavier, but the two remain the same. In Piaget's opinión (1980), children use two basic mental operations to carry out conservation tasks: denial, compensation and identity. These operations are reflected in the way an 8-year-old child could explain why the amount of water in two glasses remains unchanged: "You can empty it again and it will be the same" (denial). "The water rises higher but it is because the glass is thinner" (compensation) You just emptied it, nothing was added or removed" (identity) (Miller, 1993, p. 57). Between the ages of 7 and 11, the child learns the mental operations necessary to reflect on the transformations represented in conservation problems (García-Peñalvo, 2016). He will be able to carry out reflective abstraction, when he needs logically reasoning respect to number and volume without being confused by physical appearances. You can then distinguish between the invariable characteristics of the stimuli (weight, number or volume, for example) and the shape of the object before your eyes. Acquisition of mental operations for conservation tasks is not carried out at the same time in all areas. Understanding conservation problems follows a gradual sequence. As a peculiar feature, the child acquires the capacity to conserve numbers between the ages of 5 and 7. The conservation of the area and weight appears between 8 and 10 years. Between the ages of 10 and 11, nearly all children can perform tasks related to volume conservation. Horizontal misalignment is the name Piaget (1980) gave to this lack of uniformity in children's thinking at one stage.

4. **Logical-Formal Stage (12-16 Years):** Operations on the possible configuring the real as one of its cases. In the face of a problem, all possible validity relationships are foreseen and then it is determined by experimentation and logical analysis which of the possible relationships have real validity. It is a hypothetical-deductive, propositional and probabilistic thought (García-Peñalvo, 2016).

Once the ability to solve problems such as serialization, classification and conservation has been achieved, the 11-12 year-old child begins to form a coherent system of formal logic (García-Peñalvo, 2016). By the end of the specific operations period, you already have the cognitive tools that allow you to solve many types of logic problems, understand the conceptual relationships between mathematical operations (e. g. $15 + 8 = 10 + 13$), sort and classify knowledge sets. During adolescence, mental operations that arose in the previous stages are organized into a more complex system of logic and abstract ideas (Rey, 2006). This is where we work in four important areas.

Propositional logic is the ability to draw a logical inference from the relationship between two statements or premises. In everyday language it can be expressed in a series of hypothetical propositions (García-Peñalvo, 2016).

Scientific reasoning. It formulates hypotheses, determines how to compare them with facts and excludes false ones. Piaget gave the name of hypothetical-deductive thinking to the ability to generate and test hypotheses in a logical and systematic way. deductive hypothetical thinking is the ability to generate and test hypotheses in a logical and systematic way (García-Peñalvo, 2016).

Combinatorial Reasoning. Teenagers can come up with a way to represent all possible combinations, including three and four tokens. They are also more likely to generate combinations in a systematic way.

Reasoning about probabilities and proportions. It requires more abstract forms of reasoning.

According to Piaget (1980), the main features of the operations are:

- Transformations of the symbols in an internalised and reversible form, configuring structures and processing systems (reasoning) (García-Peñalvo, 2016).
- Simultaneous symbolic and internal apprehension of a unique synthesis of a complete series of separate facts.
- Reflection on the organization of one's own acts: contemplative character, not only active.
- Independence beyond the present acts and concrete objects of the real environment: transition to symbolically manipulating non-tangible entities.

Knowledge

Castro, Olmo & Castro (2012) refers that according to Piaget, knowledge is organized into a structured and coherent whole where no single concept can exist. This author considers that there are four factors that influence the development of intelligence:

- Maturation
- Experience with objects
- Social transmission
- Balancing

It explains development in terms of abstraction processes and distinguishes between:

- **Simple Abstraction:** It abstracts what is seen and observed in objects.
- **Reflective Abstraction:** The relationships between objects are abstracted.

It distinguishes three types of knowledge according to Kamii (1981):

1. **Physical:** Physical knowledge is acquired by acting on objects and the discovery of their behavior is produced through the senses.
2. **Social:** Social knowledge is obtained by oral transmission.

3. **Logical-Mathematical:** Logical-mathematical knowledge is constructed by reflective abstraction. In addition, physical and social knowledge have in common the fact that they both need information of origin external to the child, physical knowledge is based on the regularity of the reactions of the objects while social information is arbitrary, originates from agreements and consensus and cannot be logically inferred. These three types of knowledge have in common the demand for activity on the part of the subject in order to achieve it. There are also strong bonds of union between them, so that physical knowledge cannot be built outside a logical-mathematical framework, since one cannot interpret any fact of the external world except through a framework of relationships.

All the actions carried out by an individual have two aspects, one physical and observable in which the attention of the subject is in the specifics of the fact and the other logical-mathematical in which are taken into account, above all, what is general of the action that produced the fact.

Logical-mathematical knowledge, which is what we are now dealing with, has the following characteristics (Luceño, 2012):

- It is not directly teachable
- It is always moving in the same direction and this is towards greater coherence
- Once it's built, you never forget

Mental Representations of Computational Thought Linked to Piaget Theory

The Active Representation

It consists in manipulating objects. Knowledge is represented in actions (Day, C. 2011). This performance is the only one that takes place in young children, and it corresponds to Piaget's sensorimotor stadium (Moreno-León, Robles, & Román, 2016).

Representative Representation

Mental representation is given through the use of materials. appears when the child is able to imagine objects without having to act on them: he is able to replace the action with an image or a spatial scheme. Although it is limited to the perceptual field, it is already a way of representing information and facilitates the execution of certain tasks. This corresponds to Piaget's pre-operational stadium.

The Symbolic Representation

The representations are given through symbolic forms using mathematical language. Appears when the child is able to express his or her experiences in linguistic terms. It corresponds to the thinking of Piaget's concrete operations and formal operations.

When the child begins to internalize, an interpersonal process is transformed into another intrapersonal one. Therefore, in the construction of thought, a function appears twice first on the social, (interpersonal) and then on the personal (intrapersonal) level (Yadav, Mayfield, Zhou, Hambrusch & Korb, 2014). These processes are applied in any situation that the subject performs. Attention, memory, the formulation of concepts are first a social phenomenon and then progressively, they are transformed into a property of the individual (Pinto-Llorente, Casillas-Martín, Cabezas-González & García-Peñalvo, 2017), (Seoane Pardo, 2016).

Everything mentioned above becomes the basis of Computational Thought, they would form the basic or essential elements to develop it later, below is a concrete approximation of the components of Computational Thought.

Taking into account the King's model of competence (2006), we consider computational thinking to be an authentic competence observed in context. We identified six components, or computer thinking skills. These components are as follows:

1. **Analysis and Abstraction:** Make decisions to interpret the situation, for example, with emphasis on one element and less on others.
2. **Organization and Modelling:** Develop a plan of the solution to be developed identifying the actors, objects, processes.
3. **Code**. Choose an appropriate language, know how to read an algorithm, know elements of the digital culture that allow to discuss these concepts.
4. **Technological Systems:** We must take into account the computer and technological systems that will support our software. Do we need a network, a site, or a robotic kit?
5. **Programming:** Once we have defined the language, tool and computer systems that meet the needs of steps 1 and 2, we will create a software program.
6. **Evaluation and Repetitive Improvement:** The final step in the process is a commitment to evaluate the iterative process to improve the program.

The development of the components of computational thinking in learning activities is often achieved in a non-linear manner and by coming and going between them, in a process that maintains close links with collaborative problem solving (Calder, 2010), (Wing, 2006) and the design thinking approach in education (Kafai, 1996).

Figure 2. Six components of computational thinking

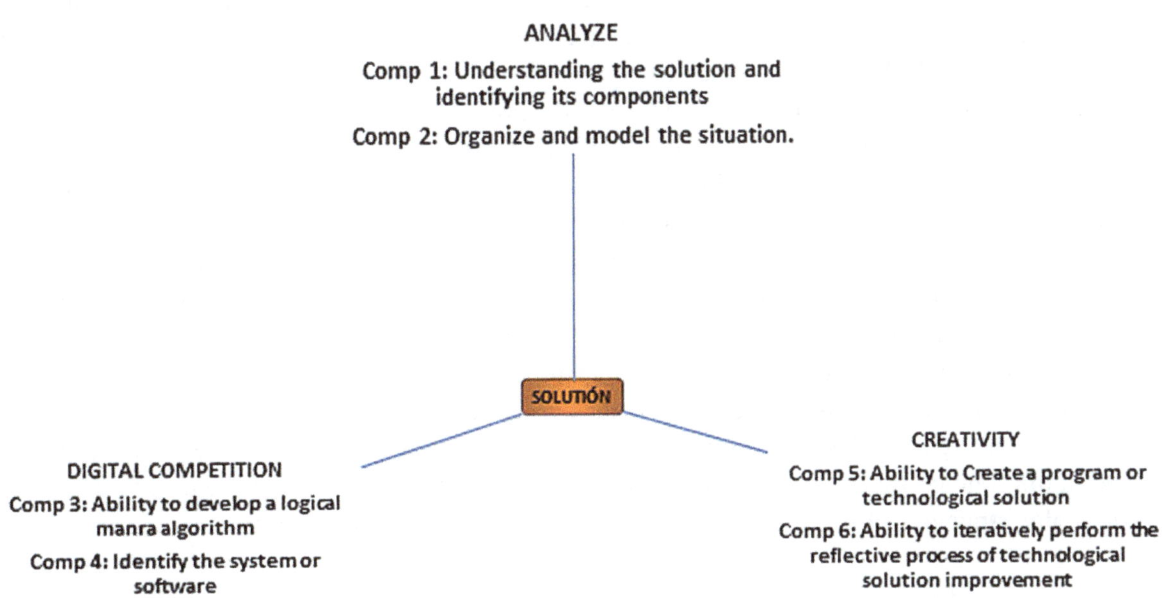

This training development is part of the basic or general training, then we outline the level of deepening already in the specialized training of the teacher in training. These competences can be worked on by incorporating them in the discipline of Mathematics (Wing, 2008), (García, Deco & Collazos, 2017).

About Teaching Mathematics

A subject of constant study and questioning in the teaching of mathematics, lies in how to teach and what to teach in mathematics. Regarding how to teach, the preceding decades show a pedagogical practice in which an instrumental teaching predominates, prioritizing the learning of mathematical concepts without connections to real or immediate contexts to the students'social practice, the application of formulas, the memoristic and algorithmic reproduction of operational processes. In contrast to the practice of an instrumental mathematician and looking for meaningful learning, we are currently promoting the implementation of a relational mathematics, in which the student is seeking to establish relationships between objects from the action on them, for example, the increase of mass in relation to life time.

The distinction of an instrumental and relational mathematician was proposed by Skemp (1978), based on the type of conception that each one reflects, classified relational understanding as knowing what to do and why to do, and instrumental understanding as having rules without a reason.

The contributions of Skemp (1978) that we will present progressively, support the proposal of a teaching based on the solution of mathematical problems, which in essence seeks to develop mathematical competencies and the construction of concepts from situations that imply simulating facts, recording data, organizing them and reaching mathematical models to understand and abstract concepts, which in turn are related to other concepts. In turn, this scenario presents us with different aspects of analysis, such as the complexity of the problem or mathematical situation to be worked on, the steps or stages to reach the solution of the problem, the mental processes expected to develop mathematical abilities that allow the student to look at reality with mathematical eyes, understand, interpret, symbolize and communicate it.

Mathematical Modeling

The Seminar on Formative Assessment in Mathematics (SEAM) defines Mathematical Modeling as the process of constructing a mathematical model to help explain and study a phenomenon. This phenomenon can be natural or social. (Flores & Gómez, 2013).

Hans Freudenthal (1991) on the other hand, from the point of view of MRE (Realistic Mathematical Education) argues that "the model is like an intermediary, often indispensable, through which a complex reality or theory is idealized or simplified in order to make it susceptible to formal mathematical treatment".

It also argues that the models "are representations of problematic situations that necessarily reflect fundamental aspects of concepts and mathematical structures relevant to the problem situation, but that they may have different manifestations. This means that the term model is not taken literally. Materials, visual sketches, paradigmatic situations, diagrams, diagrams and even symbols become models. For example, a paradigmatic situation that functions as a model is the repetitive subtraction.

Mathematical modeling as a strategy represents a potential learning opportunity that also promotes processes of high cognitive demand. Stein (1996), defines them as types of cognitive processes that are involved in the solution of a mathematical problem, from the comprehension of the task to its solution using complex strategies of a "mathematical thought"; therefore, working on mathematical modeling implies the implementation of higher order mental processes, such as differentiating, inferring, relating, arguing, comparing, classifying, etc.

The knowledge generated through the history of humanity, accounts for innumerable mathematical relationships between variables, which constituted mathematical models have served to fundamentally modify our living conditions. This process has not yet finished, on the contrary, we must assume that it is a characteristic of our social praxis, look around us, questioning and investigating the reason for processes, phenomena or facts arriving, mathematicize, explanation and interpretation (D'Abreu, & Condori, 2017).

Stages

1. **Identify the Mathematical Situation:** At this point, it is recommended to present situations of interest to the student, taking into account their context and more specifically their social practice. According to the socio-epistemological approach, (Cantoral & Farfán, 2006); it is assumed that knowledge is the result of adapting theoretical explanations with empirical evidence, ignoring, in some sense, the role that historical, cultural and institutional scenarios play in human activity, placing emphasis on the role of social practices for the construction of knowledge (Lavonen, Lattu, Juuti, & Meisalo, 2006). This premise leads us to think that any situation surrounding the student is possible to be mathematically mathematically, the challenge is to take these scenarios to learning spaces suitable for students.
2. **Raise the Issue:** Based on the aspects that interest us about the situation presented, we pose a problem to be solved by posing questions or challenges.
3. **Mathematicize the Situation Presented:** According to PISA 2006, the fundamental process that students employ to solve real-life problems is called Mathematization.

It can also be understood that mathematicizing is translating a real situation into mathematical language, which would imply, in the first place, understanding the problem. In this process it is important to consider the contribution of Polya (2014), with the proposed steps to solve a problem, a first step is to understand the problem, for this reason it is advisable to ask some questions, such as: What is the mathematical situation presented, what should we solve? do we understand the conditions? Are they enough? What do we need to figure out? What data do we have? What data we don't know?, it is also very important to paraphrase the problem.

4. **Definition of Variables:** The questions posed in the previous stage will allow us to identify attributes that are relevant in the solution of the mathematical situation, thus defining the variables involved.
5. **Design of the Mathematical Model:** The design of the mathematical model requires the implementation of different strategies and students' abilities, as well as the teacher's ability to propose activities that guide the modeling process (Seone, 2016), (Barcelos, & Silveira, 2012). At this stage, it is valuable to consider the algebrization levels proposed by Godino (2014), with the ontosemiotic approach, which emphasizes the role of algebra as a modeling instrument that goes beyond considering it as a "generalized arithmetic" in the modeling process.

Algebraic reasoning involves identifying patterns, regularities, relationships and their generalization, providing new capabilities to analyze solutions, generalize them and justify their scope. It also allows reducing the types of problems and unifying the techniques of solution. (Godino & Wilhelmi, 2014).

- At this stage, the following actions are considered:
- Data recording in a table, from the identification of variables and the simulation of data to find a numerical solution, in some mathematical situations the data recording will correspond to propose quantities per trial and error to satisfy conditions.
- Identification of patterns, regularities and recursive data, from observation and analysis of data recorded in a table.
- Identification of the mathematical relationship.
- Generalization or symbolic representation of the identified mathematical relationship.
- The following is an example of a succession.

Identifying the Mathematical Situation

We put forward a situation of interest for the student, in this case the logical game of jumping frogs, available at: http://acertijosymascosas.com/el-juego-de-las-ranas-juego-de-logica/

Whose rules are:

- Pass the male and female frogs to the opposite side of where they are.
- Keeping in mind that they can only move forward to an empty stone or jump over another frog to the next stone, if they are of different genres.
- At first, students play by the rules and find the solution by trial and error, involving and motivating all students.

Raising the Issue

After finding the solution,they are asked if, in order to reach the solution there is a fixed number of movements? Or does it vary?, They play again several times, identifying that it will always be the same number, in this case, 15 moves. Then we ask another question, what would be the number of moves if we had two frogs on each side? If it were four? Would it increase and decrease the amount of movements? If it was 10 frogs on either side? Could we find a formula or a fast track?

Mathematizing

To translate the situation into symbolic language, we ask: "What should we solve? Does the number of frogs on each side influence the number of movements to find the solution? What is the number of movements dependent on? What data do we have? What data can we symbolize?

Defining Variables

Identify the relevant data, starting from questions about which data are maintained or varied, in this case, the amount of movements varies if the number of frogs on each side is increased or decreased, so they are attributed the term of variables.

Designing the Mathematical Model

From the identification of variables, we symbolize them and organize the information in Table 1 to identify patterns, regularities and relationships between the variables.

Once the data has been recorded, by trial and error it is possible to determine the mathematical relationship that calculates the number of movements given n number of frogs on each side. Alternatively, we can find the mathematical model, giving an algebraic treatment to the data recorded in the table.

For algebraic processing, we start from identifying that since it is a second-degree succession, its general algebraic expression would be Ax^2+Bx+C,, and for this particular case of succession, the expression would be An^2+Bn+C

By replacing the algebraic expression with numerical data, of the first three data, we would have:

$A(1)^2+B(1)+C=3$; $A(2)^2+B(2)+C=8$; $A(3)^2+B(3)+C=15$

Solving the system of three equations, we have to A=1, B=2 and C=0, ratifying the model found n^2+2*n.

The situation presented offers the opportunity to identify a recursive situation, an aspect that is today important in the development of computational thinking.

CONCLUSION

- Computational Thought considers aspects related to mathematical thought.
- Cross-cutting strategies exist to strengthen the development of computational thinking
- There are three components for computational thinking, analysis, creativity and digital competencies.

Table 1.

N° of Frogs on Each Side	N° of Motion	Algorithm (We Relate Numericamante N° of Frogs on Each Side and N° of Movements)
1	3	1^2+2*1
2	8	2^2+2*2
3	15	3^2+2*3
4	24	4^2+2*4
5		
n		n^2+2*n

REFERENCES

Barcelos, T. S., & Silveira, I. F. (2012). Pensamento computacional e educação matemática: Relações para o ensino de computação na educação básica. In *XX Workshop sobre Educação em Computação, Curitiba. Anais do XXXII CSBC* (*Vol. 2*, p. 23). Academic Press.

Calder, N. (2010). Using Scratch: An Integrated Problem-Solving Approach to Mathematical Thinking. *Australian Primary Mathematics Classroom, 15*(4), 9–14.

Cantoral, R., Farfán, R. M., Lezama, J., & Martínez-Sierra, G. (2006). Socioepistemología y representación: algunos ejemplos. *Revista Latinoamericana de Investigación en Matemática Educativa.*

Castro, M. E., Olmo, R. M. A., & Castro, M. E. (2012). *Desarrollo del pensamiento matemàtico infantil.* Granada: Depósito legal: GR-1173-2002.

D'Abreu, J. V. V., & Condori, K. O. V. (2017). Educación y Robótica Educativa. *Revista de Educación a Distancia*, (54).

Day, C. (2011). Computational Thinking Is Becoming One Of The Three Rs. *Computing in Science & Engineering*, *13*(1), 88–88. doi:10.1109/MCSE.2011.5

Flores, Á., & Gómez, A. (2013). *La modelación matemática y la enseñanza de las cónicas*. Academic Press.

Freudenthal, H. (2006). *Revisiting mathematics education: China lectures* (Vol. 9). Springer Science & Business Media.

García, M. A., Deco, C., & Collazos, C. A. (2017). Robotics Based Strategies to Support Computational Thinking: The Case of the Pascual Bravo Industrial Technical Institute. *Journal of Computer Science and Technology*, 17.

García-Peñalvo, F. J. (2016). What Computational Thinking Is. *Journal of Information Technology Research*, *9*(3), v–viii.

García-Peñalvo, F. J. (2016). A brief introduction to TACCLE 3 – Coding European Project. In F. J. García-Peñalvo, & J. A. Mendes (Eds.), *2016 International Symposium on Computers in Education (SIIE 16)*. IEEE. doi:10.1109/SIIE.2016.7751876

García-Peñalvo, F. J. (2016). Proyecto TACCLE3 – Coding. In F. J. García-Peñalvo & J. A. Mendes (Eds.), *XVIII Simposio Internacional de Informática Educativa, SIIE 2016* (pp. 187–189). Salamanca: Ediciones Universidad de Salamanca.

García-Peñalvo, F. J., & Cruz-Benito, J. (2016). Computational thinking in pre-university education. In F. J. García-Peñalvo (Ed.), *Proceedings of the Fourth International Conference on Technological Ecosystems for Enhancing Multiculturality (TEEM'16)* (pp. 13-17). New York, NY: ACM. doi:10.1145/3012430.3012490

García-Peñalvo, F. J., Rees, A. M., Hughes, J., Jormanainen, I., Toivonen, T., & Vermeersch, J. (2016). A survey of resources for introducing coding into schools. In F. J. García-Peñalvo (Ed.), *Proceedings of the Fourth International Conference on Technological Ecosystems for Enhancing Multiculturality (TEEM'16)* (pp. 19-26). New York, NY: ACM. doi:10.1145/3012430.3012491

García-Peñalvo, F. J., Reimann, D., Tuul, M., Rees, A., & Jormanainen, I. (2016). *An overview of the most relevant literature on coding and computational thinking with emphasis on the relevant issues for teachers.* Belgium: TACCLE3 Consortium. doi:10.5281/zenodo.165123

Godino, J. D., Aké, L. P., Gonzato, M., & Wilhelmi, M. R. (2014). Niveles de algebrización de la actividad matemática escolar. Implicaciones para la formación de maestros. *Enseñanza de las ciencias: revista de investigación y experiencias didácticas, 32*(1), 199-219.

Gonzales, M. R. (1998). Psicología educacional de las Matemáticas. *Revista de Investigación en Psicología*, 17.

Grgurina, N., Barendsen, E., Zwaneveld, B., Van Veen, K., & Stoker, I. (2014). *Computational thinking skills in Dutch secondary education: Exploring pedagogical content knowledge.* Paper presented at the ACM International Conference Proceeding Series. doi:10.1145/2674683.2674704

Gross, S., Kim, M., Schlosser, J., Lluch, D., Mohtadi, C., Schneider, D., & Ieee. (2014). Fostering Computational Thinking in Engineering Education Challenges, Examples, and Best Practices. *2014 IEEE Global Engineering Education Conference*, 450-459.

Kafai, Y. B. (1996). Learning design by making games. *Constructionism in Practice: Designing, Thinking and Learning in a Digital World*, 71–96.

Kamii. (1981). *La teoría de Piaget y la educación preescolar.* Madrid: Visor.

Lavonen, J., Lattu, M., Juuti, K., & Meisalo, V. (2006). Strategy-based development of teacher educators' ICT competence through a co-operative staff development project. *European Journal of Teacher Education, 29*(2), 241–265. doi:10.1080/02619760600617433

Li, Y., Liu, Y., & Shu, P. (2015). *Teaching Research and Practice of Blended Leaning Model Based on Computational Thinking Frontiers in Education Conference*. IEEE.

Llorens-Largo, F. (2015). Dicen por ahí … que la nueva alfabetización pasa por la programación. *Re-Visión, 8*(2), 11–14.

Luceño C., J. (2012). *Enseñanza, aprendizaje de las competencias matemáticas.* libros en Red.

Lye, S. Y., & Koh, J. H. L. (2014). Review on teaching and learning of computational thinking through programming: What is next for K-12? *Computers in Human Behavior, 41*, 51–61. doi:10.1016/j.chb.2014.09.012

Moreno-León, J., Robles, G., & Román-González, M. (2016). Comparing Computational Thinking Development Assessment Scores with Software Complexity Metrics. *Proceedings of the IEEE Global Engineering Education Conference*.

Piaget, J., Beth, E. W., & de Zavala, V. S. (1980). *Epistemología matemática y psicología: relaciones entre la lógica formal y el pensamiento real.* Editorial Critica.

Pinto-Llorente, A. M., Casillas-Martín, S., Cabezas-González, M., & García-Peñalvo, F. J. (2017). Building, coding and programming 3D models via a visual programming environment. *Quality & Quantity*. doi:10.1007/s11135-017-0509-4

Polya, G. (2014). *How to solve it: A new aspect of mathematical method.* Princeton university press.

Rey, B. (2006). *Les compétences à l'école: apprentissage et évaluation* (2nd ed.). De Boeck. Retrieved from http://ariane.ulaval.ca/cgi-bin/recherche.cgi?qu=i2804150046

Rojas López, A., & García-Peñalvo, F. J. (2016). Relationship of knowledge to learn in programming methodology and evaluation of computational thinking. In F. J. García-Peñalvo (Ed.), *Proceedings of the Fourth International Conference on Technological Ecosystems for Enhancing Multiculturality (TEEM'16)* (pp. 73-77). New York, NY: ACM. doi:10.1145/3012430.3012499

Román-González, M. (2015). Computational Thinking Test: Design guidelines and content validation. *Proceedings of EDULEARN15 Conference*, 2436-2444.

Román-González, M., Pérez-González, J.-C., & Jiménez-Fernández, C. (2017). Which cognitive abilities underlie computational thinking? Criterion validity of the Computational Thinking Test. *Computers in Human Behavior, 72*, 678–691. doi:10.1016/j.chb.2016.08.047

Romero, M., Usart, M., & Ott, M. (2015). Can serious games contribute to developing and sustaining 21st century skills? *Games and Culture, 10*(2), 148–177. doi:10.1177/1555412014548919

Seoane Pardo, A. M. (2016). Computational thinking beyond STEM: an introduction to "moral machines" and programming decision making in Ethics classroom. In F. J. García-Peñalvo (Ed.), *Proceedings of the Fourth International Conference on Technological Ecosystems for Enhancing Multiculturality (TEEM'16)* (pp. 37-44). New York, NY: ACM. doi:10.1145/3012430.3012494

Skemp, R. (1978). Relational understanding and instrumental understanding. *Arithmetic Teacher, 26*, 9-15.

Snalune, P. (2015). The Benefits of Computational Thinking. *ITNOW, 57*(4), 58–59. doi:10.1093/itnow/bwv111

Wing, J. M. (2006). Computational Thinking. *Communications of the ACM, 49*(3), 33–35. doi:10.1145/1118178.1118215

Wing, J. M. (2008). Computational thinking and thinking about computing. *Philosophical Transactions of the Royal Society a-Mathematical Physical and Engineering Sciences, 366*(1881), 3717-3725. doi:10.1098/rsta.2008.0118

Yadav, A., Mayfield, C., Zhou, N. E., Hambrusch, S., & Korb, J. T. (2014). Computational Thinking in Elementary and Secondary Teacher Education. *Acm Transactions on Computing Education, 14*(1), 1–16. doi:10.1145/2576872

Zapata-Ros, M. (2015). Pensamiento computacional: Una nueva alfabetización digital. *Revista de Educación a Distancia*, (46).

Section 2

Educational and Learning Technologies

Chapter 5
Addressing Knowledge Management and Virtual Learning Communities in MOOC Using Open Resources and Gamification

Luis Martin-Fernandez
Universidad Politecnica de Madrid, Spain

Margarita Martinez-Nuñez
Universidad Politecnica de Madrid, Spain

Oriol Borras-Gene
Universidad Politecnica de Madrid, Spain

Angel Fidalgo-Blanco
Universidad Politecnica de Madrid, Spain

ABSTRACT

The confluence of thousands of students in a MOOC is an opportunity to manage all the knowledge generated through the creation of open educational resources (OER), especially when a connectivist approach is applied and the MOOC makes use of virtual learning communities. The challenge is transferring the flow of knowledge, activity, and interactions of the course to the community and making that transference sustainable and ongoing over time. For this purpose, the use of elements of gamification to train and retain the knowledge creators of the community along with the use of social networking platforms is proposed. This chapter analyses several editions of a MOOC and the opportunity offered by the use of different types of learning (formal, non-formal, and informal) that occur in them, thus characterizing patterns to train the open content and knowledge generation through gamification. From the results, indicators for managing successful and sustainable knowledge communities are proposed along with indicators for persistence and interaction between participants.

DOI: 10.4018/978-1-5225-4944-4.ch005

1. INTRODUCTION

The emergence of MOOC (Massive Open Online Courses) had an impact on the educational world, with very different positions. These are free online courses with open access, a publicly shared curriculum and open academic results (McAuley, Stewart, Siemens, & Cormier, 2010). Among these positions, there are two streams, conductist approach (xMOOCs) with a behavioural methodology and like traditional online courses approach, and the conectivist approach (cMOOCs) (Fidalgo, Sein-Echaluce, Borrás, & García, 2014) based on virtual learning communities (VLC), this vision becomes an opportunity to take advantage of the collective intelligence of all participants (Borrás-Gené, Martinez-Nuñez & Fidalgo-Blanco, 2015) in a collaborative way (Martínez-Núñez, Borrás-Gené & Fidalgo-Blanco, 2014). According to Downes (2012), Connectivism is "the thesis that knowledge is expanded through the connections and where learning is the ability to build that knowledge for yourself."

In general, MOOC platforms offer a number of technical limitations to apply different models or pedagogical elements, especially in cMOOC (Borrás-Gené et al., 2015). One of the most common problems found in the world of e-learning and amplified in MOOC is this lack of interaction between students because there isn't a physical space or classroom which favours relationships to motivate and help each other (Huang, Dasgupta, Ghosh, Manning & Sanders, 2014). However, although one of the main features of MOOC is that are open, its contents are really only accessible to registered users (Borrás-Gené, Martínez-Núñez & Blázquez-Sevilla, 2016). It is necessary to choose external solutions such as VLC that allow access at any time regardless of being enrolled in the course or on specific dates.

MOOCs offer new opportunities for learning because of their intrinsic characteristics: the massiveness of participants, peer-to-peer interactions, free-of-charge, openness and scalability that lead to a large heterogeneity of participants.(Martínez-Núñez et al, 2016) The incorporation of virtual learning communities (VLC) may provide greater interaction between participants, support and guidance to people with difficulties and may increase collaborative processes between participants (García-Peñalvo, Cruz-Benito, Borrás-Gené and Fidalgo Blanco, 2015). Enhancing interactions in MOOCs could mean more knowledge generated and shared

Knowledge management processes must be able to support the transfer of knowledge (García-Holgado et al, 2015) that occurs in MOOCs. Virtual learning communities (VLC) may become as an essential tool for managing knowledge in Technology Enhanced Learning Systems. There is a clear need to design MOOCs that allow the creation of learning communities that complement the contents of the course, providing new resources and making this community last even after the course ends. (Fidalgo Blanco, Sein-Echaluce Lacleta & García-Peñalvo, 2013). MOOCs make available to many users (thousands in many cases) virtual classes as knowledge containers that provide open learning resources for all users enrolled in the course, as well as methods and systems to reinforce that knowledge acquisition (Cruz-Benito et al, 2017). A MOOC is not only another online course; it has different and specific features: its teachings must be completed with knowledge management, informal and social learning and lifelong learning principles (Borras-Gene at al, 2016)

The completion rate of a MOOC is one of the most widely indicators used to measure the success of a MOOC (Breslow, Pritchard, DeBoer, Stump & Seaton, 2013; Fidalgo Blanco, Sein-Echaluce Lacleta & García-Peñalvo, 2015). While in the xMOOC the specific event in time can be the most representative success indicator, in the case of cMOOC developed in different editions as the one analysed in this work, new items need to be taken into account in order to consider a xMOOC as successful. The identification of those most active and expert participants that will be defined as group leaders and may become facilitators of the VLC is one of the key factors for the MOOC success. (Martínez-Núñez et al, 2016)

This chapter presents the incorporation of VLC in a MOOC and its sustainability over time as a new element to study, which leads to new possibilities for improving success rate indicators. Some new elements from the VLCs as open contents generation or the analysis of the types of non-formal and informal learning can be obtained to take into account when understanding the development of MOOC courses from a knowledge management perspective. In addition, VLCs allow the inclusion of gamification activities during the MOOC and in the period between different editions of the same MOOC.

This chapter aims to go one step further of the MOOC models conducted so far and seeks how to link the contents and the type of knowledge and learning generated over time with the perception of participants about motivation, utility, commitment and collaboration over time. Finally, the evolution of the different resources that are incorporated to the MOOC in order to cause an improvement of the course and a higher level of commitment and usefulness of the participants is also presented. This chapter also investigates to which extent the incorporation of gamified elements for generating open resources is key to enhance the sustainability of the VLC associated with these elements.

In the next section the theoretical principles on which the proposal is based will be reviewed. The methodology section explains how the assessment is applied in the different editions of a MOOC. The results are shown in the next section, finishing with a discussion of these results and a conclusions section.

2. THEORETICAL BACKGROUND

The main pedagogical principle behind a MOOC proposal should be that participants would be able to create new knowledge in a social and collaborative way, allowing that knowledge may be openly used both to improve the MOOC itself and to give continuity to the MOOC learning community. .(Martínez-Núñez et al, 2016)

2.1. Open Educational Resources and MOOC

Closely related to movements such as free software or open source, within the world of computing where resources are shared and teamwork is encouraged, arise another movement with a focus on the educational world, known as Open Educational Resources (OER) (UNESCO, 2012). These OER seek the involvement of students in the development and reuse of open content, the innovation in processes and people, the creation of learning communities (Liddo, Buckingham Shum, McAndrew & Farrow, 2012) and the exchange of educational resources for reuse and sustainability of these resources (Fidalgo, Lacleta, Borras & Peñalvo, 2014).

Other initiatives that have taken a step further in the use and consolidation of knowledge are open MOOC (Downes, 2008; Markoff, 2011). The development of initiatives such as OpenCourseWare can be considered as the first OER initiative that offered open courses (Abelson, 2008). Since its appearance in 2011, MOOC platforms like Miriada X, Edx, Coursera or Udacity have been constantly evolving and adapting, incorporating new features, but with restrictions on design and management (Martínez-Núñez et al., 2014), that sometimes hinder teachers to take advantage of the benefits offered by the massiveness or to include new elements adapted to the educational needs for the proposed instructional design (Fidalgo Blanco et al., 2015).

2.2. Formal, Non-Formal, and Informal Learning

It is possible to identify three types of learning (Eshach, 2008; García-Peñalvo, Cruz-Benito, Borrás-Gené & Fidalgo-Blanco, 2015), formal learning is described as learning that occurs in an organized and structured environment (in an education or training institution or on-the-job), it typically leads to validation and certification. The non-formal learning is included in planned activities related to formal learning, and can be validated and even lead to certification. Informal learning resulting from daily activities related to job, family or leisure; starts from the learners (Cross, 2011) and is not organized or structured in terms of objectives, time or learning support. This informal learning accounts for 80% of an individual's knowledge in job (Dabbagh & Kitsantas, 2012).

The informal learning emerges in MOOC from connections between students such a spontaneous way against non-formal that is proposed by teachers to start a discussion or conversation out of the MOOC (García-Peñalvo et al., 2015). This kind of learning out of the MOOC (non-formal and informal) can influence the results and achievements of the students within the MOOC (Fidalgo Blanco, Lacleta & García Peñalvo, 2013). The conversations and interactions in the social networks could reveal some aspects and results of learning activities, offline activities, etc. (Aramo-Immonen, Jussila, & Huhtamäki, 2014). In this type of learning, the student chooses the way to acquire the knowledge, the learning is given everywhere and applies to any situations in common life (Eshach, 2007) and is not evaluated.

2.3. Gamification, Social Networks, and Learning Communities

Gamification is understood as the use of mechanical elements and game design techniques within contexts that are not games, looking to motivate, involve users or change behavior (Zichermann & Cunningham, 2011; Werbach & Hunter, 2012; Deterding, Dixon, Khaled, & Nacke, 2011).

Related to motivation and from a behaviorist point of view, "extrinsic motivation" can be seen as based on material aspects as money, punctuation or degree of completion, associated with rewards that reinforce behaviors (Groh, 2012). On the other hand, there are cognitive theories which highlights the self-determination theory (Deci E. & Ryan, 1985), according to which individuals do not need rewards to motivate, actions are rewarding and fun for themselves, it speaks of "intrinsic motivation". Both types of motivation (intrinsic and extrinsic) could be related to gamification (Hamari et al, 2014, Hanus and Fox, 2015, Muntean, 2011, Seaborn and Fels, 2015)

There are a number of recurring mechanisms in the literature (Werbach & Hunter, 2012; Zichermann & Cunningham, 2011; Vaibhav & Gupta, 2014) directly related to game design, such as points, badges, leaderboards, levels, challenges, competition or cooperation, feedback, time pressure, etc.

Social media channels are one of the main ways to generate loyalty, engagement or commitment (Tardón & de la Vega, 2014, Laroche et al, 2012, Paine 2011). Social networks as Twitter, Google+ or Facebook can accommodate learning communities, given that are focused on interaction and on building relationships, which are essential in models based on connectivism (Martínez, Borrás & Fidalgo, 2014). Leveraging its gamified tool and adding other external tool, it is possible to achieve higher levels of loyalty from students in the community and therefore greater commitment (Borrás-Gené et al, 2015).

2.4. Knowledge Management

Nonaka and Takeuchi (1995) claim that the knowledge creation begins with socialization, which is described as a phase that begins with individuals sharing their knowledge and experiences with other individuals in their group. Knowledge management depends on several core competencies, including the acquisition of knowledge and storage, knowledge application, knowledge sharing, and knowledge creation (Alavi & Leidner, 2001). Llorens and Capdeferro (2011) showed that social networks promote informal learning, in turn enabling knowledge construction and skills development.

The Virtual Learning Community of a MOOC as the space where participants share common values and interests, and that are connect via different communication tools that such networks offer, whether synchronous or asynchronous (Cabrero, 2006). These communities offer new learning values from the traditional learning. The VLC is characterized by the exchange of information in different formats, and by the generation and construction of new knowledge; usually flexible in time and a multidirectional communication, ranging from one to one or one to all is done.

Little is known about the structure of knowledge exchange through forums in online courses although it is known that only a small fraction of registered participants on a large MOOC or discussion forums are really engaged over to complete course, completing all course activities (Clow, 2013)

The operation of these learning communities is related to the behavior of the participants of it. The sustainability of the same time be an indicator of success associated MOOC. The generation of knowledge and the ability to share content can define the level of viability of the same. To this end, we must seek what level of persistence and interaction between individuals of the community, at the same time to know what their activity and the perception of their ability to participate as motivation in it.

According to Cabrero (2007) "... learn in VLC is a group learning, and learning in a collaborative and non-competitive environment, where all community members contribute with their knowledge to achieve common goals, which can range from solving a problem or project, to the development of a single activity". These communities are sources of important knowledge management and represent an important model of social learning (Cruz-Benito et al, 2015a).

3. METHODOLOGY

3.1. MOOC Context

In November 2013 the Miriada X platform published the first edition of "Application of social networks to teaching" MOOC (RSEMOOC). This course aimed to improve the social networks skills of teachers of all educational levels (Martínez Núñez et al., 2014), establishing a framework for teachers to open the classroom to social networks. Three editions of this MOOC have been launched in 2013, 2014 and 2015.

The "Application of social networks to teaching" MOOC consists of 5 modules focused on teaching and on the application of Facebook and Twitter in education along with other more specific social networks. Each module has a number of short videos accompanied by texts with additional information of links, exercises and discussions. The course evaluation is based on a self-assessment questionnaire for each module and a final evaluation activity peer (P2P).

The MOOC follows a cooperative model that combines the features of xMOOC and cMOOC (Fidalgo Blanco et al., 2013), which since the second edition was added a layer based on elements of gamification, called model gcMOOC (Borrás-Gené et al., 2016). The main differences between editions are the incorporation of new gamification elements and other mechanics that look for the improvement of student engagement and motivation both in MOOC and in the associated learning community.

3.2. Learning Community

There is a virtual learning community, using Google+, as workspace and common thread binding among editions, in which all connectivist part of the MOOC takes place. In this learning community, students can interact, present, ask or answer questions, post the results of the activities proposed in the MOOC and share content that generate or are in the network. The Google + community is where both holistic and empirical knowledge generated during the MOOC is organized.

The community has 8 post categories (Member presentations, announcements, discussions, doubts, activities and exercises, resources, various and contests). There were 3 most used categories (activities, exercises and resources) whose publications give rise to useful resources for the community and the course

Within the MOOC they have been proposed in different editions, a number of predefined hashtags and associated with an oriented and non-formal learning, were used to characterize specific publications (questions, debates, etc.) or proposed exercises; compared with more informal proposed by students. Table 1 summarizes the number of publications, replies, and members. From this information the community can measure the persistence and interaction thereof.

3.3. Motivational Activities

Since the second edition of MOOC activities engagement associated with the videos were incorporated, which consisted of voluntary exercises of practical application of the lesson, using a specific hashtag and within the category "Activities and exercises", students could share in the community, showing their results and becoming examples or resources. In addition, the students shared informal, spontaneous and voluntary assessment activities since the first edition: P2P-end-activity and also in P2P-activity 1 (only first edition).

From the third edition gamified activities were added in the form of challenges (modules 2 and 3) sent by e-mail without notice. These challenges were a total of 6 and were to perform some kind of task from MOOC, earning badges as a reward for the first ones. Highlight gamification elements such as time pressure, points, rankings or badges; using elements of Google+ and external tools (e.g. OpenBadge-Factory). In this edition, for publication of the final activity students will be recommended a specific hashtag (#RSEActividadFinal) to recover the resources.

Table 1. Community information

	Google + Community			
	New Members	**New Publications**	**Replies**	**Publications (Total)**
1st edition	1181	1199	460	1659
2nd edition	2646	2106	745	2851
3rd edition	2392	2168	523	2691

Figure 1 summarizes the interactions between students and community through motivational activities along MOOC editions. In the "Activities and exercises" living informal resources as a result of P2P (1st and 2nd edition) and non-formal from exercises and challenges activities.

In Figure 2, can be seen in more detail both the activities of motivation as specific hashtags used on different issues and that kind of learning was associated with the resources generated in the community.

3.4. Data Analysis

To measure the sustainability of community-based method is proposed two characteristics: interaction and persistence. These indicators can be measured by analysing the behaviour of participants and generated activities, for example, the number of open resources in the form of publication, and if they are shared with others, read and followed.

Applying own concepts of learning analytics (Tabaa & Medouri, 2013) and analysis of learning communities (Rovai, 2002), persistence indicates the stability of the user concentration in a course in temporal terms. Although, it is a complex phenomenon, it is possible to measure the persistence of students through an important indicator: the number of materials generated and viewed. Moreover, the indicator of the interaction is used to measure the interaction of students in the community. The participant-participant

Figure 1. Students' interactions from different MOOC editions to learning community

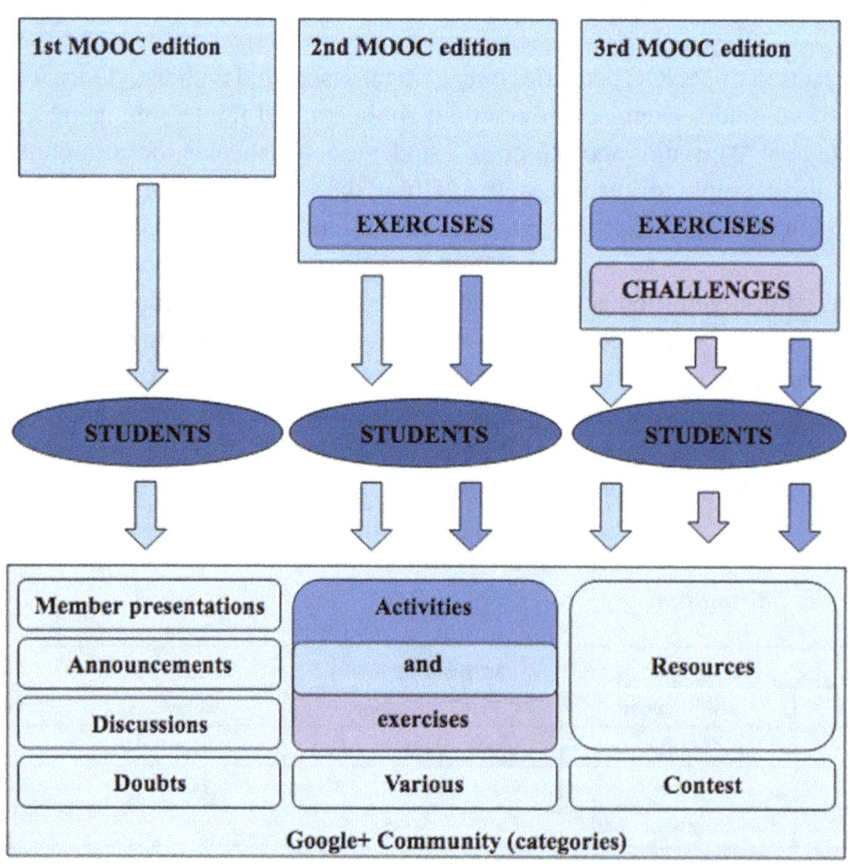

Figure 2. Set of non-formal (blue arrows) and informal (brown arrows) actions in MOOC towards learning community

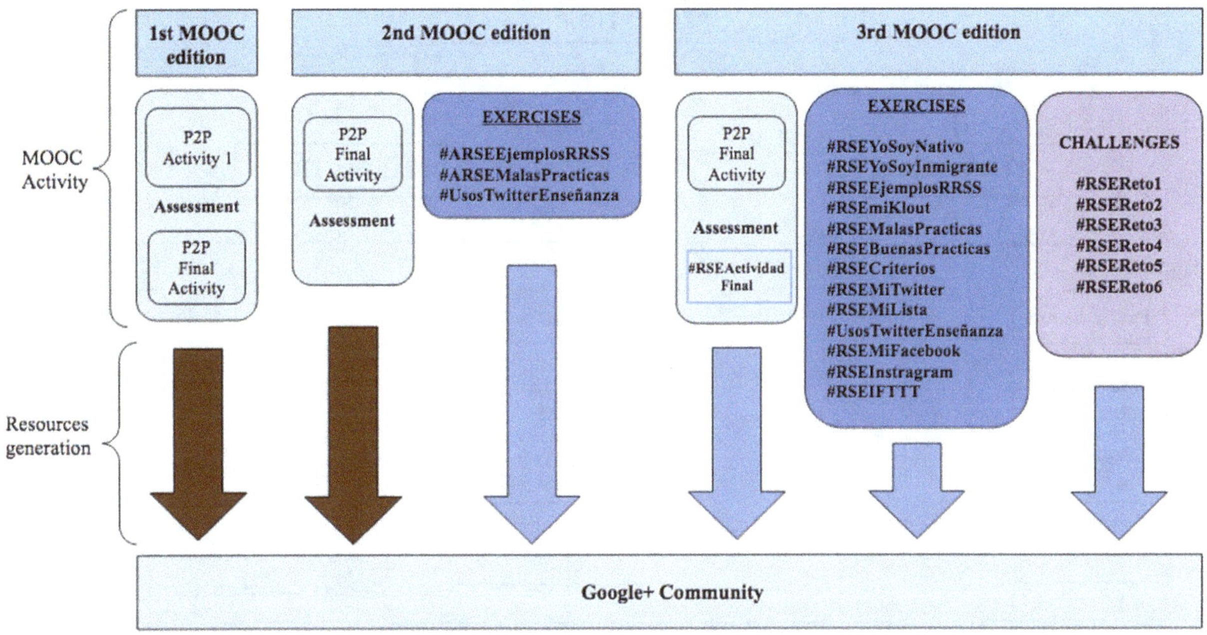

specific relationship is carried out based on two assumptions: (1) use of +1 and comments to other users indicate more interest than he/she has in the community publications and (2) use of reshare assumes quality publications and giving greater visibility to the authors of these.

Given the level of participation and the degree of motivation, it is not only important which is the extent of this activity but also the knowledge of what is the perception of the participants in the persistence and the ability to interact.

For the analysis of Google+ activity used two web tools, for AllMyPlus side for hashtags in the community and interactions; on the other hand the development tool itself GILCA (Borrás-Gené et al., 2015) for the development of the community. With these tools have been able to obtain data related publications, such as the evolution over time, interaction between users from these (+ 1s, comments and reshares), hashtags analysis used and the type of content or resources associated with the purpose of characterizing the community associated with the MOOC.

To obtain the resources generated by students informally, without being associated with specific hashtags, it was necessary to study the community manually in the categories "Resources" and "Activities and exercises".

As for the perception of the participants, it has worked with the questionnaire motivation and participation, through a survey at the end of the course partly based on the SEEQ (Marsh, 1982), adding new questions focused on Google+ community and activities added, as are the challenges. Figure 3 shows the determination of learning behaviour perception, as specified learning community.

Figure 3. Learning behaviour perception

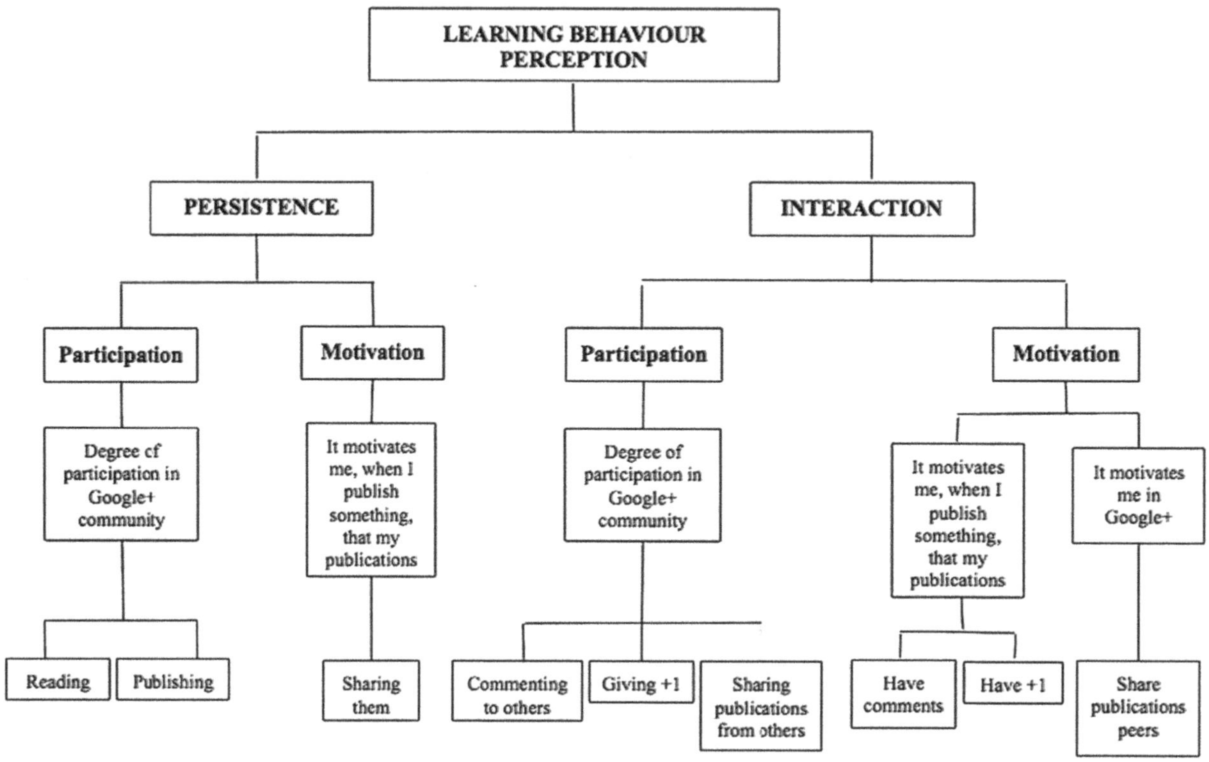

4. RESULTS AND DISCUSSION

4.1. Open Educational Resources Generated

Starting with the first objective of this article, the results obtained in the course community in terms of both revenue generation from non-formal activities and informal initiatives are offered Figure 4 shows the number of publications with non-formal (proposed by the teaching staff) and informal hashtags, chosen by members of the community, along three editions of the MOOC taking into account the periods between editions. It is found that the number of publications has increased and following the community alive even among editions.

Motivational activities, proposals from the second edition, leading to alternation in such publications between non-formal or informal depending on whether the hashtags were associated with MOOC or activities were proposed by students. This last type of knowledge generated by students who provide one of the highest values of the community to become genuine learning resources (Sein-Echaluce, Fidalgo-Blanco & García-Peñalvo, 2016), allowing reuse by either being in open community.

4.1.1 Resources From Non-Formal Activities

Focusing view in activities led by the teaching team and proposals from the MOOC, Table 2 shows the publications and interactions in community related to hashtags (non-formal) for each exercise (second

Figure 4. Switching between resource generation induced by non-formal and informal activities

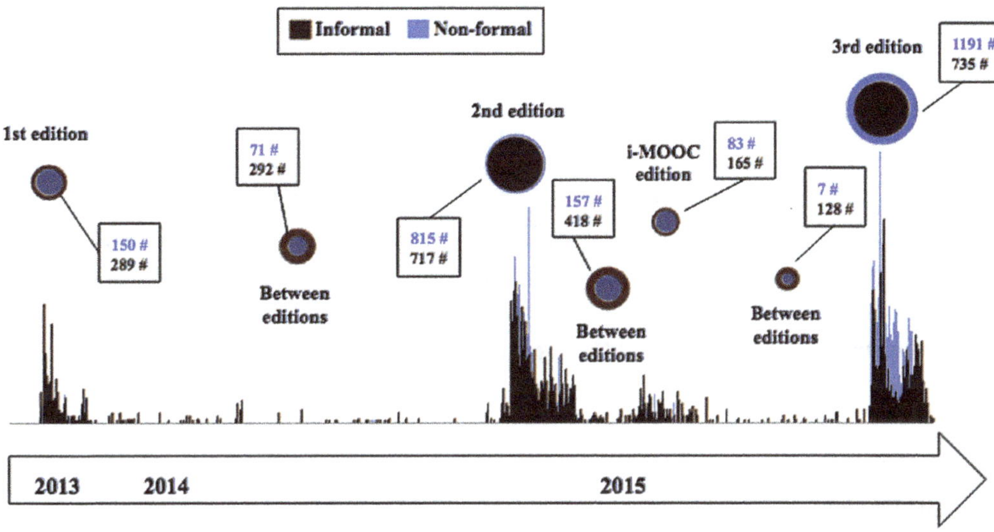

Table 2. Hashtags from exercises analysis

	Stadistic	Views	Publications	+1s	Comments	Reshares
2nd edition	Min	3132	148	272	28	7
	Max	4251	288	701	56	19
	Mean	3770,33	240,00	446,00	38,33	11,67
	SD	575,920	79,699	225,679	15,373	6,429
3rd edition	Min	1635,0	80,0	184,0	13,0	1,0
	Max	3347,0	325,0	481,0	50,0	26,0
	Mean	2721,833	204,833	286,333	29,000	7,167
	SD	537,5473	64,7777	82,5583	11,4256	7,3957
	Final Activity					
	Total	744	136	615	156	13

and third edition), figure 2 also shows the different hashtags used by exercise. “Views” column shows the number of unique users who have seen the video corresponding to the lesson where each exercise is proposed, in the case of the final activity refers to students who successfully completed this. If the number of students who have seen videos associated with exercises with the number of publications using each specific hashtag on average is compared there is an increase of 6.2% to 8.3% between the second and third edition in terms of publications in the community regarding viewings unique, regardless of the final evaluation activity which represented 18.2% of publications.

The use of metadata, such as hashtags in the learning process provides greater interoperability between different systems used to automate the process of knowledge creation and structure and unify educational resources to finally share and reuse content created (Ghenname et al., 2013). The following graphs (Figure 5) show, for publications containing hashtags (non-Formal) proposed, those of non-

Figure 5. Relationship between total hashtag publications, non-formal and which had a wrong hashtag

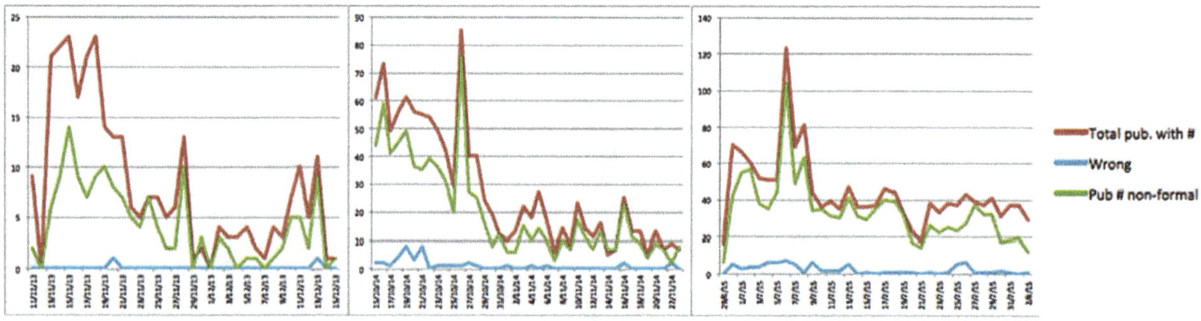

(a) 1st MOOC edition (b) 2nd MOOC edition (c) 3rd MOOC edition

formally misspelled and publications associated with these hashtags. As activities increase and spread over time greater persistence seen in publications, once past the initial classic peak observed in MOOC and saved the initial drop-outs course (Fidalgo Blanco et al., 2015). The Google+ network has created a differentiated space for the course community by using the Communities tool; thanks to this differentiation can be easily recovered the conversations even when new hashtags appear (not defined in the course) or they are not tagged.

The result in terms of publications and interactions for the challenges proposed in the third edition of MOOC can be seen in Table 3. Of the six proposed, the first three only used Twitter and not the community so they are not the subject of this study and they have only been analysed the following three using Google+. Great interaction between course participants, without forgetting that ran in among the second and third week, so we must keep in mind that students drop is already important is checked.

4.1.2. Resources From Informal Interaction

Related to those actions associated with a more informal learning, Figure 6 shows the result of analysing the category "Resources" in the Google+ community, where each publication coincides with an appeal brought by a student and provides information what sort of content. This analysis extends to each of the editions and periods between them.

As for informal learning, there are two categories that have generated most community resources. In "Resources" any member could upload content found on the web, highlighting the variety of these (Figure 6) and especially the continuity in the periods between editions noting numerous interactions between members. In the case of evaluation activities "Activities and exercises" (Table 4) were rises

Table 3. Data related to the challenges of the 3rd edition

Challenges (Google+)				
Hashtag	**Publications**	**+1s**	**Comments**	**Reshares**
#RSEReto4	120	395	33	11
#RSEReto5	89	155	31	4
#RSEReto6	68	219	40	12

Figure 6. Analysis of the resources generated in learning community

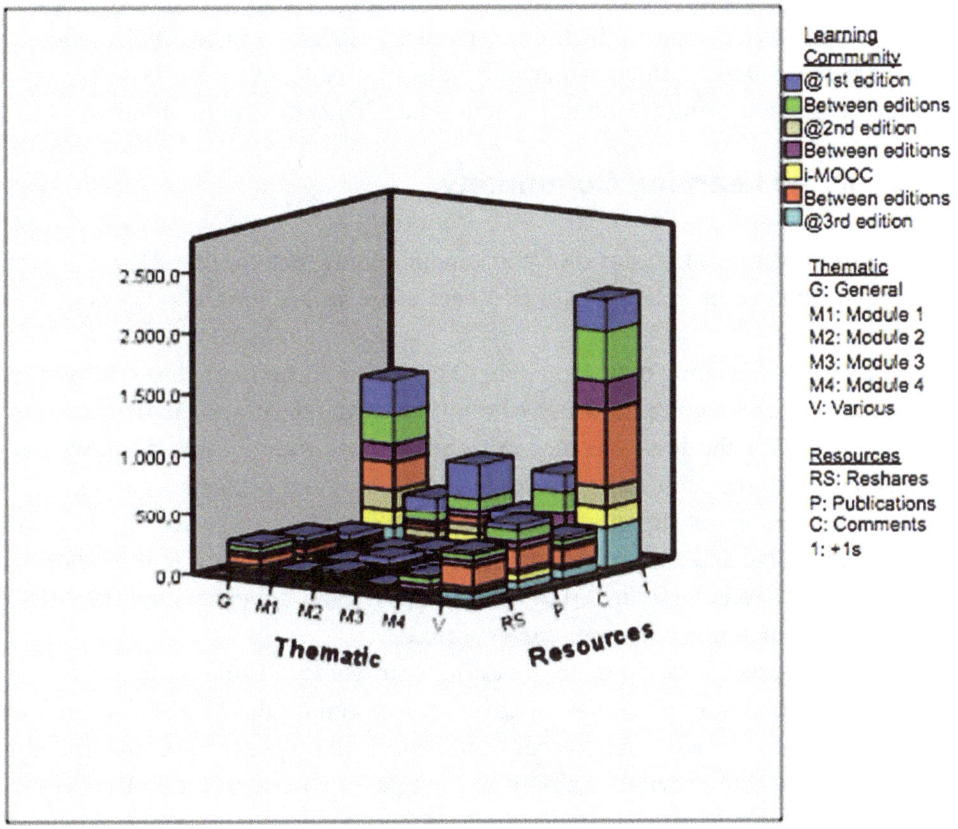

spontaneously and altruistically by students during the first two editions assuming resources of great interest to the other fellow, as they are complete works uploaded in open.

Particularly, according to the results, there were some significant informal conversations in the community started by the students, which is endorsed to provide new content to other students. This can be identified as a result of informal activity initiated or even created by the mechanics of the MOOC course. It is possible to detect, for example, how informal activities are revealed as publications of students, and how it leads to a appreciable interaction of other students through indicators of approval such as "+ 1s" with more than 875, "Comments" or "Publications" with more than 495, throughout the several course editions associated with those activities or informal comments associated.

Table 4. Publications and interactions result of evaluation activities

MOOC Edition		Publications	+1s	Comments	Reshares
1st	P2P-activity 1	99	366	129	2
	P2P-final-activity	69	284	61	6
2nd	P2P-final-activity	57	225	80	25

To characterize the course participants and what their perception to generate open resources against the activities introduced in recent editions, such as gamification approach, gives us information to improve the design of them. Proactivity seems to be the key element especially in cMOOC, where the development of content That can lead to continuous learning and sustainable over time, based on casual features, imitating That can be given in a job (Dabbagh & Kitsantas, 2012) or in school hallways with peers.

4.2 Perception in the Learning Community

Knowing how students perceived their level of participation and what was their level of motivation will indicate as much as if the course activities studied here are working for the success of the course and the learning community.

At the time to know what is being operation of activities introduced in the last edition seems interesting to know whether there is a feature for users who influence the development of them and the course, taking them into account for the improvement in training those users. Table 5 shows that the age of the participants is related to the achievement of both certificates and performing to gamified activities. People from 26 to 35 years are at the greatest motivation. The challenges activity is also linked with the overcoming of the course and the perception of proactivity of the user. On the other hand, work in learning community seems to be only linked to the proactivity and not exceeding the MOOC or in this case teaching profile to which it is addressed, as expected.

If we delve into the characterization of the learning community (Table 6) shows that both interaction and persistence have a similar perception, except in overcoming the MOOC, where there seems to be related to reading content, or with share other publications that appear to be specific to community. Finally, it should be noted that previous training in Google+ or performance of previous editions of

Table 5. Main results of the application of statistical test for MOOC and motivation in generating resources open

		MOOC Characterization and Activities			
		Certificate	**Interaction**	**Challenges**	**Google+ Community**
Characterization of Participants	**Gender**	*NO*	*NO*	*NO*	*NO*
		0,183	0,953	0,156	0,082
	MOOC Succeed	*YES*	*YES*	*YES*	*NO*
		0,000	0,008	0,008	0,974
	Age	*YES*	*NO*	*YES*	*NO*
		0,007	0,473	0,004	0,726
	Teacher	*NO*	*YES*	*NO*	*NO*
		0,128	0,000	0,383	0,726
	Perception of Proactivity in MOOC	*YES*	*YES*	*YES*	*YES*
		0,000	0,000	0,000	0,000

NO: There is no functional relationship; YES: There is a functional relationship; Number in parentheses corresponds to critical level of Chi-square test

Table 6. Key outcomes of the application of statistical test in the learning community

<table>
<tr><th rowspan="4" colspan="2"></th><th colspan="9">Perception Learning Behaviour</th></tr>
<tr><th colspan="3">Persistence</th><th colspan="6">Interaction</th></tr>
<tr><th>[Reading]</th><th colspan="2">[Publishing]</th><th colspan="2">[Comment]</th><th colspan="2">[+1]</th><th colspan="2">[Sharing Publication From Others]</th></tr>
<tr><th>Partic.</th><th>Motiv.</th><th>Partic.</th><th>Motiv.</th><th>Partic.</th><th>Motiv.</th><th>Partic.</th><th>Motiv.</th><th>Partic.</th></tr>
<tr><td rowspan="8">Training Learning Community</td><td rowspan="2">Prior Knowledge [Google+]</td><td>NO</td><td>NO</td><td>NO</td><td>NO</td><td>NO</td><td>NO</td><td>NO</td><td>NO</td><td>NO</td></tr>
<tr><td>0,384</td><td>0,505</td><td>0,055</td><td>0,956</td><td>0,067</td><td>0,623</td><td>0,520</td><td>0,143</td><td>0,199</td></tr>
<tr><td rowspan="2">Participation in Other Editions</td><td>NO</td><td>NO</td><td>NO</td><td>NO</td><td>NO</td><td>NO</td><td>NO</td><td>NO</td><td>NO</td></tr>
<tr><td>0,335</td><td>0,636</td><td>0,164</td><td>0,692</td><td>0,692</td><td>0,810</td><td>0,163</td><td>0,708</td><td>0,279</td></tr>
<tr><td rowspan="2">Familiarization With Hashtags</td><td>YES</td><td>YES</td><td>YES</td><td>YES</td><td>YES</td><td>YES</td><td>YES</td><td>YES</td><td>YES</td></tr>
<tr><td>0,000</td><td>0,000</td><td>0,000</td><td>0,000</td><td>0,000</td><td>0,000</td><td>0,000</td><td>0,000</td><td>0,000</td></tr>
<tr><td rowspan="2">MOOC Succeed</td><td>NO</td><td>NO</td><td>YES</td><td>YES</td><td>YES</td><td>YES</td><td>YES</td><td>YES</td><td>NO</td></tr>
<tr><td>0,058</td><td>0,059</td><td>0,001</td><td>0,006</td><td>0,006</td><td>0,008</td><td>0,004</td><td>0,041</td><td>0,069</td></tr>
</table>

NO: There is no functional relationship; YES: There is a functional relationship; Number in parentheses corresponds to critical level of Chi-square test

MOOC not imply any upgrading in perception of participants for work in the community relationship, but improve familiarity with the work of hashtags itself, which it is important for the subsequent editions of the course.

The activities discussed here generate scaffolding, which maintains and engages the student in the course (Tang & Kay, 2014), using elements of gamification as challenges promoting competitiveness and rewards in form of badges. It has been found that identification with this type of resource is greater in the young participants and actively use is linked to overcoming MOOC. Elements such as +1s (Google+), which promote feelings of engagement, reputation or visibility and cooperation, are linked to the capacity of interaction in the community, but not with the overcoming of course but proactivity participant. It seems essential that in order to have links with it, the user is familiar with hashtags. While it was known until now that emotional comfort and the usability features of social media are the main motivators for using social media in education (Mao, 2014) a good learning in the use of hashtags increases digital skills or even care when choosing a correct category to post (García-Peñalvo et al., 2015) which would solve another of the limitations identified in this study, which is to arise in collecting the information generated and analysed, being mislabelled. Consequently, the knowledge of the MOOC users' perception about motivation, utility and learning, is essential to improve the deployment of MOOC courses. (Martinez-Nuñez et al, 2016)

As a result of our analysis is consistent with the resources generated by users over time they are a more valid indicator of success and quality of MOOC, given the lack of consensus in measuring this success (Besteiro, 2015), generalization in so far that the ratio of end depends on various factors (Fidalgo Blanco et al., 2015) and not just the instructional MOOC design or resources provided. This paper describes a method and tools necessary to study the community and their perceptions, based on two indicators such as perception and interaction.

5. CONCLUSION

The strength of gamification allow, through its elements and mechanics, involve the student from MOOC through different motivational activities, to learning community, seeking to create a habit at the same train him as a digital citizen developing the necessary skills to be useful in these spaces. The participation of thousands of students in these learning community, through their ideas or content in form of publication, serve as a source of open educational resources (OER) and will feed the next editions of the MOOC.

Evolution and sustainability of learning community associated with MOOC are related, on the one hand, with its ability to generate useful resources for members of the same and the perception of participants regarding usefulness, comprehension and learning. Regarding the results of this work, and agreeing with Cruz-Benito et al (2015b), it can be concluded that it is possible to obtain and classify non-formal and informal learning activities in environments like social networks and virtual learning communities.

This work has detected the existence of an improvement in the learning process as well as in the quality, quantity and variety of resources that appear in a MOOC through the cooperation in the generation of knowledge. Learning communities are technological resources where knowledge is generated, stored and organized. Finally, an analysis of type of user would be important to adapt methodologies as not all participants in these courses seek socialization or even go into games as challenges, which can also lead to dropouts.

To the future there is a need to improve methods of collecting resources to give them greater visibility and look for ways that the community is not only sustainable in terms of content generation, but it is self-regulate from members, reviewing from wrong or duplicate publications. In addition, according to Fidalgo-Blanco et al, (2013) it is necessary to progress in the management of the contents and the knowledge generated in virtual learning communities.

REFERENCES

Abelson, H. (2008). The creation of OpenCourseWare at MIT. *Journal of Science Education and Technology*, *17*(2), 164–174. doi:10.1007/s10956-007-9060-8

Alavi, M., & Leidner, D. E. (2001). Knowledge management and knowledge management systems: Conceptual foundations and research issues. *Management Information Systems Quarterly*, *25*(1), 107–136. doi:10.2307/3250961

Aramo-Immonen, H., Jussila, J., & Huhtamäki, J. (2014, October). Visualizing informal learning behavior from conference participants Twitter data. *Proceedings of the Second International Conference on Technological Ecosystems for Enhancing Multiculturality*, 603-610. doi:10.1145/2669711.2669962

Besteiro Martínez, B. (2015). *Herramienta para el análisis y visualización de las características principales de un MOOC*. Academic Press.

Blanco, Á. F., Lacleta, M. L. S., Gené, O. B., & Peñalvo, F. J. G. (2014). Educación en abierto: Integración de un MOOC con una asignatura académica. Teoría de la Educación. *Educación y Cultura en la Sociedad de la Información*, *15*(3), 233.

Borrás-Gené, O., Martinez-Nuñez, M., & Blázquez-Sevilla, A. (2016). Gamified open Q&A forum model as a platform for a CMOOCs. *INTED2016 Proceedings*, 4903-4910.

Borrás-Gené, O., Martinez-Nuñez, M., & Fidalgo-Blanco, A. (2015). Gamificación de un MOOC y su comunidad de aprendizaje a través de actividades. *Actas del III Congreso Internacional sobre Aprendizaje, Innovación y Competitividad.* CINAIC 2015.

Borrás-Gené, O., Martínez-Nuñez, M., & Fidalgo-Blanco, Á. (2016). New Challenges for the Motivation and Learning in Engineering Education Using Gamification in MOOC. *International Journal of Engineering Education*, *32*(1B), 501–512.

Breslow, L., Pritchard, D. E., DeBoer, J., Stump, G. S., Ho, A. D., & Seaton, D. T. (2013). Studying learning in the worldwide classroom: Research into edX's first MOOC. *Research & Practice in Assessment, 8.*

Cabrero, J. (2006). Las TIC y las inteligencias múltiples. Infobit. *Revista para la difusión y el uso educativo de las TIC, 13*, 8-9.

Cabrero, J. (2007). Comunidades virtuales para el aprendizaje. Su utilización en la enseñanza. Eduweb. *Revista de Tecnología de la Información y Comunicación en Educación, 1*(1), 5-22.

Clow, D. (2013). MOOCs and the funnel of participation. In *Proceedings of the Third International Conference on Learning Analytics and Knowledge* (pp. 185–189). ACM. doi:10.1145/2460296.2460332

Cross, J. (2011). *Informal learning: Rediscovering the natural pathways that inspire innovation and performance.* John Wiley & Sons.

Cruz-Benito, J., Borrás-Gené, O., García-Peñalvo, F. J., Fidalgo Blanco, Á., & Therón, R. (2015a). Extending MOOC ecosystems using web services and software architectures. [*Proceedings of the Interacción, 2015*, XV.

Cruz-Benito, J., Borrás-Gené, O., García-Peñalvo, F. J., Fidalgo-Blanco, A., & Therón, R. (2015b). Detection of non-formal and informal learning in Learning Communities supported by social networks in the context of a cooperative MOOC. In *2015 International Symposium on Computers in Education (SIIE)* (pp. 195-200). IEEE. doi:10.1109/SIIE.2015.7451675

Cruz-Benito, J., Borrás-Gené, O., García-Peñalvo, F. J., Fidalgo-Blanco, A., & Therón, R. (2017). *Learning communities in social networks and their relationship with the MOOCs.* IEEE Revista Iberoamericana de Tecnologias del Aprendizaje.

Dabbagh, N., & Kitsantas, A. (2012). Personal Learning Environments, social media, and self-regulated learning: A natural formula for connecting formal and informal learning. *The Internet and higher education, 15*(1), 3–8. doi:10.1016/j.iheduc.2011.06.002

Deterding, S., Khaled, R., Nacke, L. E., & Dixon, D. (2011, May). Gamification: Toward a definition. In *CHI 2011 Gamification Workshop Proceedings* (pp. 12-15). ACM.

Downes, S. (2008). *MOOC and Mookies: The Connectivism & Connective Knowledge Online Course.* Retrieved June 02, 2017, from: http://www.slideshare.net/Downes/mooc-and-mookiesthe-connectivism-connective-knowledge-online-course-presentation

Downes, S. (2012). *Connectivism and connective knowledge: Essays on meaning and learning networks*. National Research Council Canada. Retrieved June 10, 2017, from http://www.downes.ca/files/books/Connective_Knowledge-19May2012.pdf

Eshach, H. (2007). Bridging in-school and out-of-school learning: Formal, non-formal, and informal education. *Journal of Science Education and Technology*, *16*(2), 171–190. doi:10.1007/s10956-006-9027-1

Fidalgo, A., Sein-Echaluce, M. L., & Garcíapeñalvo, F. J. (2013). Coperative MOOC. An integration between cMOOC and xMOOC. *II Congreso internacional sobre Aprendizaje (CINAIC 2013)*, 481-486.

Fidalgo, A., Sein-Echaluce, M. L., Lerís, D., & Castañeda, O. (2013). Teaching Innova Project: The incorporation of adaptable outcomes in order to grade training adaptability. *Journal of Universal Computer Science*, *19*(11), 1500–1521.

Fidalgo Blanco, Á., Sein-Echaluce Lacleta, M. L., Borrás Gené, O., & García Peñalvo, F. J. (2014). Educación en abierto: Integración de un MOOC con una asignatura académica. *Education in the Knowledge Society*, *15*(3), 233–255.

Fidalgo Blanco, Á., Sein-Echaluce Lacleta, M. L., & García-Peñalvo, F. J. (2015). Methodological Approach and Technological Framework to break the current limitations of MOOC model. *Journal of Universal Computer Science*, *21*(5), 712–734.

García-Holgado, A., Cruz-Benito, J., & García-Peñalvo, F. J. (2015). Análisis comparativo de la gestión del conocimiento en la administración pública española. In *La Sociedad del Aprendizaje. Actas del III Congreso Internacional sobre Aprendizaje, Innovación y Competitividad*. Madrid: Fundación General de la Universidad Politécnica de Madrid. Retrieved from http://www.dmami.upm.es/dmami/documentos/liti/Actas_CINAIC_2015.pdf

García-Peñalvo, F. J., Cruz-Benito, J., Borrás-Gené, O., & Blanco, Á. F. (2015). Evolution of the Conversation and Knowledge Acquisition in Social Networks related to a MOOC Course. In *Learning and Collaboration Technologies* (pp. 470–481). Springer International Publishing. doi:10.1007/978-3-319-20609-7_44

Ghenname, M., Abik, M., Ajhoun, R., Subercaze, J., Gravier, C., & Laforest, F. (2013). Personalized recommendation based hashtags on e-learning systems. *ISKO-Maghreb. 3rd International Symposium*, 8-9. doi:10.1109/ISKO-Maghreb.2013.6728109

Groh, F. (2012). *Gamification: State of the art definition and utilization*. Institute of Media Informatics Ulm University.

Hamari, J., Koivisto, J., & Sarsa, H. (2014, January). Does gamification work?--a literature review of empirical studies on gamification. In *System Sciences (HICSS), 2014 47th Hawaii International Conference on* (pp. 3025-3034). IEEE.

Hanus, M. D., & Fox, J. (2015). Assessing the effects of gamification in the classroom: A longitudinal study on intrinsic motivation, social comparison, satisfaction, effort, and academic performance. *Computers & Education*, *80*, 152–161. doi:10.1016/j.compedu.2014.08.019

Huang, J., Dasgupta, A., Ghosh, A., Manning, J., & Sanders, M. (2014, March). Superposter behavior in MOOC forums. In *Proceedings of the first ACM conference on Learning@ scale conference* (pp. 117-126). ACM. doi:10.1145/2556325.2566249

Laroche, M., Habibi, M. R., Richard, M. O., & Sankaranarayanan, R. (2012). The effects of social media based brand communities on brand community markers, value creation practices, brand trust and brand loyalty. *Computers in Human Behavior*, *28*(5), 1755–1767. doi:10.1016/j.chb.2012.04.016

Liddo, A., Buckingham Shum, S., McAndrew, P., & Farrow, R. (2012). The open education evidence hub: A collective intelligence tool for evidence based policy. In *Joint OER12 and OpenCourseWare Consortium Global 2012*. Cambridge, UK: Conference.

Llorens, F., & Capdeferro, N. (2011). Facebook's Potential for Collaborative e-Learning. *Revista de Universidad y Sociedad del Conocimiento (RUSC), 8*(2), 31–45. Retrieved from http://rusc.uoc.edu/ojs/index.php/rusc/article/view/v8n2-llorens-capdeferro/v8n2-llorens-capdeferro

Mao, J. (2014). Social media for learning: A mixed methods study on high school students' technology affordances and perspectives. *Computers in Human Behavior*, *33*, 213–223. doi:10.1016/j.chb.2014.01.002

Markoff, J. (2011). Virtual and artificial, but 58,000 want course. *The New York Times*, p. 15.

Marsh, H. W. (1982). SEEQ: A reliable valid and useful instrument for collecting students evaluations of university teaching. *The British Journal of Educational Psychology*, *52*(1), 77–95. doi:10.1111/j.2044-8279.1982.tb02505.x

Martínez -Núñez, M., Borrás-Gené, O., & Fidalgo- Blanco, Á. (2014). Social community in MOOCs: practical implications and outcomes. In *Proceedings of the Second International Conference on Technological Ecosystems for Enhancing Multiculturality* (pp. 147-154). ACM.

Martínez-Núñez, M., Borrás-Gene, O., & Fidalgo-Blanco, Á. (2016). Virtual Learning Communities in Google Plus, implications and sustainability in MOOCs. *Journal of Information Technology Research*, *9*(3), 18–36. doi:10.4018/JITR.2016070102

McAuley, A., Stewart, B., Siemens, G., & Cormier, D. (2010). *The MOOC Model for Digital Practice*. SSHRC Knowledge Synthesis Grant on the Digital Economy.

Muntean, C. I. (2011, October). Raising engagement in e-learning through gamification. In *Proc. 6th International Conference on Virtual Learning ICVL* (pp. 323-329). Academic Press.

Nonaka, I., & Takeuchi, H. (1995). *The knowledge-creating company: How Japanese companies create the dynamics of innovation*. Oxford University Press.

Paine, K. D. (2011). *Measure what matters: Online tools for understanding customers, social media, engagement, and key relationships*. John Wiley & Sons.

Rovai, A. P. (2002). Sense of community, perceived cognitive learning, and persistence in asynchronous learning networks. *The Internet and Higher Education*, *5*(4), 319–332. doi:10.1016/S1096-7516(02)00130-6

Seaborn, K., & Fels, D. I. (2015). Gamification in theory and action: A survey. *International Journal of Human-Computer Studies*, *74*, 14–31. doi:10.1016/j.ijhcs.2014.09.006

Sein-Echaluce, M. L., Fidalgo-Blanco, Á., & García-Peñalvo, F. J. (2016). Students' Knowledge Sharing to improve Learning in Engineering Academic Courses. *International Journal of Engineering Education*.

Tabaa, Y., & Medouri, A. (2013). LASyM: A learning analytics system for MOOCs. *International Journal of Advanced Computer Science and Applications*, *4*(5), 113–119. doi:10.14569/IJACSA.2013.040516

Tang, L. M., & Kay, J. (2014). Gamification: Metacognitive scaffolding towards long term goals? *Personalization Approaches in Learning Environments, 63*.

Tardón, C. G., & de la Vega, R. A. (2014). #dametuits: Programa de formación y comunicación corporativa gamificada. *Obra digital: revista de comunicación*, (7), 47-68.

UNESCO. (2012). *2012 Paris OER Declaration*. Retrieved April 20, 2016, from: http://www.unesco.org/new/fileadmin/MULTIMEDIA/HQ/CI/CI/pdf/Events/English_Paris_OER_Declaration.pdf

Vaibhav, A., & Gupta, P. (2014). Gamification of MOOCs for increasing user engagement. *MOOC, Innovation and Technology in Education (MITE). 2014 IEEE International Conference*, 19-20. doi:10.1109/MITE.2014.7020290

Werbach, K., & Hunter, D. (2012). *For the win: How game thinking can revolutionize your business*. Wharton Digital Press.

Zichermann, G., & Cunningham, C. (2011). *Gamification by design: Implementing game mechanics in web and mobile apps*. O'Reilly Media, Inc.

Chapter 6
Technology-Enhanced Learning:
Good Educational Practices

David Fonseca
Universitat Ramon Llull, Spain

Ricardo Torres Kompen
Universitat Ramon Llull, Spain

Emiliano Labrador
Universitat Ramon Llull, Spain

Eva Villegas
Universitat Ramon Llull, Spain

ABSTRACT

There are numerous examples from recent years of the incorporation of all types of applications and technological systems into the classroom at all educational levels, with the aim to improve both student motivation and academic performance; we can group these initiatives under the term technology-enhanced learning, TEL. The TEL research field has been profoundly involved with the development and application of collaboration apps. Computers, mobile devices, and applications play diverse roles at different times along the project lifecycle. The most common lifecycle comprises four distinct phases: design, implementation, approval, and final assessment. In this chapter, the authors discuss key concepts of these TEL phases as well as some different approaches that can be defined as "good technological practices" and their main results in order to implement technologies in the formative process.

INTRODUCTION

One of the key issues in all formative levels is the management of student motivation and the best way to acquire the competences and skills defined in the academic plans (Fonseca, Redondo & Villagrasa, 2015). The change in the educational systems of the past decade (which involved a reduction in the number of hours of teaching classes), along with a reduction in the number of university students, has caused a

DOI: 10.4018/978-1-5225-4944-4.ch006

shift in the teaching paradigm, which is currently focused on how to improve student ability and skills. As previously, proposed Biggs (1999), the evaluation and control of the quality of student education and development is a key concept in the implementation of technologies utilized in the classroom (Fonseca, Martí, Redondo, Navarro & Sánchez, 2014). The interest, need, and urgency of implementing new technologies in education and universities in particular is a relatively new situation (Rogers, 2000). However, technological innovation, which is intended to improve the student learning process, must be capable of providing support to address difficulties that could arise with the student in the use of and interaction with technological elements. These elements must not obstruct the auto-learning process, which is altered by this technology, and the students must be motivated with the new educational methodology.

Hence, the use of Personal Learning Environments (PLE), Augmented and Virtual Environments and devices (AR/VR), gamification of educational activities, or Robotics appear to have the potential to increase student motivation, while also motivating them to attain better scores in practical assignments (as well as higher overall score). It also has the capability to generate new virtual and embedded methods for assessing student learning. In contrast to traditional programs (passive and focused on subject matter), the new methods based on Project Based Learning (PBL) offer an experience equivalent to learning a trade, as learners must face a well-planned series of real situations (scenarios) in a significant and motivating role, something knows as Scenario Centered Curriculum, SCC.

To incorporate a new IT-based methodology into a specific teaching environment, some recommendations for avoiding student rejection must be considered. The literature defines so-called "good educational practices" that are primarily focused on virtual rooms, distance education (or e-learning), and semi-present teaching (Area, San Nicolás & Fariña, 2010). These studies have focused on maximizing profit from web service content, alternative methods using the intranets of each university, and auto-evaluation systems of information (Chickering & Gamson, 1987; Epper & Bates, 2004; González & Rodriguez, 2010).

In our chapter, we will discuss about what we can understand under the definition of "Good Practices", and some interesting possibilities that new technologies can add to the educative framework. After that, we will present three educational cases using different technologies and different assessment approaches that we can be defined as a set of proposed "Good Practices", an easily adapted in other academic fields (Navarro & Fonseca, 2017):

- The first one is part of a project based on the application of gamification techniques in a first-year course of the Degree in Multimedia Engineering at La Salle-Universitat Ramon Llull. The work discusses the advantages of the applying of immersive gamification techniques in the acquisition of role-based learning. This project aims to improve teaching quality and student learning experience. Our objective was two-fold: firstly, to meet the expectations of users, and second to guarantee that the user experience is memorable and satisfactory (Villagrasa, Fonseca, Redondo & Duran, 2014).
- The second experience started with a pilot study based on a framework for designing PLEs using Web 2.0 services, which was proposed by Torres, Edirisingha and Mobbs (Torres et al, 2009). The framework outlined four types of Web 2.0 services and how they could serve as "hubs" for the construction of PLEs. Two projects emerged based on this framework, in two different settings: secondary education and higher education. Their goal was to explore and understand how high school and university students used Web 2.0 services in their learning, thus helping in the devel-

opment of good practices for the adoption and implementation of these tools in the classroom, which could in turn be extended to other levels (primary) and types (informal, lifelong learning) of education. The lessons learned and the methodology that was developed were then successfully applied in other environments and settings.

- Finally, we present a case study where the aim of the project is focused on evaluating both the intrinsic motivation and satisfaction of the student in the process of incorporating VR systems for visualization and interaction in 3D complex projects related with Architectural and Urban projects. In this case the main goal is the relation established between the Multimedia students, who are developing the interface and the interaction of the 3D environment and the Architecture students, who are developing the 3D models and are one the potential final users of the project (in addition to the citizenship) (Fonseca, Villagrasa, Navarro, Redondo, Valls & Sánchez, 2017).

BACKGROUND

TEL focus its research challenges and strategic objectives on the most common lifecycle of technological implementation that comprises four distinct phases: design, implementation, approval and final assessment. Focusing in these four main phases, we the students' learning experience may be improved by:

- Supporting student engagement, motivation, satisfaction and retention;
- Helping to produce enterprising graduates with the skills required to compete in the global business environment;
- Encouraging inspirational and innovative teaching;
- Personalizing learning that promotes reflection;
- Delivering and supporting Continuing Professional Development and Internationalization.

All these activities fit inside the TPACK model (Technological Pedagogical Content Knowledge; Koehler, M., & Mishra, P., 2009). This model describes the different types of knowledge that teachers must integrate into their teaching in order to fulfil an effective learning methodology, and it is the basic theoretic framework of TEL work and research. Based on the different intersections of the TPACK model, as well as on the benefits of implementing ICTs in learning and teaching environments, the challenges and difficulties, and the potential benefits, we can identified the following scientific challenges for the upcoming work and research:

- Research, implement and evaluate new methods and systems (both technological and non-technological) that may help generate both Best Practices and more efficient educational experiences. For this reason, we will study the relationships that define this efficiency in technological fields (both inside and outside the university), with ICTs. The main challenge is to characterize aspects such as subject comprehension, competency assessment, motivation and usability studies, preparation and training of teachers, analysis of proposals, and the implementation environment, always taking into account how they complement each other, in order to help in the improvement in educational efficiency.
- Creation of new methodologies (both for teaching and for application in other professional fields), that incorporate new technical approaches (with an emphasis on those related to augmented and

virtual visualization, robotics, and gamification or serious games). Due to the current trend of a society with a high use of social networks and very heterogeneous learning environments, it is necessary to democratize the use of information and technology.
- Improve the impact and dissemination of technological good practices.

Gamification

Games are created by designers/teams of developers and are consumed by players. They are purchased, used and eventually cast away like most other consumable goods. The difference between games and other entertainment products (such as books, music, movies and plays) is that their consumption is relatively unpredictable. The string of events that occur during gameplay and the outcome of those events are unknown at the time the product is finished. The MDA framework formalizes the consumption of games by breaking them into their distinct components (Hunicke, Leblanc & Zubek, 2004), i.e., Rules => System => Fun, and establishing their design counterparts:

- Mechanics describes the components of the game at the level of data representation and algorithms.
- Dynamics refers to the run-time behavior of the mechanics acting on player inputs and outputs.
- Aesthetics pertains to the desirable emotional responses evoked in the player through game interaction.

Gamification in classes helps to improve the connection between the material and the student, offers the opportunity to reflect on a topic in depth and allows positive changes in behavior (Kapp, 2012). In this approach, learning through gamming is achieved by aligning the game mechanics with Bloom's taxonomy of learning (Zichermann, Gabe & Cunningham, 2011), allowing learning to be classified into three domains (Bloom, 1953):

- Cognitive, which is taught in traditional education and implies understanding and synthesis of knowledge.
- Affective (involving emotions), which reflects the attitude toward a situation.
- Psychomotor (the physical), which is activated by requiring a union of mental and physical activity.

To encourage the use of games in learning beyond simulations and puzzles, it is essential to develop a better understanding of the tasks, activities, skills and operations that different game types can offer and to examine how these might correspond to the desired learning outcomes (Cronk, 2012). Extant studies in education/learning contexts considered the learning outcomes of gamification as mostly positive. For example, outcomes were assessed in terms of increased motivation and engagement in the learning tasks, as well as enjoyment derived from partaking in the tasks (Hamari, Koivisto & Sarsa, 2014). However, these studies revealed some negative outcomes that need to be addressed, such as the effects of increased competition, task evaluation difficulties, and design features (Connolly, Boyle, MacArthur, Hainey & Boyle, 2011).

Game playing is associated with trial, error, failure and eventual success through practice, experience, reflection and learning. A key objective of most games is not to prevent failure but to develop a positive relationship with it. Failure is not seen as an end but as a step in the journey to mastery. Gamified learning interventions seek to maintain a positive relationship with failure by creating rapid feedback cycles and keeping the stakes for individual learning episodes low. In many ways, the paradigm that governs current educational systems has many game-like elements. Most assessments strive for objectivity, and continuous assessment is seen as desirable. Students earn points for completing assignments correctly. These translate into comparable rewards—grades. If they perform well, students "level up" by proceeding to a more advanced course of study at the end of every academic year (Nicholson, 2012).

What distinguishes gamification most distinctly from more traditional approaches is the explicit use of competition as a motivational tool. This competitive element is a source of motivation (Domínguez, Saenz-de-Navarrete, De-Marcos, Fernández-Sanz, Pagés & Martínez-Herráiz, 2013). These ranking systems serve as motivators because users see their efforts instantly recognized (Brophy, 2013). In education, motivation is considered a key determinant of learning. It is used to explain the attention and effort students dedicate to particular learning activities (Buckley & Doyle, 2014). In our approach to gamifying learning, we employ common gamification design elements (Harlen & Deakin, 2003):

- **Objective:** Gamified activities have rules that predetermine the actions a player can or cannot take.
- **Rewards:** Gamified learning activities have a reward system that provides users with SAPS (Status, Access, Power and Stuff) for interacting with the game successfully.
- **Quick Feedback:** By receiving timely feedback, the users can quickly learn how to improve at the game.
- **Cycles:** Incorporating competitive elements into gamified activities presents users with a challenge, while the objective outcome associated with the games allows for ranking of users.

Intrinsic motivation is another key element because it involves learners that are interested in what they learn and in the learning process itself. In contrast, individuals possessing extrinsic motivation engage in learning because it is a means to an end. Thus, their approach is relatively disassociated from the content and subject of learning (Piaget, 1952). Intrinsic motivation, conceptually, it is closely associated with cognitive behavioral theories and the work of Piaget. The author posits that, when individuals experience discrepancy between their experienced knowledge of the world and their private, internally held knowledge, they are driven to eliminate this discrepancy (Skinner, 1976). On the other hand, extrinsic motivation is associated with B. F. Skinner's behavioral theories of human learning and focuses on the provision of rewards to direct and control learning behavior (Vallerand, Pelletier, Blais, Briere, Senecal & Vallieres,1992).

Prompting and mediating positive learning behaviors is seen as the key advantage of gamification. However, learning motivation differs among individuals. Some learn for pleasure or to satisfy curiosity, whereas others learn to obtain rewards (e.g., a high-status job and/or financial rewards). Student motivation is an important factor in their reaction to learning activities. While some students may be motivated by having their learning gamified, this approach may demotivate others. Even if gamified learning impacts individuals differently, this fact does not necessarily deny its utility. Rather, gamified learning interventions can be included as part of a range of learning interventions. However, these interventions must be chosen in a manner that ensures no type of learner is systematically disadvantaged. For example, it is argued that common forms of academic assessment, such as essays and reflective pieces, favor learners with relatively high intrinsic motivation. A key research question raised in the pertinent literature is how individuals with different motivations for learning are impacted by gamified learning activities. A game is an activity related to the resolution of a problem, following an approach based on a clearly playful attitude. Continuing with some of the lectures by Jesse Schell (2008), we posit that gamified learning requires a very well-implemented background story with "complex" goals, along with a set of rather "simple" (Usable/Playable) rules.

We should not forget that the audience of any "game" has a transcendent interest. Thus, the "story" should orbit around the four axes of Game Design. This is the definition of the Aesthetic Criteria, which require creating a set of rules or mechanics for the game, a good and dense history, and a good technological approach. The considerations in terms of the game space include the Objects (enigmas and challenges that relate to inventory Player Characters), the actions to undertake (as in conversational Adventure graphs, these enable the relationship between objects and the game space), pre-defined rules (same for all players/students), and a proper balance between dexterity and luck. One of the most important concepts to take into account is the flow. In this context, flow is defined as the state in which we find ourselves when we are focused on a task and nothing can distract us from it. A game, a movie, a book, a concert or any kind of pleasurable and engaging experience should have a good flow, and we can be so immersed in it to the point that we lose our sense of time and space. The flow should not be linear but should take the form of a rocking curve. In this way, the user is allowed a break after a period of time in which he/she experienced intense action or emotion.

Personal Learning Environments

Personal Learning Environments, or PLEs, first emerged as a new concept in the educational filed as early as 2001 (JISC, 2007). Being a relatively new field, several definitions have been proposed through the years, and they vary from those firmly rooted in the technological to those with a more social orientation (Fiedler, 2010). Currently, there is no standard definition of PLEs, as this in turn depends on how are PLEs conceptualised. The two main approaches are PLEs as objects versus PLEs as a concept or approach; platforms and services have been proposed, with various degrees of success (Van Harmelen, 2008) while other research efforts have been directed to the application of PLEs, and the process of developing and maintaining them, to learning and teaching methodologies (Castañeda and Adell, 2011).

Adell and Castañeda propose that PLEs are "a set of tools, sources of information, connections and activities which each individual uses on a regular basis to learn [and is created by the users themselves]" (Adell and Castañeda, 2010). The "personal" element is central to the definition of PLEs.

The personalization of learning is based on the idea that learning technologies should enable the various aspects of learning (the content, the mode of delivery and access) to be offered, according to the personal characteristics of the learner, thus providing the learner with greater flexibility and options for learning. As shown in The 2009 Horizon Report (Johnson et al, 2009) the personal web was one of the trends in the second horizon of adoption (Time-to-Adoption: Two to Three Years). The ubiquity of technology and the increased access to it, along with the ability to carry it with us wherever we go. Younger generations spend more and more time online, immersed in their gadgets and devices, and depending on connectivity, apps and online services. In 2009, Torres, Edirisingha and Mobbs (Torres et al, 2009) proposed that the future shape of PLEs would probably be mobile, and that seems to have become a reality.

One of the elements to be considered in any pedagogical implementation involving PLEs is the profile of learners; the speed at which technology has been evolving for the last decade means that there is no standard profile for which to aim when adapting or personalizing teaching and learning approaches. Much has been said and written about Digital Natives (Prensky, 2001) to establish a difference between generations, but this view has been contested by proposals such as Digital Visitors and Residents (White, 2011). We support this approach, based on the fact that age is not necessarily the main element that marks the difference between users, but how they use and appropriate technology-based tools and services, especially those considered as Web 2.0, in their own learning process.

The question remains, how can teachers and learners appropriate these technologies in a way that enhances the learning process but without being intrusive or overly disruptive, and blends in with the current methodologies and approaches?

New Visual Tools for Graphic Expression (GE)

The evolution of the Computer Aided Design (CAD), and Building Modelling Modelling (BIM) methods at university level has been gradual since the late 1980s. The curricula has progressively integrated instrumental subjects with CAD/BIM applications. Engineering and Architecture have adopted the computers in subjects of Graphic Expression (GE) in different levels.

Industrial engineering initially adopted the implementation of computer systems to improve the teaching development of subjects (virtual platforms, web spaces, etc.), as well as the use of these tools in 3D modeling of projects. Multimedia and interactive applications have favored the performance and speed of learning as well as personal and intrapersonal skills of students. Multimedia systems can improve the acquisition of knowledge, motivation and participation in the process, as demonstrated by the experiment of the University School of Industrial Technical Engineering of San Sebastián (Garmendia et al., 2001). In this way, the use of multimedia systems in the area of GE was an opportunity to improve the quality of teaching. Another example of the application of the methodologies based on on-line platforms was the one undertaken in the University of Las Palmas of Gran Canarias by the teachers (García, Martín, Suárez, Pérez, & Suárez, 2002). They developed the DSS tool (Drawing Support System) to teach the subject of EG at the same time as it was used as a study resource outside the classroom by the students.

The need to get as close as possible to knowledge from the new digital platforms, has forced to adapt the theoretical material to multimedia formats. The study of matter outside the classroom is promoted through multimedia systems, interactive tutorials, animation, hypermedia systems, visualization, etc. Some examples show favorable results in the application of these methodologies, as is the case of teachers (Álvarez Peñín, Del Canto, Charro, Rubio, García, & Suárez, 2001) in the implementation of multimedia systems as a reduction to traditional teaching systems. Other proposals focus more on the subject of EG from multimedia formats and interactive tutorials for drawing views such as the work of teachers (Álvarez Peñín, Pando, García, & Pérez, 2006) and (Álvarez Cuervo, Roces, & Alonso, 2006). The delivery of teaching in technical drawing is one of the most suitable for these media as proven by (Fernández, Reina, Mateo, & Sánchez, 1998) and the dihedral system, (Blanco, et al, 2006). In addition to methodologies, tools have been created to improve visualization (Ramos, García, Baños, & Melgosa, 2003) and as a means to problem solving (Garmendia, Gisasola, & Gorozika, 2004)

Support for technical modeling software allows adding specific modules on more powerful graphics platforms such as AutoCAD for manipulating 3D figures and spatial comprehension of your projections (Mora, Palomo, García, & Marín 2001). The ability in spatial representation capacity requires a learning of spatial perception. Following this assumption, new technologies like Augmented and Virtual Reality (AR/VR) help to improve this spatial learning and perception.

The AR consists of the vision through a technological device, directly or indirectly, of a physical environment of the real world, whose elements are combined with virtual elements for the creation of a mixed reality in real time. The term coined by Caudell & Mizell (1992) was initially oriented towards the visualization of digital information in manufacturing processes. The AR introduced a new variable in the field of the development of spatial skills, it is about the virtual objects that can be manipulated by the user interacting with the contents from a mobile device. Some references of the applications in the field of spatial understanding of geometry is the constructions as Construc3D, by professors Kaufmann, Steinbügl, Dünser, & Glück, (2003), or the AR-DEHAES app of Professor Martín-Gutiérrez, et al, (2010).

On the other hand, the VR refers to an environment generated by computer technology, which creates in the user the feeling of being immersed in it. The space is normally visualized by the user through a device known as glasses or helmet of virtual reality. The term VR is difficult to date since it can be considered that the origin can be panoramic images or stereoscopic photography. Possibly, the first example of VR was the flight simulator "Link Trainer" by Professor Mills (1941). The basis of the VR is to create an immersive experience and allow the user to interact with objects. This concept is enriching from the point of view of training to operate in space. Some studies show that training in a virtual environment in which three-dimensional objects can be manipulated and rotated to become familiar with them from any angle allows better recognition of objects than if they are taught on paper (James, Humpherey, & Vilis, 2002).

CASE 1: GAMIFICATION APPROACH IN MULTIMEDIA EDUCATION

Gamification has been introduced as an incentive to encourage students to study more and acquire more knowledge, and thus obtaining higher marks. All dynamics, mechanics and aesthetics are applied to improve the perception of the subjects towards the student. User Experience techniques have been applied to improve the system and to provide a solution that meets the real problem as effectively as possible for all involved.

To solve the problem of students' motivation, gamification techniques are very attractive and it is obvious to think about them as a good solution to the problem at hand. However, as researchers from the Research Group on Media Technologies (GTM) from La Salle-Universitat Ramon Llull, we had the need to validate this claim. Thus, we created a system that could ensure that the use of these techniques was to be fully justified and would effectively serve to our purposes. To that effect, we created the methodology Fun Experience Design (FED). FED is a methodology based on the User-Centered Design (UCD) (Abras, Maloney-Frichmar & Preece, 2004). Knowing the actual needs, concerns, behavior and motivations of our users and their perception of the subject was the key point in developing a gamified system that worked properly. Although the methodology includes iterative design, i.e., the end of the implementation re-start the process to identify problems that might have occurred and correct them in the implementation of the next version, we wanted the user experience in this first version to be pleasant as possible.

The evaluation techniques used in this methodology can be classified in two groups:

- **Inquiry or Query Techniques:** These techniques are to identify requirements, both from users and product, at an early stage of the development. Among other things, these techniques provide information about the profile of potential users, the context of use, product competition, the market in which the product is included, motivations, interests, needs and opinions about the new

Figure 1. Basic scheme of fun experience design methodology

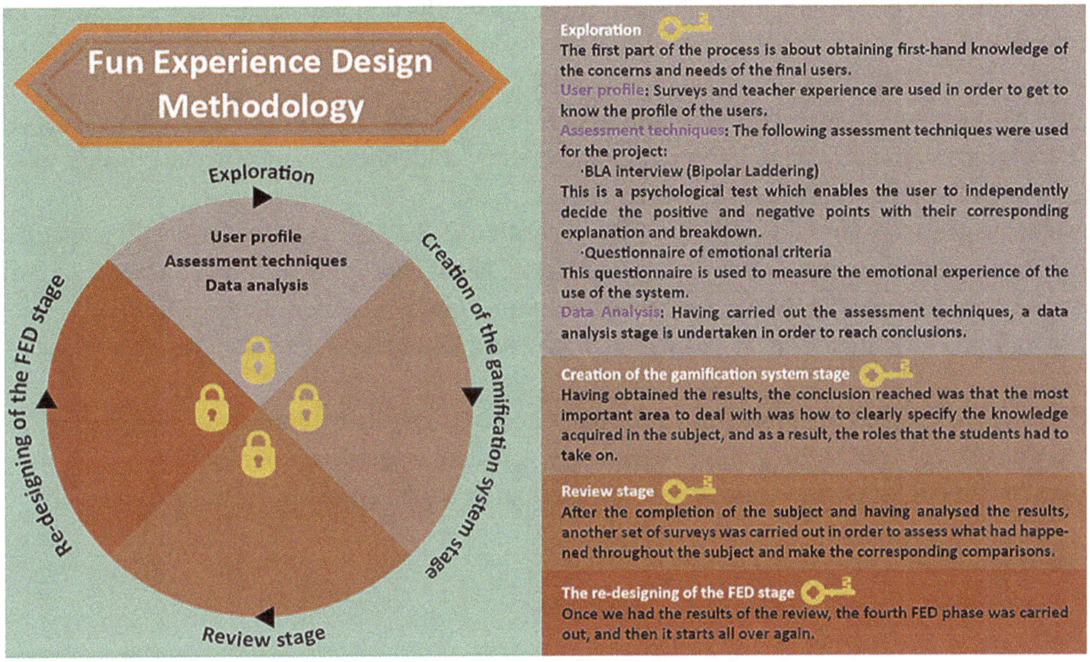

application. This group is composed of techniques such as interviews, focus groups, card sorting, contextual inquiry, ethnographic studies, and surveys or guided sessions.

- **Test or Evaluation Techniques:** These techniques are used to check the efficiency and effectiveness of users to find specific usability problems and to compare possible solutions through the evaluation of prototypes or similar interfaces. Given that the inquiry or inspection techniques are used to define requirements and identify certain aspects of the usability of the product, it is with these test techniques that we can find the specific usability problems that users have. The evaluation techniques are composed of general tests with users, thinking aloud protocol, the method set discovery, the protocol of questions, the measure of performance, the pencil test, eye-tracking, Bipolar Laddering interview (BLA; Pifarré & Tomico, 2007) and the task test.

In order to conduct this specific project the following techniques were used:

- **BLA Interview:** Bipolar Laddering methodology is a psychological exploration technique that highlights key points of user experience. The objective is to find which specific points of the interface cause user frustration or confidence, among others. Each finding is scored between 0 and 10 for the level of satisfaction of each of them. The goal of a laddering interview is to discover how they link in the mind of the user, the attributes of the product, the consequences of using it and the transmitted personal values.
- **Emotional Appraisal Questionnaire:** The emotional appraisal questionnaire is used to calibrate the emotional experience of use of the system. Obtaining the pairs of feelings are based on the research from (Schmidt-Atzert, 1985). Sixty words that express emotion were the basis to ask 20 users to order in columns from the dimensional point of view between "pleasant" vs. "unpleas-

ant" in a 5 scale range, mapped in to a percentage value afterwards (from 0% to 100% in steps of 25%). Once the results are grouped, we average the results to obtain an average score of emotional appraisal.

In order to achieve a clearly defined immersion in which the student can feel identified, we have created customized images, named every role and created a hierarchy of profiles according to student performance in the different areas (see Figure 2).

The user profiles are: Initial Profile, profiles according to the Level of Knowledge of a specific area (design, usability and communication skills) and User Clan according to the profile level acquired (mix of types of knowledge profile). An example of the initial "clans" is shown in the Table 1.

Following this example, we have defined new profiles, and images, based on different concepts:

- User profiles related to knowledge in design: Creative, recycled, conformist.
- User profiles related to knowledge in usability: Critical, adaptable, tolerant.
- User profiles at knowledge related to communication skills: communicator, sociable, shy.

Figure 2. Examples of some image profiles created

	TECHNICAL
	ARTISTIC
	COMMUNICATOR
	SPHINIX: Technic + Artist + Communicator
	CENTAUR: Technic + Communicator
	FENIX: Artist
	HUMAN: Nothing

Table 1. The initial clans

Technical: You like working on the technological part of design, albeit technological, physical, telematics or the programming of a project.
Artistic: You like working on the visual, narrative and/or aesthetic parts of a project.
Communicator: You like to share, report on and document a Project, either verbally or in writing.

Table 2. User Clan according to the profile level acquired

Sphinx: Technical + Artist + Communicator	**Harpy:** Artist + Communicator	**Hydra:** Comm.
Hippogriff: Technical + Artist	**Minotaur:** Technical	**Human:** Nothing
Centaur: Technical + Communicator	**Fenix:** Artist	

As the course progresses and students acquire a higher or lower level of knowledge according to each profile (evaluation), they will belong to one clan or another (Table 2).

The results obtained from the questionnaire allow a series of evaluations about what students thought about the subject, thereby facilitating the identification of those aspects that should be empowered -since they are valued positively-, and what needs to be improved as it considered negative. After analyzing the results, it was observed that the emotional assessment of students towards the subject was positive. Around 60% of the users considered the subject fun, useful, innovative and close to reality. Therefore, we can conclude that it is a well-received subject. Still, we need this perception be transmitted to a higher percentage of students and improve the motivation for the study of the subject (Labrador & Villegas, 2014).

At the end of the course the roles that were achieved in a satisfactory way were the Communicator with 23%, followed by the Artist with 9.9% and Technical with 7.9%. The percentage of acquisition of a specific role is very low, less than 10% in the case of the Technical and Artist, and although it rises considerably in the case of communication, up to 23%, it remains an unsatisfactory value. In this regard, it should be noted that a student might have more than one profile.

The results showed that more than 75% of the class did not acquire any of the profiles with sufficient quality (humans). Only 10% have acquired the profile of Communicator, and a very low number achieved unique Artist or Technical profiles. The values of the combination of two of the profiles are also very low. However, a larger number of students acquired a combination of the three profiles when compared to a combination of just two. After checking the data and analyzing the results it was verified that, although the final grades were satisfactory, the acquisition of concrete profiles was not being achieved.

That is, the students had an average grade of all the exercises that allowed them to pass the subject, but their knowledge was of isolated elements, without coherence amongst them. In this first iteration of the course, a high dropout rate of 14.4% was still obtained, which led to a re-calibration of the system for the following course, modifying the ranges according to which students obtain the game medals, and assessing the exercises presented much more gamified, in order to get students more engaged and involved. The results of this adaptation showed an increase with respect to the previous course, with rates of 35.5% for the Communicator case, and above 15% for both the Artist and Technical profile. These increases were reflected at all levels (both academic and game), and resulted in a decrease of the dropout rate to 12%.

CASE 2: PLES BASED ON WEB 2.0 SERVICES IN SECONDARY AND HIGHER EDUCATION

The project, known as PELICANS (Personal E-Learning In Communities And Networking Spaces) was a joint effort between the Beyond Distance Research Alliance, University of Leicester, UK and the Universitat Politècnica de Catalunya, i2Cat Foundation and Citilab, in Catalonia, Spain and was carried out at university level, at the Escuela Superior de Estudios Internacionales, ESEI (Torres et al, 2010). Web 2.0 tools and services were introduced in a specific subject, which was focused on Business and Information Systems, as part either of the syllabus, or as tools to achieve specific learning outcomes. The students were asked to create their own personal learning environments, and then provide feedback and comments on their usefulness and advantages, as well as obstacles and problems found. The methodology used was design-based research, which allowed for several modifications in the successive iterations of the project.

The project was carried out with the participation of three groups of students, in three successive academic years. Students were gradually introduced to several Web 2.0 services, in the context of the subject, by means of activities and e-tivities (Salmon, 2004), and were then asked to reflect on the impact and role these could have in their learning process. Piece by piece, they built what they knew as "Web 2.0 diagrams", and only once they had finished this process, they were introduced to the concept of PLEs. At that point, a series of questionnaires and interviews were conducted, in order to gather data about the experience. A follow-up interview was done five years later, in order to explore the long-term impact of this exercise on their professional lives.

The main outcomes of this project were the evidences of the impact of working with PLEs in the classroom; gaining insight on the perception of students about the role of technology in their learning process; and the identification of obstacles and difficulties. PLEs were shown as potential enablers for organizing and managing content, acquiring and developing skills, and strengthening social interactions. At the same time, issues such as interoperability, learning curves, fragmentation, distraction and technological limitations were flagged as potential problems.

The experience acquired in this project led to the second idea, a training space/course for secondary education teachers, with the aim of providing the tools and methodologies so they could apply the PLE idea in their specific subjects and schools. This, project, known as Hort Digital (loosely translated as Digital Orchard) was ran at Citilab-Cornellà (http://en.citilab.eu) with the support of the i2CAT foundation (http://www.i2cat.cat). Its goal was to help high school teachers implement and use technology in innovative ways in the classroom. It was offered in a course format during the three academic years, an initiative supported by the Department of Education of the regional government in Catalonia, Spain. In the course, teachers and facilitators worked together to propose, discuss and test innovative approaches to using technology in the classroom, with a focus on Web 2.0 tools and the personalization of learning. One of the main goals was that the collaborative and social approach that was followed, supported by the creation of a community of practice, would have a positive impact in closing the digital gap between teachers and students, help teachers realize the potential of new technologies, and support the new policies that at the time were being issued by both the regional and national governments in Spain. Hort Digital was designed as an open space, in which both new and traditional technologies were showcased, so users could test them, find new applications in the classroom, develop innovative projects in a collaborative environment, and follow up the implementation of those.

By using a Living Lab approach, the teachers participated not only during the training sessions, but also took part in the proposal, design and execution of two new levels of this training program.

As a Living Lab, the project complied with all the main characteristics of one: it took into account the needs of final users, was designed in a way that encouraged user-driven innovation, users were co-creators, and specific methodologies were used in order to involving them in the process, it had local impact, was sustainable and the technology needed was easily available.

The project impacted hundreds of students and other teachers (besides the 80 that formally registered in it) that were not directly participating in the training sessions. The success of the project, evidenced by the diversity of educational initiatives taken by the participants (Torres et al, 2011) showed that the idea of PLEs could be turned into a teaching and learning resource without the need for expensive or complicated technological set-ups. Variations on this approach were applied in a diversity of settings, such as ICT courses for senior citizens, PLEs for researchers in the healthcare field, digital marketing approaches and creativity and innovation workshops.

CASE 3: INTERACTIVE 3D ENVIRONMENTS IN ARCHITECTURAL EDUCATION

The study described in this section focuses on the description of the implementation process of gamified RV systems in architecture teaching and the perception of utility and motivation of the student in front of these proposals. The overall aim of the project is to virtually recreate urban areas of the city of Barcelona to establish a user-friendly environment. In this system, the users (students, professionals, architects and citizens of all types and ages) can interact (play) with, in order to recreate new spaces centered in pedestrian uses, instead of the present ones that do not contemplate predominantly this typology of space. The virtual three-dimensional space is also intended to have a photographic quality (i.e. maximum realism), incorporating the movements, materials, textures and even sounds of the environment. We have defined these following phases:

- Definition of the urban environment, initial proposal of uses, modeling and texturization of buildings using modeling tools like 3DMax, Rhyno, Sketchup, Photoshop.
- Definition of game mechanics, interaction, constraints and implementation system, programming and integration of models and interface. Unreal is being used for this process.
- Definition of metrics and analysis evaluation.

For the study, we used an initial questionnaire to obtain both the technological profile of the students and the degree of interest in the topics and approach of the course. For the section of the profile, we have asked for statistical data such as age, gender, place of birth, completed, ongoing studies, and possible work experience. They have also asked about the devices and access to Internet services both at the level of use / membership and frequency of use.

In a second block, we have included 18 questions divided into three blocks (see Table 3). The first one (consisting of eight questions identified as A1-A8), is focused on obtaining the student's perception of the degree of utility of technologies such as RA / RV for visualization and understanding of architectural space. The second block of three questions (G1-G3), is focused on evaluating the perception of utility of games in the understanding of space and its educational use. Finally, through seven questions (identified as IM1-IM7), we evaluated the subjective response of students referred to IMI.

The sample of work was a total of 75 students, 30 men and 35 women, with a mean age of 23.8 years (Typical Deviation, DT: 5.47). 56% are Spanish and the rest of multiple nationalities. From the first block of general issues analyzed [A], the average obtained was 3.5, having weighed down the first two questions (A1: 2.5 and A2: 2.4), which correspond to the statements of how important visualization 3D architectural projects for your understanding is and in the same line, that the use of VR is useful for such understanding. Centered on the gamification block [G], the average of the three variables studied was 3.2.

Table 3. Pre-test control variables

Question	#
3D digital visualization of architectural and urban projects is of vital importance for their understanding	A1
The use of virtual reality to visualize projects is useful for understanding	A2
The use of augmented reality to visualize projects is useful for understanding	A3
I am motivated to use AR / VR in the project presentation phase	A4
The materials, textures and lighting of a virtual scene should always be as realistic as possible	A5
Environmental sounds should be as realistic as possible	A6
The existence of ambient music satisfies me in the visualization and interaction with virtual proposals	A7
The display device (mobile, tablet, computer, glasses) greatly influences the perception of virtual quality	A8
The use of a crowded environment (with missions and achievements) is better than free navigation by virtual zones	G1
In the case of gaming systems, I am motivated more those who are single-user regarding the multi-player	G2
I consider that the use of games in educational environments provide a better understanding of the subjects	G3
I enjoy playing in virtual environments. I consider these kind of experiences very funny and entertaining	IM1
Through the 3D visualization I acquire better skills related to the architecture compared to traditional systems	IM2
I believe that the use of virtual and gamified proposals need less effort than traditional systems	IM3
The use of virtual proposals generates less tension, stress, nervousness than models based systems and panels	IM4
Do you consider that by using virtual proposals you change the way you work in the future with architectural presentations?	IM5
This type of activities (games and virtual navigation), are useful for my future and can benefit me	IM5
These systems help me interrelate with other users / partners / friends by expanding my relationships	IM7

This value reflects an average interest, highlighting in the lower zone with a 2.9 the possible utility that students see in games in educational settings. The last block focused on IMI measurement has averaged 3.0. The best-rated aspect with a 3.5 has been the IMI-7 (the ICT systems helps the personal relationship with my colleagues), and the smaller the IMI-1 (enjoy the game as a fun and entertaining experience).

Once the course has been completed, the students have had to make the urban proposals for the "pedestrianization" of specific streets in Barcelona (see Fig. 3). Finally, they had to integrate their proposals into the interactive system for visualization (Fig. 4). After that, a Post-Test was performed in which, in addition to repeating the initial variables, they were asked about certain aspects related to usability and satisfaction with the proposed method. Fig. 5 shows the results obtained from the Post-Test, and their comparison with the data obtained from the Pre-test. Globally, the indices have increased in value from an overall average of 3.30 to an average of 3.75.

CONCLUSION

Technologies hold the potential to help and enhance learning processes, at different levels of education and in a variety of settings, covering formal, informal and lifelong learning. The cases presented here show some of the efforts of our research group and the lessons learned and demonstrate that can be adapted easily to other environments and situations.

Figure 3. Examples of the student's proposals and the programmed environment

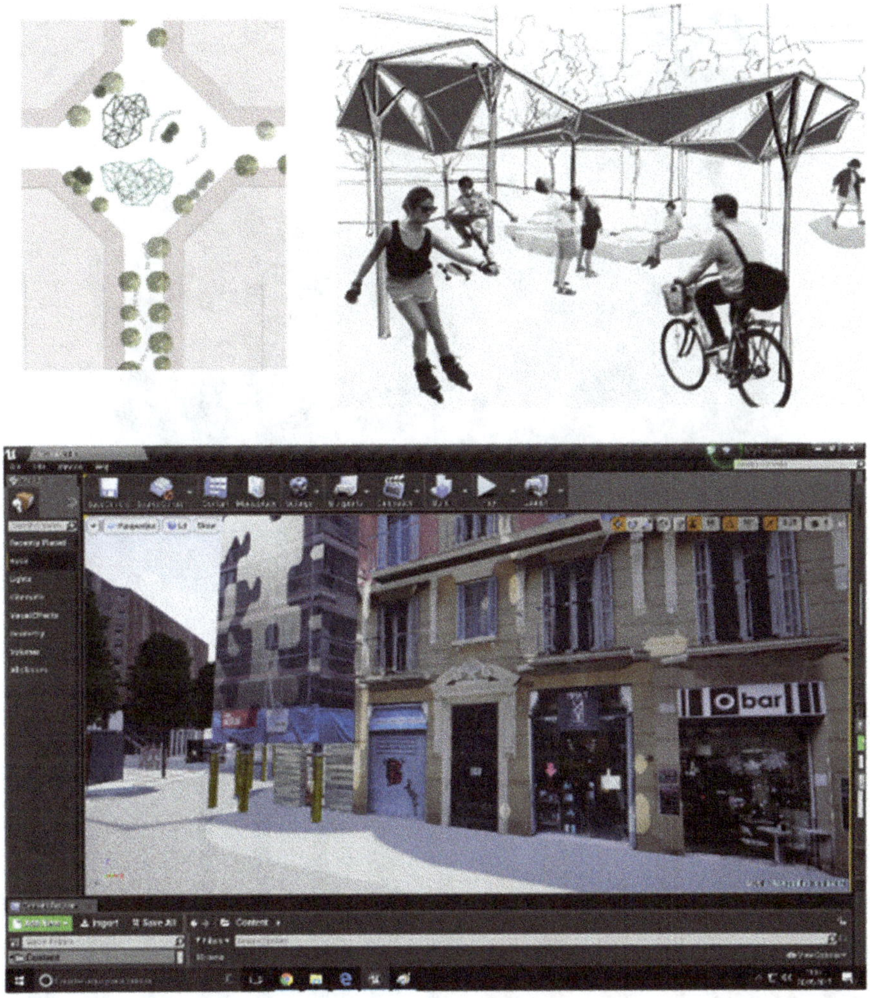

The analysis of the data gathered in the multimedia gamification study revealed that the implementation of the methodology based on gamification had the desired effect in terms of increasing student performance (Labrador & Villegas, 2014). Taking into account the various profiles defined in the game, 90% have increased their level, some of them significantly. These changes evidence that the practical exercises were perceived as more motivating, which in turn shows in better results in the practical exercises, especially considering that the adjustment of the points allocation was not so exaggerated as to have justified such a major change alone. Based on the data analyzed, and still considering a high dropout rate, it can be concluded that the implementation of the gamification-based methodology has worked well with the students who are predisposed to study, but does not work with those who are not so predisposed. Gamification has helped students learn more effectively the roles of the industry by increasing their awareness of their strengths and weaknesses, far beyond a mere final numerical note, which gives no idea of the actual acquisition of knowledge.

In the specific case of PLEs, the projects allowed us to identify evidences for the impact of PLEs in acquiring skills and knowledge, strengthen social interactions, and their usefulness as management and

Figure 4. Final practice and interaction at class

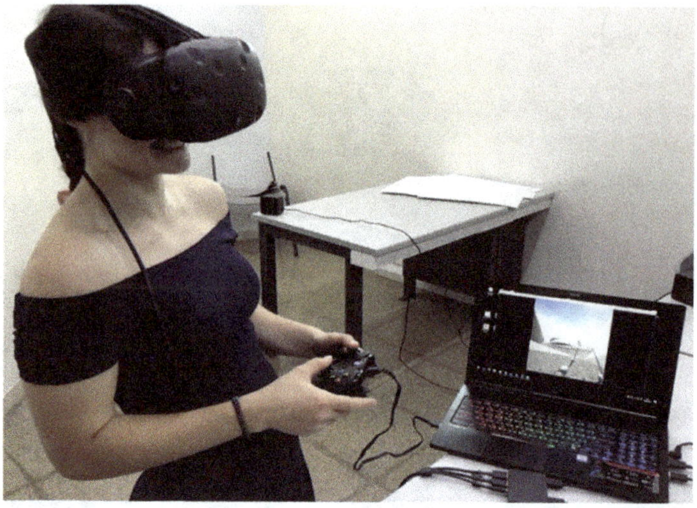

Figure 5. Pre-test vs. post-test data analysis

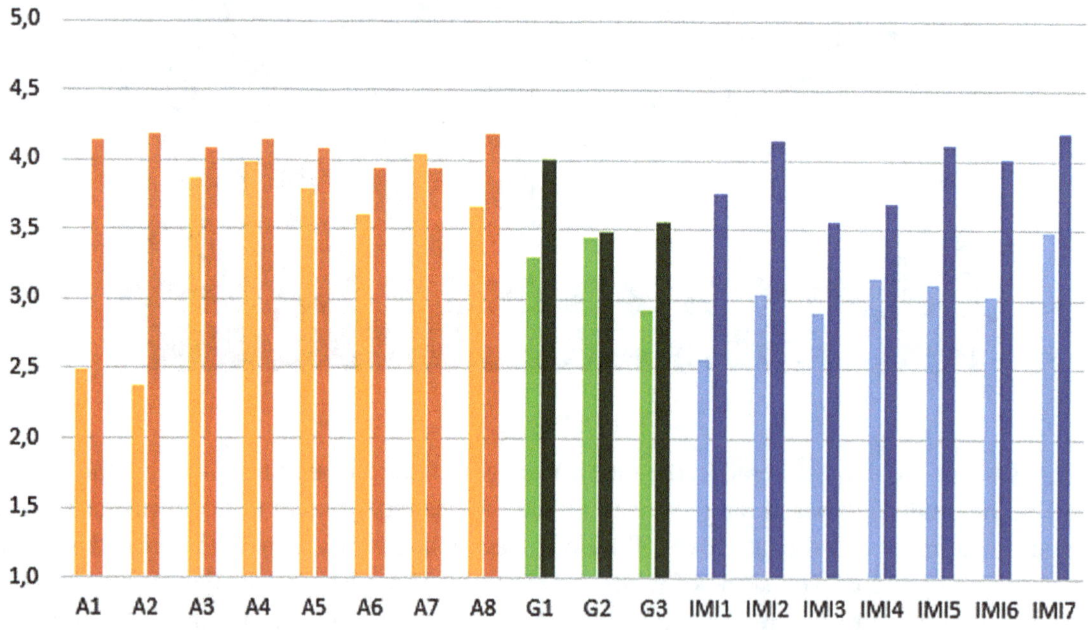

organization tools. It also showed some of the perceived disadvantages and obstacles they presented, and their dynamic nature. PLEs are not only difficult to define as a concept, but their ever-changing nature, and the fact that they depend on the learner's choices, makes it difficult to come up with guidelines or fixed approaches to their implementation, development and use. Nevertheless, a few of the lessons learned from these experiences may be summed up in a few basic best practices, which can be grouped in three main areas:

- **Social:** The Personal Learning Networks (PLNs) built during the process of developing PLEs are of great help and support for learners. Communities and a culture of collaboration and cooperation allow for peer-support and foster participation. Creating activities that require teamwork and active involvement with other learners could help some learners with the difficulties and obstacles they encounter during their learning process.
- **Technological:** Not focusing on specific tools and services, but rather on their functions and how to apply them works best in the long term, especially in the current dynamic and changing environment. As services disappear, stop working or are replaced by other applications, learners are forced to keep up with these changes. Showing them alternatives takes away the focus from the tool itself and puts it back on its value and usefulness. Having options increases the learners' confidence in managing and dealing with problems and difficulties.
- **Methodological:** Including activities and e-tivities that allow learners to explore different paths and services, and that fit within the broader context of the topics they are learning, helps making sense of the usefulness and applicability of the tools. A "learn to learn" approach, where self-learning is fostered, prepares them for learning new tools on their own and helps with transitions between services.

Allowing learners to follow their own approach and pace helps with making the experience more personal, but it does not require teachers and facilitators to personalize it for each one of the participants. By offering options and alternatives, learning objectives may be achieved in different ways.

Finally, in our example related with the educational framework of Architecture studies, we have analyzed the impact of including new interactive and gamified visualization systems in the development of the urban project as an educational tool. The continuous advancement of Information and Communication Technologies and all the different teaching approaches may allow systems such as games, interaction or real-time rendering to be used in areas such as urban design or project design. The initial reluctance logic is overcome when the results can be seen. The case study confirms this evolution of the student, especially in terms of motivation and the perception that the systems used can have in the representation of both architectural and urban projects. While initially the motivation could be considered medium / low, after the completion of the case study, it has increased significantly, an aspect that not only reflects the usefulness of the method, but also the potential in the academic and competence improvement of the student, which previously had already been referenced is linked to the student's motivation. The project in sub-following phases will improve the navigation and interaction of the system to adapt it to the needs of each urban zone, all of which will be introduced cyclically in the contents of the subjects to involve the student in real projects.

ACKNOWLEDGMENT

This research was supported by the National Program of Research, Development and Innovation aimed to the Society Challenges with the references BIA2016-77464-C2-1-R & BIA2016-77464-C2-2-R, both of the National Plan for Scientific Research, Development and Technological Innovation 2013-2016, Government of Spain, titled "*Gamificación para la enseñanza del diseño urbano y la integración en ella de la participación ciudadana* (ArchGAME4CITY)", & "*Diseño Gamificado de visualización 3D con sistemas de realidad virtual para el estudio de la mejora de competencias motivacionales, sociales y espaciales del usuario* (EduGAME4CITY)". (AEI/FEDER, UE).

REFERENCES

Abras, C., Maloney-Krichmar, D., & Preece, J. (2004). User-centered design. In Encyclopedia of Human-Computer Interaction. Thousand Oaks, CA: Sage Publications.

Adell, J., & Castañeda, L. (2010). Los entornos personales de aprendizaje (PLEs): una nueva manera de entender el aprendizaje. In Claves para la investigación en Innovación y calidad educativas. La integración de las Tecnologías de la Información y la Comunicación y la Interculturalidad en las aulas (pp. 19-30) Alcoy: Marfil.

Álvarez Cuervo, R., Roces, J., & Alonso, J. (2006). *Sistema tutorado interactivo para el dibujo de vistas SDV®. XVIII Congreso Internacional Ingeniería Gráfica*, Sitges, Barcelona.

Álvarez Peñín, P., Del Canto, A., Charro, M., Rubio, R., García, R., & Suárez, J. (2001). *Propuesta de un sistema multimedia para minimizar el efecto de la reducción de docencia tradicional en Expresión Gráfica. XIII Congreso Internacional de Ingeniería Gráfica*, Badajoz.

Álvarez Peñín, P., Pando, P., García, R., & Pérez, M. (2006). *Entorno multimedia para uso docente en materias de Expresión Gráfica. XVIII Congreso Internacional de Ingeniería Gráfica*, Sitges, Barcelona.

Área, M. M. A., San Nicolás, M. B., & Fariña, E. (2010). Buenas prácticas de aulas virtuales en la docencia universitaria semipresencial. *Revista de Teoría de la Educación Sociedad de la Información, 11*(3), 7–31.

Attwell, G. (2006). *Personal Learning Environments – a position paper*. Retrieved from http://www.knownet.com/writing/weblogs/Graham_Attwell/entries/6521819364

Biggs, J. (1999). What the Student Does: Teaching for enhanced learning. *Higher Education Research & Development, 18*(1), 57–75. doi:10.1080/0729436990180105

Blanco, M. (2006). Estudio del sistema diédrico mediante un tutorial multimedia. *XVIII Congreso Internacional Ingeniería Gráfica*, Sitges, Barcelona.

Bloom, B. S. (1953). Thought-processes in lectures and discussions. *The Journal of General Education, 7*(3), 160–169.

Brophy, J. E. (2013). *Motivating students to learn*. Routledge.

Buckley, P., & Doyle, E. (2016). Gamification and student motivation. *Interactive Learning Environments, 24*(6), 1–14. doi:10.1080/10494820.2014.964263

Castañeda, L., & Adell, J. (2011). El desarrollo profesional de los docentes en entornos personales de aprendizaje (PLE). In R. Roig Vila & C. Laneve (Eds.), *La práctica educativa en la Sociedad de la Información: Innovación a través de la investigación / La pratica educativa nella Società dell'informazione: L'innovazione attraverso la ricerca* (pp. 83–95). Alcoy: Marfil.

Caudell, T. P., & Mizell. (1992). Augmented reality: An application of heads-up display technology to manual manufacturing processes. "System Sciences, 1992. *Proceedings of the Twenty-Fifth Hawaii International Conference on.* IEEE. doi:10.1109/HICSS.1992.183317

Chickering, A. W. & Gamson, Z. F. (1987, March). Seven Principles for Good Practice in Undergraduate Education. *The American Association for Higher Education Bulletin.*

Connolly, T. M., Boyle, E. A., MacArthur, E., Hainey, T., & Boyle, J. M. (2012). A systematic literature review of empirical evidence on computer games and serious games. *Computers & Education, 59*(2), 661–686. doi:10.1016/j.compedu.2012.03.004

Cronk, M. (2012, June). Using gamification to increase student engagement and participation in class discussion. *World conference on educational multimedia, hypermedia and telecommunications*, 2012(1), 311-315.

Domínguez, A., Saenz-de-Navarrete, J., De-Marcos, L., Fernández-Sanz, L., Pagés, C., & Martínez-Herráiz, J.-J. (2013). Gamifying learning experiences: Practical implications and outcomes. *Computers & Education, 63*, 380–392. doi:10.1016/j.compedu.2012.12.020

Epper, R., & Bates, A.W. (2004). *Enseñar al profesorado cómo utilizar la tecnología. Buenas prácticas de instituciones líderes.* Barcelona: Editorial UOC. Colección Educación y Sociedad Red.

Fernández, A., Reina, R., Mateo, F., & Sánchez, J. (1998). Docencia de dibujo técnico a través de herramientas de animación asistida por ordenador. *X Congreso Internacional Ingeniería Gráfica*, Málaga

Fiedler, S. (2010). *Personal Learning Environments: concept or technology?* The PLE Conference, Barcelona, Spain. Retrieved on September 23rd 2015 from: http://pleconference.citilab.eu/wp-content/uploads/2010/07/ple2010_submission_45.pdf

Fonseca, D., Martí, N., Redondo, E., Navarro, I., & Sánchez, A. (2014). Relationship between student profile, tool use, participation, and academic performance with the use of Augmented Reality technology for visualized architecture models. *Computers in Human Behavior, 31*, 434–445. doi:10.1016/j.chb.2013.03.006

Fonseca, D., Redondo, E., & Villagrasa, S. (2015). Mixed-methods research: A new approach to evaluating the motivation and satisfaction of university students using advanced visual technologies. *Universal Access in the Information Society, 14*(3), 311–332. doi:10.1007/s10209-014-0361-4

Fonseca, D., Villagrasa, S., Navarro, I., Redondo, E., Valls, F., & Sánchez, A. (2017, April). Urban Gamification in Architecture Education. In *World Conference on Information Systems and Technologies* (pp. 335-341). Springer.

García, M., Martín, G., Suárez, J., Pérez, J., & Suárez, F. (2002). *S.A.D. (Sistema de apoyo al dibujo): Una herramienta integrada en la enseñanza de la expresión gráfica. XIV Congreso Internacional de Ingeniería Gráfica*, Santander.

Garmendia, M., Gisasola, J., & Gorozika, J. (2004). *Enseñanza de la visualización de piezas como resolución de problemas. XVI*. Zaragoza, Huesca: Congreso Internacional de Ingeniería Gráfica.

González, T., & Rodríguez, L. (2010). El valor añadido de las buenas prácticas con TIC en los centros educativos. *Revista de Teoría de la Educación Sociedad de la Información, 11*(3), 262–282.

Hamari, J., Koivisto, J., & Sarsa, H. (2014, January). Does gamification work?--a literature review of empirical studies on gamification. In *System Sciences (HICSS), 2014 47th Hawaii International Conference on*. IEEE.

Harlen, W., & Deakin Crick, R. (2003). Testing and motivation for learning. *Assessment in Education: Principles, Policy & Practice, 10*(2), 169–207. doi:10.1080/0969594032000121270

Hunicke, R., LeBlanc, M., & Zubek, R. (2004, July). MDA: A formal approach to game design and game research. *Proceedings of the AAAI Workshop on Challenges in Game AI*, 4.

James, K., Humpherey, G., & Vilis, T. (2002). "Active" and "Passive" Learning of Three-Dimensional Object Structure Within an Immersive Virtual Reality Environment. *Behavior Research Methods, Instruments, & Computers, 34*(3), 383–390. doi:10.3758/BF03195466 PMID:12395554

JISC. (2007). *A report on the JISC CETIS PLE project*. Available online at http://wiki.cetis.ac.uk/Ple/Report

Johnson, L., Levine, A., & Smith, R. (2009). *The 2009 Horizon Report*. Austin, TX: The New Media Consortium.

Kapp, K. M. (2012). *The gamification of learning and instruction: game-based methods and strategies for training and education*. John Wiley & Sons.

Kaufmann, H., Steinbügl, K., Dünser, A., & Glück, J. (2003). Improving Spatial Abilities by Geometry Education in Augmented Reality - Application and Evaluation Design. *Proceeding of the 2005 VRIC Laval Virtual*, 25-34.

Koehler, M., & Mishra, P. (2009). What is technological pedagogical content knowledge (TPACK)? *Contemporary Issues in Technology & Teacher Education, 9*(1), 60–70.

Labrador, E., & Villegas, E. (2014). Sistema Fun Experience Design (FED) aplicado en el aula. *Re-Visión, 7*(2).

Martín-Gutiérrez, J. (2010). AR_Dehaes: an educational toolkit based on augmented reality technology for learning engineering graphics. *2010 10th IEEE International Conference on Advanced Learning Technologies*.

Mills, E. K., Jr. (1941). *U.S. Patent No. 2,243,973*. Washington, DC: U.S. Patent and Trademark Office.

Mora, P., García, & Marín. (2001). *Módulo de AutoCAD orientado a la enseñanza de piezas*. XIII Congreso Internacional de Expresión Gráfica en la Ingeniería, Badajoz.

Navarro, I., & Fonseca, D. (2017). Nuevas tecnologías de visualización para mejorar la representación de arquitectura en la educación. *ACE: architecture, city and environment, 34*, 218-237. DOI:10.5821/ace.12.34.5290

Nicholson, S. (2012). A user-centered theoretical framework for meaningful gamification. *Games+ Learning+ Society, 8*(1).

Piaget, J. (1952). The origins of intelligence in children. New York: International Universities Press. doi:10.1037/11494-000

Pifarré, M., & Tomico, O. (2007). Bipolar laddering (BLA): A participatory subjective exploration method on user experience. In *Proc. Conference on designing for user eXperiences*. ACM, doi:10.1145/1389908.1389911

Prensky, M. (2001). Digital natives, digital immigrants part 1. *On the Horizon, 9*(5), 1–6. doi:10.1108/10748120110424816

Ramos, G., Baños, & Melgosa. (2003). *Aprendizaje innovador en la visualización de piezas y dispositivos, en la formación de dibujo técnico mediante aplicación hipermedia*. XV Congreso Internacional de Ingeniería Gráfica, Nápoles, Italia.

Rogers, D. L. (2000). A paradigm shift: Technology Integration for Higher Education in the New Millennium. *Educational Technology Review*, 19–33.

Salmon, G. (2004). *E-tivities: The key to active online learning*. London, UK: Routledge.

Schell, J. (2014). *The Art of Game Design: A book of lenses*. CRC Press. doi:10.1201/b17723

Schmidt-Atzert, L. (1985). *Psicología de las emociones*. Herder.

Skinner, B. F. (2011). *About behaviorism*. Vintage.

Torres, R., & Edirisingha, P. (2010). Flight of the PELICANS. *Proceedings of The PLE Conference 2010*. Retrieved from http://pleconference.citilab.eu

Torres, R., Edirisingha, P., & Mobbs, R. (2009). Putting the pieces together: Conceptual frameworks for building PLEs with Web 2.0 Tools. In *Distance and e-learning in transition*. Wiley-ISTE.

Torres, R., Martín, R., & Ortiz, A. (2011). *New Technologies, New Environments: developing Personal Learning Environments in Secondary Education*. Presented at McLuhan Galaxy Conference, Barcelona, Spain.

Vallerand, R. J., Pelletier, L. G., Blais, M. R., Briere, N. M., Senecal, C., & Vallieres, E. F. (1992). The academic motivation scale: A measure of intrinsic, extrinsic, and amotivation in education. *Educational and Psychological Measurement, 52*(4), 1003–1017. doi:10.1177/0013164492052004025

Van Harmelen, M. (2008). Design trajectories: Four experiments in PLE implementation. *Interactive Learning Environments, 16*(1), 35–46. doi:10.1080/10494820701772686

Villagrasa, S., Fonseca, D., Redondo, E., & Duran, J. (2014). Teaching case of gamification and visual technologies for education. *Journal of Cases on Information Technology*, *16*(4), 38–57. doi:10.4018/jcit.2014100104

White, D., & Le Cornu, A. (2011). Visitors and Residents: A new typology for online engagement. *First Monday, 16*(9).

Zichermann, G., & Cunningham, C. (2011). *Gamification by design: Implementing game mechanics in web and mobile apps*. O'Reilly Media, Inc.

KEY TERMS AND DEFINITIONS

Computer Skills/Competences: The level of knowledge regarding the use of IT systems that a student has.

Engagement: The participation of learners in the activities offered in the course.

Gamification: Serious games applied to improve the motivation of the student and the interaction with the educational content.

Good Educational Practices: Digital educational models produced specifically with the purpose of supporting teaching and learning in a more productive environment.

Personal Learning Environment: An environment that allows the individual to take control of managing the learning process, facilitating his/her learning.

Technology-Enhanced Learning: New models of learning based on the use of technology that can improve the engagement of the student.

Virtual and Augmented Reality: New type of digital content with a set of resources and tools that favor the interaction with 3D contents among students and teachers.

Chapter 7
Current Situation and Appraisal Tendencies of M-Learning

Laura Briz-Ponce
University of Salamanca, Spain

Juan Antonio Juanes-Méndez
University of Salamanca, Spain

Francisco José García-Peñalvo
University of Salamanca, Spain

ABSTRACT

Mobile devices and apps are placed in a prominent position in the daily routine of all people. The fast evolution of these devices and their main advantages have caused a real impact in the society. Currently, they are essential tools to be totally connected anywhere and to consult and access information of any field. The use of these devices has been increasing since the last decade. The emergence of new and sophisticated devices and new services has contributed to this sparkling uptrend. One of the fields that society is using these mobile technologies is in learning. The aim of this chapter is to describe the current situation of these technologies and to make an approach of the future tendencies of these tools. To achieve this goal, it was necessary to conduct a survey and involve different undergraduate students of the university and different professionals. Results reveal that students are using more and more apps and mobile devices but there is an important gap between students and professionals so it is still necessary to boost their relevance to improve their potential use.

INTRODUCTION

Currently, there are 51,298 million of mobile lines in Spain (CNMC, 2017). Step by step, people have got used to these devices, which have provided new ways of communication, interacting each other, getting information or even learning.

The number of these lines has been increasing dramatically since the last decade. Not only that, this trend has sparked the appearance of mobile devices more sophisticated as Smartphone and tablets where

DOI: 10.4018/978-1-5225-4944-4.ch007

it is possible to run mobile applications or apps on them. In fact, according to the study conducted by *Fundación Telefónica* (Fundación Telefónica, 2017), there is roughly 27,7 million of active users and 3,8 million of daily downloads in Spain. Smartphones have, on average, 30 apps installed on it per user, whereas tablets have around 24 apps.

These apps can be found in the marketplace, available since different mobile devices. In fact, according the reports, there are currently almost 3 million of apps in Google Play (Statista, 2017a) and more than two million in App Store (Statista, 2017b).

On the other hand, there is a study conducted by Educause Center for Applied Research (ECAR) (Eden Dahlstrom, 2012) that has published an article about the use of mobile technology in higher education. One of the results claimed that around 67% of surveyed students reported that mobile devices are very important in their academic success and their activities in the University. In fact, one of the upward trends of the last years is precisely the application of new technologies within the educational context and specifically of mobile devices and its educational use, which is known as mobile learning. Therefore, higher education and consequently and specifically, medical education, are highly impacted by this trend.

There are many reports related with the use of mobile devices (Mobile Marketing Association, 2017; Nielsen, 2014; *The State of Mobile World 2014*, 2014) in general and the use of students in particular (Eden Dahlstrom, 2012). One of the reports (Fundación Telefónica, 2014) affirmed that around 90% of medical professionals has accessed Internet during 2013 and 51% of them has used Smartphones to access medical information. Recently, it has published another report that reveals that 29,9% of Internet users have used them for training courses (Fundación Telefónica, 2017). Besides, the percentage of Internet users from 14 to 19 years old, that have chosen education as first reason to use Internet, has been increased in 8 percentage points.

The use of mobile devices and its inclusion as learning tools in the classrooms of Medical Schools is becoming a reality. Some Schools, as the Medical School of University of Stanford have already adopted the new technologies (mainly tablets) for leaning, forcing their use providing one of them to all undergraduate students (Dolan, 2011; Gallegos, 2013)

Students want to be more and more prepared for the work, so University and Medical Schools must adopt the cutting-edge technologies for providing the students the best education and allow them to be part of the new digital era. Because of that, they must solve some pitfalls in order to modify their curriculum. It is important to notice that the Medical Schools that have implanted these devices, they are still using them (Briz-Ponce, Juanes-Méndez, & García-Peñalvo, 2014b)

However, in spite of these advantages, the use of medical apps for medical education is still limited or unknown. This paper wants to analyse these benefits and the behaviour of a specific group such as students and medical professionals. Besides, it wants to provide some insights about the main factors to consider when it is necessary to promote the use of apps in an academic environment. This outreach includes an up-to-date of different participants' characteristics and how they could influence on the use of mobile devices for learning. It is important to notice that this paper also tries to analysis the relationships between these characteristics and how they may have an effect on the participants' behaviour. Most part of them were analysed previously (Briz-Ponce & Juanes-Méndez, 2015) but this paper completes the information and also includes some current tendencies in the field of m-learning.

The study is divided in four sections. The first section describes the methodology used to perform the survey in the University of Salamanca and provides a review of the main characteristics of the participants' profile. The second section explains the results obtained focusing on the main uses of mobile devices and the different types of apps utilized and the time employed. The third part is focused on the discussion and finally it describes the main conclusion drawn of this research.

METHODOLOGY

Method

The method used for the current investigation is a cross-sectional survey distributed in two ways: face to face and on line.

The face-to-face survey was conducted in the Medical School of the University of Salamanca, distributing it directly to undergraduate students ten minutes before a class. In case of medical professionals, the survey was giving directly to personal contacts.

The online survey was developed using GoogleDocs. The participants received a link to the survey. The first page described in detail the goal of the survey as part of the study of a doctorate programme. The participants answered the survey anonymously.

The conducted survey was formed by 19 questions which covered the main participant's characteristics: gender, age, ownership of mobile devices, daily use of Smartphone and tablets, the most important issues for downloading apps and the type of apps downloaded by the participants.

The data were collected for a month since March 2014 to April of 2014 and the final number of participants was 124. The data were computerized in the program SPSS v21 (Vinacua, 2007) to analyse the different variables obtained from the data (Lee & Wang, 2003)

Participants

The survey was conducted in the University of Salamanca and the participants were undergraduate students of Medical School and professionals. It involved medical residents, medical specialists and medical teachers in the group of medical professionals. The Table 1 provides information of the survey participant's characteristics, such as frequency and the percentage of each dimension.

As for the data, the dominant age of the participants falls within the range from 18 to 35 years old with 71% of the participants.

Besides, half of the participants were students (n=61, 49,2%), medical residents were 19,4% (n=24), medical specialists 16,9% (n=21) and finally medical teachers 14,5% (n=18) which it means that medical professionals represent 50,8% of total. More information about the characteristics of participants could be found on the published articles (Briz-Ponce, Juanes-Méndez, & García-Peñalvo, 2014a, 2014c)

It is important to emphasize that most part of the survey respondents (94,4%) owned a mobile device.

RESULTS

Use of Mobile Devices

This section analyses the daily use of participants with their smartphones, tablets or both. The obtained results reported that most part of the participants used the mobile devices daily and only 9% of them

Table 1. Descriptive statistics on participant characteristics

Variable	Participants' Characteristics		
	Descriptive	**Frequency**	**%**
Gender	Male	45	36,3%
	Women	79	63,7%
Profile	Student	61	49,2%
	Medical Residents	24	19,4%
	Medical Specialists	21	16,9%
	Medical teachers	18	14,5%
Range of age	From 18 to 25 years	60	48,4%
	From 26 to 35 years	28	22,6%
	From 36 to 45 years	6	4,8%
	From 46 to 55 years old	23	18,5%
	+ 55 years	7	5,6%
Ownership	Only Smartphone	58	46,8%
	Only Tablet	1	0,8%
	Smartphone and Tablet	58	46,8%
	None	7	5,6%

that owns a smartphone, did not use it to download apps. This data reveals the importance that the user gives to the apps and how the apps are widely spread in the Society, as it was commented before.

Frequency of Use

In the survey, all the participants had to answer their frequency of use of apps on Smartphone and/or tablets. The results differentiate between the data obtained for students and the one obtained for professionals, overall to analyse if there is any difference between them. The results claimed that medical professionals (medical residents, specialists and teachers) used Smartphones between 1 and 2 hours per day, whereas students used them between 3 and 4 hours per day. In addition, analysing more in depth the tablets, the results obtained are completely different. In this case, students, medical specialists and medical teachers on average do not use tablets. On the contrary, medical residents are very active users using the tablets on average between 3 and 4 hours per day.

It is interesting to know the dependency between two variables: daily use of apps on Smartphone and the profile of participants. The null hypothesis indicates that there is no dependency between them and the alternative hypothesis is there is a dependency between them. As they are nominal variables, it was necessary to use the non-parametric Chi-Square method (Chakravarti, Laha, & Roy, 1967, pp. 392–394). In this case, the SPSS program was used. The variable selected as row is: daily apps for Smartphone and as column, the variable is: the profile of the participants (students, medical residents, medical specialists and teachers). The values of Phi and Cramer's V (Alan Agresti, 2002) were calculated as well. The results were: Chi-Square=71,012, ρ=0,000, Phi=0,757 ρ=0,000 and Cramer's V =0,437, ρ=0,000.

At the α=0,05 level of significance, there exists enough evidence to reject the null hypothesis that considers both variables independents (p=0,000, p<0,05); besides, the values of Phi and Cramer's V falls within the range from 0,3 to 0,7 so it can be said that there is a moderate relationship between the daily use of the smartphones and the profile of the participants.

Then, the same process was performed again, but in this case, between these two variables: the profile of the participants and the use of tablets. The results obtained were Chi-Square=14,054, ρ=0,120, Phi=0,347 ρ=0,120 and Cramer's V=0,200, ρ=0,120.

At the α=0,05, level of significance reveals that there is no evidence enough to reject the null hypothesis that both variables are independents (ρ=0,120>0,05).

In addition to this information and following with the regression analysis, it is necessary to define several criterion variables, which means the outcomes of the predictor variable (Salkind, 1998), in this case the frequency of Smartphone use and the frequency of Tablet use.

According to the survey, it is possible to identify four types of variables, as it is shown in Table 2

Analysing the results again with the SPSS program and after considering the normality and homocesticity requirements to use parametric or non-parametric techniques, the outcome data are shown in Table 3 and Table 4.

According to these results, there is a significance difference for the participant's perception of m-learning and the frequency Smartphone use (ρ<0,05), except for the variable Tools, which it means that it is not possible to predict this variable with the frequency Smartphone use.

Table 2. Criterion variables

Criterion Variable	ID	Description
Perception of the use of m-learning	PERC	This construct measuares the participants' perception of m-learning
Resources avaible for using m-learning	TOOL	This construct takes into account how the participants evaluate the available tools for using m-learning
Emotional perception of using m-learning	EMO	This construct considers also the anxiety and the trust for using m-learning
Global Evauation	VGLOB	This construct gives an overall score for the use of m-learning

Table 3. Criterion variables and frequency smartphone use

ID	Technique	Parameters
PERC	ANOVA	F=7,651
		ρ=0,000
TOOL	Kruskal-Wallis	H=7,350
		ρ=0,119
EMO	ANOVA	F=1,465
		ρ=0,043
VGLOB	ANOVA	F=196,494
		ρ=0,002

Table 4. Criterion variables and frequency table use

ID	Technique	Parameters
PERC	ANOVA	F=9,379
		ρ=0,000
TOOL	ANOVA	F=4,243
		ρ=0,007
EMO	Kruskal-Wallis	H=8,734
		ρ=0,033
VGLOB	ANOVA	F=9,199
		ρ=0,000

In this case, there is a significance difference for the participant's perception of m-learning and the Frequency Tablet use (ρ<0,05), which it means that it is possible to predict all these variables with the frequency use factor.

Characteristics

Other aspect to consider in the survey was the most important factors that participants considered to download apps. It was possible to select more than one option, without a limitation on the maximum number of answers. The Figure 1 shows the percentage of each factor considering all participants. It is important to notice that the participants could select more than one option

On average, each participant had selected 2,89 characteristics of 8 options they could choose.

Then, Figure 1 shows this information categorizing the results by profile of participants.

From here, it is possible to deduct several points. The first point is that the content is the key characteristic for all profiles to download apps. Medical specialists and residents considered that the content and the security/privacy have the same level of importance.

Recommendation of a friend is the factor most used for medical residents and teachers, whereas this characteristic is not so relevant for students.

In addition, the survey included an open question to give the participants the possibility of providing other factors not included already in the list. 83,9% of participants (n=124) considered that there are not any additional characteristics. A minority of them, 4,8% did not even answer. The total number of answers for this question was very scarce. However, 6,5% and 1,6% of participants that answered it, reported that the price and the capacity respectively, were very important to take into account when downloading any app.

Finally, the survey gave the opportunity to the participants to chose only one characteristic for downloading apps. The content factor was scored as first option (37%) for professionals, whereas security/privacy was the most important factor for students (43%) as it is possible to see in Figure 2

Figure 1. Key factors to download apps according to the profile

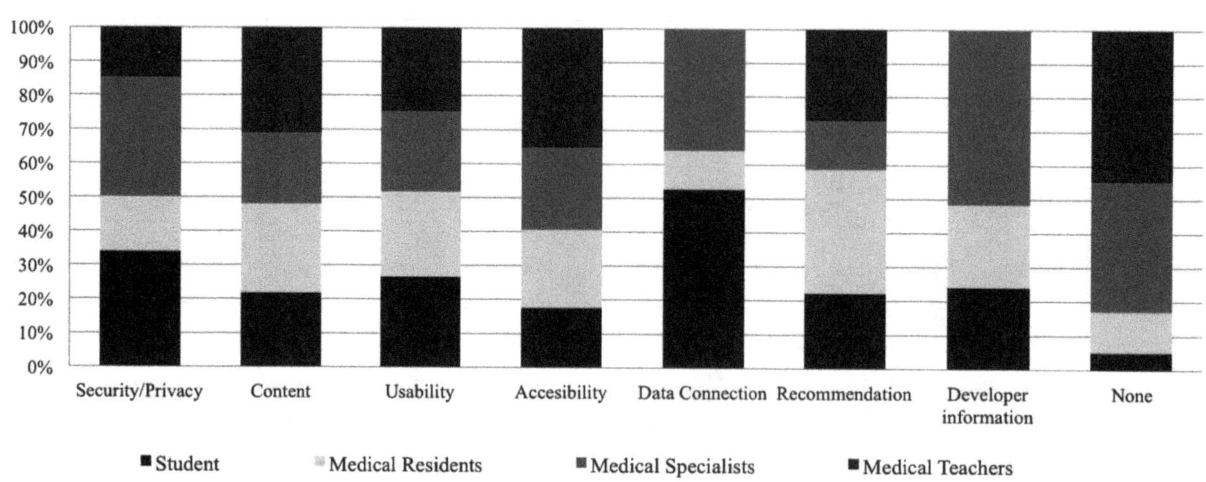

Figure 2. The most important characteristic to download apps

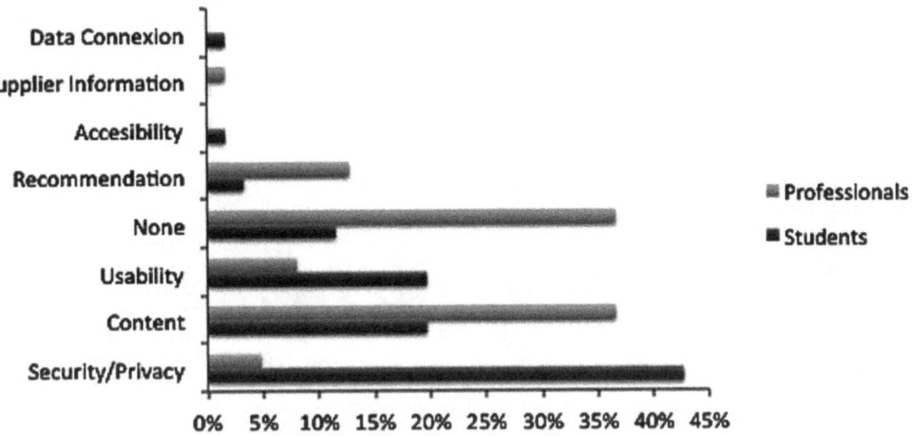

Besides, in this analysis, the gender was added to analyse which factor could be considered as the most important ones. The results revealed that for women students, the most important one is security and privacy, whereas for men students are the content factor. For professionals, both men and women consider also the content factor as the most important one to download an App.

Type of Apps

This section explains the type of apps that students and medical professionals use more frequently. This analysis applies to the type of apps, not the app itself. The most popular type of apps are related with entertainment, followed by social networks and games. The apps related with medicine: medical apps and training medical apps are placed on 5th and 7th position respectively. It is important to highlight the difference between the apps used for medical education (such as atlas, information, etc.) and the apps used for medical diagnosis (drugs dictionary, taking care of patient, medical calculator, etc.).

Besides, Figure 3 shows the type of apps selected by participants (students and professionals).

As it is possible to see in this figure, most part of the students prefer social networks apps (93%), whereas 68% of professionals chose entertainment apps as their favourite followed by medical Apps (60%). This is an important point, as it reveals the increasing use of these type of Apps among professionals.

In addition, as this study is mainly focused on apps related with medicine, specifically on two types of apps: medical apps and training medical apps. The Figure 4 depicts the detailed information of the participants' answers for these type of apps. The reasons claimed by students to explain this result is the unawareness of these types of apps or that they did not have to use them.

From the graphic of Figure 4, it is possible to observe that the number of students that had used a training medical app is higher (5%) than the one that had used a medical app (2%).

The total percentage of professionals that had used some apps related with medicine is 32%, whereas considering only the number of professionals, the percentage is increased until 63%.

Figure 3. Type of apps selected by participants

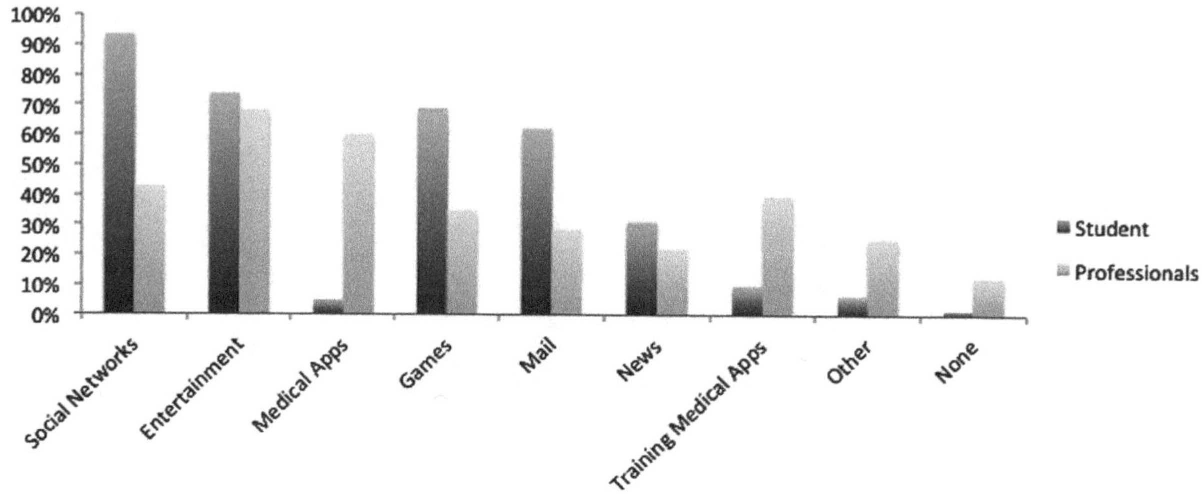

Figure 4. Percentage of participants that have selected medical apps or training medical apps according to the profile

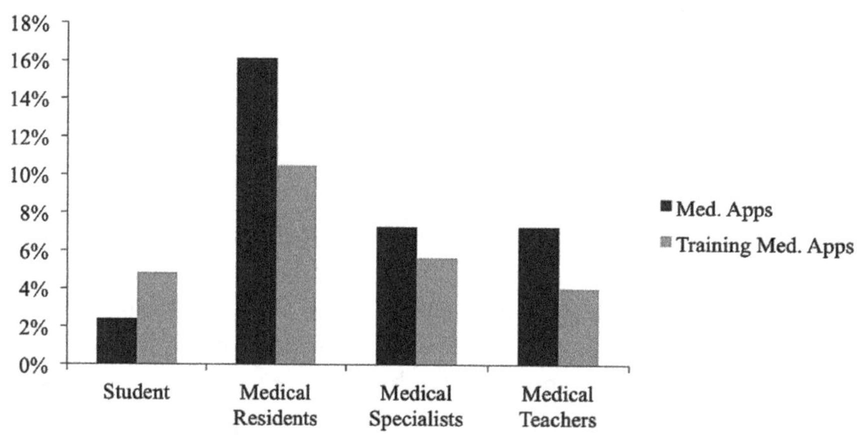

Number of Downloaded Apps

The survey also studied the number of apps that the participants had downloaded during the last month. The measurement of the participants' activity covers only a short specific period of time, as it is more relevant the activity during the last month than one or two years ago.

The mode is calculated in order to know the most frequent value of the data. The results indicate that the number of apps downloaded on Smartphone was between 1 and 10, whereas on tablets, the figures indicated that they do not use them for downloading apps.

In this case, it is calculated the grade and the correlation relationship between these two variables (download of apps for Smartphones and Tablets) and the profile of the participants (students, medical residents, medical specialists and medical teachers). In this case, the same process as it was followed in the last analysis, was performed: the Chi-square method and the Phi and Cramer´s V value were calculated as well. The results were Chi-square=13,176, ρ=0,040, Phi=0,326 ρ=0,040 and Cramer's V =0,310, ρ=0,040.

The null hypothesis indicated that there was no relationship between the number of apps with Smartphone and the profile of participants, whereas the alternative hypothesis indicated that there was a relationship between them.

At α=0,05, level of significance, the output data was ρ-value <0,05 (ρ=0,04) so it is enough evidence to reject the null hypothesis. Thus, there is a relationship between both variables. In this case, the Phi and Cramer's V values were found within the range 0 – 0,3, so there is a weak relationship between them.

The same method was used to analyse the tablets. The null hypothesis was that there is no relationship between the number of apps utilized on tablets and the profile of the participants. The results were Chi-Square=7,968, ρ=0,537, Phi=0,259 ρ=0,537 and Cramer's V=0,251, ρ=0,537.

At α=0,05, level of significance, the results reveal that there is no enough evidence to reject the null hypothesis (ρ=0,537, ρ>0,05), so both variables could be considered independents

List of Apps

The survey also included a question related with the most used App. As it was shown previously, professionals are the participants that have used training medical Apps, in fact the residents were the most active users. The Table 5 shows a list with the top Apps more used for the participants. Vademecum and apps related with Anatomy head the list

DISCUSSION

These results can provide important insights in the use of mobile devices. The use of new technologies is part of our lives and it is a reality that it is not possible to avoid.

Table 5. List of apps

App	Freq	% Participants
Vademecum	6	16%
Anatomía	5	13%
3D_Brain	4	11%
Atlas Anatómico	2	5%
Diccionario	2	5%
Gray's Anatomy	2	5%
Medmecum	2	5%
Medscape	2	5%
Prospectos	2	5%
Radiología	2	5%
Skeleton	2	5%
Visual Anatomy	2	5%

European Commission is also aware of their rapid breakthrough and their contribution to the economic growth, so their priorities are aligned with this evolution. Other organizations, as United Nations Educational, Scientific and Cultural Organization, UNESCO, claimed that the new technologies applied to the learning and teaching methods are essentials for the students' good results (UNESCO, 2009). They also recognized that the integration of mobile phones into education carries a great potential that could make important changes on the way of learning (UNESCO, 2011).

There are several publications that consider the mobile technology with a great potential in the context of higher education. However, it seems that its use it is centred on a didactic goal, more than a constructivism ecosystem (Herrington, Herrington, Mantei, Olney, & Ferry, 2009).

This analysis reveals that professionals are using training medical apps more frequently than students, which answered that the reasons for not using them is unawareness or lack of necessity. However, some authors indicated that students consider these tools very useful and they are very willing to use them (Lea & Callaghan, 2011; López-Hernández & Silva-Pérez, 2014; Sarrab, Alzahrani, Al Awan, & Alfarraj, 2014; Teri et al., 2013). On the contrary, other authors reported that it is necessary to take into account some pitfalls before adopting these new tools from institutions (Alden, 2013; Ashour, Alzghool, Iyadat, & Abu-Alruz, 2012; Handal, Macnish, & Petocz, 2013; Martinez-Perez, de la Torre-Diez, & Lopez-Coronado, 2013; Ozdalga, Ozdalga, & Ahuja, 2012). In addition, some authors do not think that the mobile technologies could bring advantages to education and can even create distractions in classrooms (Chen, Lim, & Tan, 2010; Jackson, 2013; Quicios-García, Sevillano García, & Ortega Sánchez, 2013; Rapetti, Picco, & Vannini, 2011; Visvanathan, Hamilton, & Brady, 2012).

Other identified barriers to tackle in order to adopt these new technologies as for example the direct support of institutions or a regulation of Apps (Alrasheedi, Capretz, & Raza, 2015; Khatoon, Hill, & Walmsley, 2013; Lea & Callaghan, 2011; Martínez-Pérez, de la Torre-Díez, & López-Coronado, 2015; Visvanathan et al., 2012), technical problems (Ferreira, Klein, Freitas, & Schlemmer, 2013; Nuss, Hill, Cervero, Gaines, & Middendorf, 2014; Toktarova, Blagova, Filatova, & Kuzmin, 2015) and pedagogical problems (Alden, 2013; Ferreira et al., 2013; Handal et al., 2013). Other authors present a reflection of the capacities and the resources needed to make a technology integration in the classroom (Ramírez-Montoya, & García-Peñalvo, 2017).

Besides, in 2009, the International Telecommunication Union, (Adolph, 2009) published a brief report about mobile applications. They explained the rapid growth of these devices and they claimed that the lack of interoperability between different Operation Systems (Android, IOS, etc.) could be a constraint for the development of mobile apps market. Even, it could make the process slow and may cause customer confusion about availability and features.

Currently, this prediction could be considered a reality, as there is some confusion about the number of apps in the marketplace. The developers have to follow the process of different marketplaces and sometimes it is not realistic to launch the app in all of them for two main reasons: cost and time to market.

On the other hand, there are two tendencies spreading rapidly among the educative community (Fundación Telefónica, 2017): Bring your Own Device (BYOD) and the Massive Open Online Courses (MOOCs). The first term was translated directly from the enterprise world to the education world. It consists on allowing students to use their own device as a resource for learning. There are some universities that have adopted this model in order to reduce cost and providing students electronic documents and the devices for reading it instead of printing manuals or the documents (Briz-Ponce et al., 2014b).

As for MOOCs, considered as unaccredited studies, the number of courses took 6850 from 700 universities in 2016 (Shah, 2016) and reached 4200 from over 550 universities in 2015 (Shah, 2015), so this tendency is still expanding its footprint. Even prestigious universities as Oxford University or Yale University have added MOOCs to their catalogue of courses. In fact, they monetize their business offering a paid certificate of the course.

CONCLUSION

The use of apps is gradually growing. According our results, smartphones are the devices most used for downloading apps and for looking information. The survey also suggests that the students are active users for smartphones whereas medical residents are active users for tablets. It is important to take into account that there is a relationship between the use of smartphones and the daily number of apps used and the profile (students and medical professionals).

On the other had, medical apps and training medical apps are spread among the medical industry. As it is mentioned above, the great amount of apps in the marketplace makes that the most part of students and physicians have an unawareness of them. This unawareness is precisely one of the causes that they use to justify not using them.

Surprisingly, the results obtained with the present study could suggest that around 63% of medical professionals have used apps related with Medicine, whereas only 5% of students have used them, which is against we could think at the beginning. Thus, we could suggest that medical professionals are very open-minded to the cutting-edge technologies and they consider them a tool that is worthy to use and explore.

Vademecum is the App that heads the list of most used Apps together with Anatomical Apps.

The role of these tools is still unknown, and although there is some Medical Schools that incorporate technologies in their curriculum, there is still a gap to treat. The problem is that technology goes faster that Institutions are able to adopt it. In fact, it could be possible that when a University decides to adopt it, the technology is out-dated and the leaders have to rethink the new implementation, which sometimes discourages them from putting it into practice.

This article pretends to show a brief overview of the real use of mobile devices by students and medical professionals. Besides, new tendencies are emerging in this field as for example BYOD or MOOCs. These trends give the students an opportunity to participate actively in their learning and provide new resources and facilities to develop new m-learning strategies in Universities and other higher education institutions.

Finally, due to the great amount of apps available in the marketplace and the lack of regulation of educational apps, a quality protocol should be developed, designed and implemented in order to fulfil the users' requirements. Part of this necessity has already been developed (Briz-Ponce, 2016) but it is still necessary to be commercially adopted to make it a reality.

ACKNOWLEDGMENT

This research study has been made within the PhD Programme on Education in the Knowledge Society.

REFERENCES

Adolph, M. (2009). *Mobile Applications.* International Telecommunication Union.

Agresti. (2002). *Categorical Data Analysis* (2nd ed.). New York: John Wiley & Sons.

Alden, J. (2013). Accomodating mobile learning in college programs. *Journal of Asynchronous Learning Networks, 17*(1), 109–122.

Alrasheedi, M., Capretz, L. F., & Raza, A. (2015). A Systematic Review of the Critical Factors for Success of Mobile Learning in Higher Education (University Students' Perspective). *Journal of Educational Computing, 52*(2), 252–276. doi:10.1177/0735633115571928

Ashour, R., Alzghool, H., Iyadat, Y., & Abu-Alruz, J. (2012). Mobile phone applications in the university classroom: Perceptions of undergraduate students in Jordan. *E-Learning and Digital Media, 9*(4), 419–425. doi:10.2304/elea.2012.9.4.419

Briz-Ponce, L. (2016). *Análisis de la efectividad en las Aplicaciones m-health en dispositivos móviles dentro del ámbito de la formación médica.* Salamanca: University of Salamanca.

Briz-Ponce, L., & Juanes-Méndez, J. A. (2015). Mobile Devices and Apps, Characteristics and Current Potential on Learning. *Journal of Information Technology Research, 8*(4), 26–37. doi:10.4018/JITR.2015100102

Briz-Ponce, L., Juanes-Méndez, J. A., & García-Peñalvo, F. J. (2014a). Analysis of Certificated Mobile Application for Medical Education Purposes. In F. J. García-Peñalvo (Ed.), *Proceedings of the Second International Conference on Technological Ecosystems for Enhancing Multiculturality-TEEM14* (pp. 13–17). New York: ACM. doi:10.1145/2669711.2669871

Briz-Ponce, L., Juanes-Méndez, J. A., & García-Peñalvo, F. J. (2014b). Analysis of mobile devices as a support tool for professional medical education in the university school. In *Proceedings of 6th International Conference on Education and New Learning Technologies-EDULEARN14* (pp. 4653–4658). Valencia: IATED Academy.

Briz-Ponce, L., Juanes-Méndez, J. A., & García-Peñalvo, F. J. (2014c). First Approach of Mobile Applications Study for Medical Education purposes. In *Proceedings of the Second International Conference on Technological Ecosystems for Enhancing Multiculturality-TEEM14* (pp. 647–651). New York: ACM. doi:10.1145/2669711.2669968

Chakravarti, I. M., Laha, R. G., & Roy, J. (1967). *Handbook of methods of applied statistics.* New York: John Wiley.

Chen, W., Lim, C., & Tan, A. (2010). Pre-Service Teachers' ICT Experiences and Competencies: New Generation of Teachers in Digital Age. In *Proceedings of the 18th International Conference on Computers in Education* (pp. 631–638). Putrajaya, Malaysia: Asia-Pacific Society for Computers in Education.

CNMC. (2017). *Informe Mensual Comunicaciones Móviles.* CNMC.

Dolan, B. (2011). *Nine medical schools that support mobile learning | mobihealthnews.* Retrieved May 9, 2015, from http://mobihealthnews.com/12346/nine-medical-schools-that-support-mobile-learning/

Eden Dahlstrom. (2012). *ECAR Study of undergraduate Students and Information Technology (Research Report)*. EDUCAUSE Center for Applied Research.

Ferreira, J. B., Klein, A. Z., Freitas, A., & Schlemmer, E. (2013). Mobile learning: Definition, uses and challenges. In L. A. Wankel & P. Blessinger (Eds.), Cutting-edge Technologies in Higher Education (pp. 47–82). Bingley, UK: Emerald Group Publishing Limited. doi:10.1108/S2044-9968(2013)000006D005

Fundación Telefónica. (2014). *La Sociedad de la Información en España 2013*. Barcelona, Spain: Ariel, S.A.

Fundación Telefónica. (2017). *La Sociedad de la Información en España 2016*. Barcelona, Spain: Ariel, S.A.

Gallegos, A. (2013). *Medical Schools Embrace Benefits of Tablets, Mobile Devices*. Create New Ways of Learning.

Handal, B., Macnish, J., & Petocz, P. (2013). Academics adopting mobile devices : The zone of free movement. In H. Carter, M. Gosper, & J. Hedberg (Eds.), *Proceedings 30th ascilite Conference 2013* (pp. 350–361). Sidney, Australia: Macquarie University.

Herrington, J., Herrington, A., Mantei, J., Olney, I. W., & Ferry, B. (2009). Using mobile tecnologies to develop new ways of teaching and learning. In J. Herrington, J. Mantei, I. W. Olney, B. Ferry, & A. Herrington (Eds.), *New techonologies, new pedagogies: Mobile learning in higher education* (pp. 1–14). Wollongong: University of Wollongong.

Jackson, L. D. (2013). Is mobile technology in the classroom a helpful tool or a distraction?: A report of university students' attitudes, usage practices, and suggestions for policies. *International Journal of Technology, Knowledge and Society*, *8*(5), 129–140. doi:10.18848/1832-3669/CGP/v08i05/56335

Khatoon, B., Hill, K. B., & Walmsley, A. D. (2013). Can we learn, teach and practise dentistry anywhere, anytime? *British Dental Journal*, *215*(7), 345–347. doi:10.1038/sj.bdj.2013.957 PMID:24113956

Lea, S., & Callaghan, L. (2011). Enhancing health and social care placement learning through mobile technology. *Journal of Educational Technology & Society*, *14*(1), 135–145.

Lee, E. T., & Wang, J. W. (2003). *Statistical Methods for survival Data Analysis*. John Wiley & Sons, Inc. doi:10.1002/0471458546.fmatter

López-Hernández, F. A., & Silva-Pérez, M. M. (2014). M-learning patters in the virtual classroom. *Universities and Knowledge Society Journal*, *11*(1), 208–221. doi:10.7238/rusc.v11i1.1902

Martinez-Perez, B., de la Torre-Diez, I., & Lopez-Coronado, M. (2013). Mobile Health Applications for the Most Prevalent Conditions by the World Health Organization: Review and Analysis. *Journal of Medical Internet Research*, *15*(6), e120. doi:10.2196/jmir.2600 PMID:23770578

Martínez-Pérez, B., de la Torre-Díez, I., & López-Coronado, M. (2015). Experiences and Results of Applying Tools for Assessing the Quality of a mHealth App Named Heartkeeper. *Journal of Medical Systems*, *39*(11), 1–6. doi:10.1007/s10916-015-0303-6 PMID:26345452

Mobile Marketing Association. (2017). *Estudio tendencias Mobile 2017*. Author.

Nielsen, J. (2014). The Digital. *Consumer Reports*. PMID:25350310

Nuss, M. A., Hill, J. R., Cervero, R. M., Gaines, J. K., & Middendorf, B. F. (2014). Real-time use of the iPad by third-year medical students for clinical decision support and learning: A mixed methods study. *Journal of Community Hospital Internal Medicine Perspectives*, *1*(4), 1–7. doi:10.3402/jchimp.v4.25184 PMID:25317266

Ozdalga, E., Ozdalga, A., & Ahuja, N. (2012). The Smartphone in Medicine: A Review of Current and Potential Use Among Physicians and Students. *Journal of Medical Internet Research*, *14*(5), e128. doi:10.2196/jmir.1994 PMID:23017375

Quicios-García, M. del P., Sevillano García, M. L., & Ortega Sánchez, I. (2013). Educational uses of mobile phones by university students in Spain. *The New Educational Review*, *34*(4), 151–163.

Ramírez-Montoya, M. S., & García-Peñalvo, F. J. (2017). La integración efectiva del dispositivo móvil en la educación y en el aprendizaje. *Revista Iberoamericana de Educación a Distancia*, *20*(2), 29–47. doi:10.5944/ried.20.2.18884

Rapetti, E., Picco, A., & Vannini, S. (2011). Is Mobile Learning a Resource in Higher Education? Data Evidence From an Empirical Research in Ticino. *Journal of E-Learning and Knowledge Society*, *7*(2), 47–57.

Salkind, N. J. (1998). *Métodos de investigación*. D.F, México: Prentice Hall.

Sarrab, M., Alzahrani, A., Al Awan, N., & Alfarraj, O. (2014). From traditional learning into mobile learning in education at the university level: Undergraduate students perspective. *International Journal of Mobile Learning and Organisation*, *8*(3–4), 167–186. doi:10.1504/IJMLO.2014.067014

Shah, D. (2015). *MOOCs in 2015: Breaking Down the Numbers*. EdSurge.

Shah, D. (2016). *Monetization Over Massiveness: Breaking Down MOOCs by the Numbers in 2016*. EdSurge.

Statista. (2017a). *Google Play: number of available apps 2009-2017 Number of available applications in the Google Play Store from December 2009 to March 2017*. Author.

Statista. (2017b). *Number of available apps in the Apple App Store from July 2008 to January 2017*. Author.

Teri, S., Acai, A., Griffith, D., Mahmoud, Q., Ma, D. W. L., & Newton, G. (2013). Student use and pedagogical impact of a mobile learning application. *Biochemistry and Molecular Biology Education*, *42*(2), 121–135. doi:10.1002/bmb.20771 PMID:24375862

Toktarova, V. I., Blagova, A. D., Filatova, A. V., & Kuzmin, N. V. (2015). Design and Implementation of Mobile Learning Tools and Resources in the Modern Educational Environment of University. *Review of European Studies*, *7*(8), 318–324. doi:10.5539/res.v7n8p318

UNESCO. (2009). La nueva dinámica de la educación superior y la investigación para el cambio social y el desarrollo. In Conferencia Mundial sobre la Educación Superior (pp. 1–9). París: United Nations Educational, Scientific and Cultural Organization.

UNESCO. (2011). *UNESCO Mobile Learning Week Report*. Paris.

Vinacua, B. V. (2007). *Análisis estadístico con SPSS 14: estadística básica (3rd ed.)*. McGraw-Hill Interamericana de España S.L.

Visvanathan, A., Hamilton, A., & Brady, R. R. W. (2012). Smartphone apps in microbiology-is better regulation required? *Clinical Microbiology and Infection*, *18*(7), 218–220. doi:10.1111/j.1469-0691.2012.03892.x PMID:22563840

Chapter 8
In-Service Teachers' Use of ICT for the Promotion of Collaborative Professional Learning

Ana García-Valcárcel
University of Salamanca, Spain

Juanjo Mena
University of Salamanca, Spain

ABSTRACT

Information and communication technologies (ICT) are often rendered as key tools in the promotion of teachers' collaborative learning. Their use enables teachers to complete assignments, solve problems, or create products together. The content of this chapter is based on the information published in a previous research study by the authors. In that study, they aimed at describing teachers' use of ICT towards collaboration from a triple perspective: what they believe (teachers' opinion), what they know (teachers' knowledge), and what they do (teachers' use). A questionnaire and interviews were the instruments to collect data. Some results pointed out that teachers used ICT to promote collaboration on a regular basis, but it is limited to the knowledge they have on particular tools, which is acknowledged to be intermediate. The most important implication for teacher education programs is considering the actual limitations of teachers' knowledge and use of ICT in practice to set a more accurate starting point to promote collaboration through technologies.

INTRODUCTION

Educational Research has shown how collaboration plays a key role for effective teaching (McLaughin & Talbert, 2006). Collaborative teaching implies exchanging ideas and working together in small groups to meet a specific goal. In other words, "sustained teacher collaboration about instructional strategies, curriculum, students, and assessment, as well as general collaboration is the primary vehicle for continuous improvement of teacher practice, for sharing accountability, and collective responsibility" (Killion, 2015).

DOI: 10.4018/978-1-5225-4944-4.ch008

Collaborative learning activities typically include debates, group projects, case-study discussions, role-playing exercises, personal essays, shared stories and multiple-goal activities.

Collaborative learning among teachers has its foundation in communities of practice (Lave &Wenger, 1991; Wenger, 1998) by which teachers engage in a shared enterprise over time. 'The key mechanism for individual and group learning comprises access to observing and then participating in the practices at the core of the community' (Levine &Marcus, 2010). Scharff &Brown (2004) also report the efficacy of creating Learning Communities based on socio-cultural learning principles (Vygotsky, 1978), since they provide a mutually supportive framework for teachers. Collaborative learning allows teachers to progress beyond what they are able to learn alone by observing or learning from teacher education programs.

Traditionally, four types of collaboration have been distinguished to take place in schools (Little, 1990): (1) storytelling and scanning, (2) aid and assistance, (3) sharing and (4) joint work. The first two are based on a collaboration relation among the participants (Monereo Font & Durán, 2003) and characterized by a low level of interdependency and collective autonomy (–traditional perspective) whereas the latter two are based on constructed knowledge (Scardamalia & Bereiter, 1994) and represent the highest level of collaborative work (Hernández, González & Muñoz, 2014).

Therefore, collaborative skills to be acquired need time to be interiorized (Rigelman &Ruben, 2012) as well as engaging in reflection (Wells, 2001), systematic inquiry (Ball & Forzani, 2010) and practical knowledge (Paulus, Woodside & Ziegler, 2010). But most importantly is that collaboration depends heavily on other factor: the context where it is taking place (Wenger, McDermott & Snyder, 2002). The conditions for collaborative teacher learning may differ considerably between schools that, ultimately, may lead to different types and amounts of learning activities (Bakkenes, Vermunt, &Wubbels, 2010).

The information to be presented in the following lines of the chapter basically reproduces a research study that was published in the Journal of Information Technology Research (JITR) in García-Valcárcel & Mena (2016). Please see references.

THEORETICAL FRAMEWORK

Educational research has emphasized the study of teachers' collaborative work mediated using ICT. Traditionally, this line of research has been approached under the name of Computer Supported Collaborative Learning –CSCL-(Garrison, 2006; Kirschner, 2002; Palloff & Pratt, 1999; Oakley et al., 2004; Rubia, 2010)

In the last five to eight years the vast majority of collaborative processes mediated through ICT have begun to rely on the use of the so-called web 2.0. not only in the sharing of knowledge but also in its online co-construction (Little, 1990). The ultimate promise is that web 2.0. and web 3.0 tools will increase ways of collaboration in the classroom (Pérez & Delgado, 2012) because they promote students' collaborative learning (Pérez Mateo, Romero & Romey, 2014; Domingo, Coscollola & Marquès, 2011) and create the expectation of a change in teaching practices (Crook, 2012).

In essence it implies a combination of technology (the relationship between the design of technology and the use of technology); affordances (affordances constrain the ways in which technologies can possibly be written or read); institutions (implementation of collaborative learning in the schools), and infrastructure (the provision of ICT as related to organizational and institutional factors) (Jones, Dirkinck-Holmfeld & Lindstrom, 2006).

Although it has been argued that there is a scarcity of baseline information about collaboration mediated through ICT and a lack of tools and measures to examine the processes involved in CSCL (Gress, Fior, Hadwin & Winne, 2010), many researchers have reported that using technology in collaborative professional environments promotes complex learning in in-service teachers (Dillenbourg, 2002;Pontecorvo, 2007; Livingstone, 2012; James, 2013) and better qualifies preservice teachers to learn the profession (Jirotka, Procter, Rodden & Bowker, 2006; Dobber, Akkerman, Verloop & Vermunt; 2012).

For instance, Gorghiu, Lindfors, Gorghiu and Hämäläinen (2011) created an on-line course addressing methods for experienced secondary school teachers in order to integrate face-to-face and web-based learning tools with a view to determining how participants use ICT in their teaching activity. Fifty-seven teachers assumed different on-line tutor roles for their students –pedagogical, managerial, technical and social- on a virtual platform. They mainly performed pedagogical roles -i.e., supporting learning processes by providing instructions, focusing on relevant content, and guiding and motivating learners- and managerial roles, i.e., administering the courses and tracking learners' progress. Those roles were demonstrated to be more efficient when carried out in collaboration with other teachers. Social tutoring (i.e., supporting students' feelings, motivation or integration) and technical tutoring (i.e., solving problems related to ICT) were less used by the schoolteachers.

Wegerif (2006) employed evidence from case studies to value teachers' collaborative learning around ICT in Primary Schools in the UK. In that study, they used exploratory talk with teachers, a type of talk in which questions were asked, alternatives explored and reasons given. The method was devised to assess the effectiveness of the way groups were talking together using Reven's Standard Progressive Matrices (Wegerif, 1996). The evidence suggests that dialogic theory offers a possible solution to the problem of how to conceptualize general thinking skills for collaborative learning. In other words, teachers' conversations promote the learning of general thinking skills. On the other hand, Schoor & Bannert (2010) investigated to what extent motivation influences student teachers' learning activities and knowledge acquisition during collaborative work. Two hundred graduate students participated in the study. It was hypothesized that current motivation (in terms of self-efficacy, instrumentality, beliefs, attainment value, goal orientation, interest, intrinsic task value, costs of expectancy and value components) would influence both learning activities (i.e., chats, forums, etc.) and knowledge acquisition in a positive way. The results from questionnaires indicated that only the goal orientation dimension was associated with knowledge acquisition. The expectancy and value components of current motivation were not related either to the observed learning activities or to knowledge acquisition during collaborative learning but were partly associated with learning activities and knowledge acquisition during individual learning. Dobber, Akkerman, Verloop & Vermunt (2012) also evidenced that collaborative research processes play a determinant part teachers' knowledge elaboration. Two groups of student teachers were video-recorded following a university teacher education program and two inquiry processes were analyzed: elaboration (a process of exchange, consideration, and integration of information) and decision making (making a choice about a certain matter). In one group, the participants found a balance between elaboration and decision making whereas the other group had difficulties to engage in dialogic exchanges and decision-making choices. This led to the conclusion that both inquiry processes are crucial features for collaboration. In a similar vein Simon, Anderson, Hoyer & Su (2004) investigated the use of Tablet PCs to promote active learning in computer science. They noted that the sharing of different students' (or groups of students) spontaneous attempts to solve problems 'provided an opportunity to point out common mistakes and allowed comparison of different approaches to the same problem' and that 'the instructor also received immediate feedback on whether or not students understood' (p. 215).

Finally, Tolmie and Boyle (2000) suggest that having knowledge about other participants involved in a community (i.e., knowledge about how they work and about their earlier experiences and ideas) promote collaborative work mediated (or not) by ICT.

However teachers' collaboration processes entail complex work dynamics that do not always allow for benefiting from it in terms of professional learning. According to Ann, Kim & Kim (2007), there are at least five challenges that prevent an optimal stage from being reached in collaborative working: (1) Fear of plagiarism; (2) lack of recognition within the team; (3) effort (it demands time and involvement); (4) assessment (it is difficult to accurately identify each team-worker's contribution); (5) technical support (using online approaches may be actively avoided due to unfamiliarity or because of administrative overheads and the accompanying risk of technical problems).

Strijbos and Fischer (2007) provided another five major challenges of collaborative learning research: (1) the multilevel challenge: How to deal with effects and dependencies at individual level and at group level: (2) the convergence/divergence challenge: Mutual influence of learning partners in one group but at the same time difficulties in determining whether their contributions become similar over time; (3) the segmentation and coding challenge: Since content analysis is frequently used in collaborative learning research, thorough methods must be deployed; (4) the fusion of hybrid methods challenge. Combining two frequently applied but different methods; (5) the situational challenge. Most approaches emphasize the crucial role of context in collaborative working.

Kreijns, Kirschner & Jochems (2003) added another last challenge: trust. Group participants will not share their tentative ideas unless they trust other group members and feel a sense of belonging.

Therefore, collaboration often remains shallow due to the lack of effective group support. Dirkx & Smith (2004) express this as a 'struggle with the development of a sense of interdependence and intersubjectivity ' (p. 134).

Furthermore ICT alone are unable to promote more efficient collaborative work (Engel & Onrubia, 2010) not even by adopting critical thinking or reflecting upon their use in practice.

Accordingly, collaborative networking needs support in order to prevent superficial results. More than two decades of research suggest that teacher collaboration matters for teachers' work. However, relatively little research has focused on how different types of teacher collaboration afford different types of teacher learning. Moreover, a variety of studies highlight the importance of analyzing teachers' attitudes towards ICT in improving collaborative relationships and how they represent their work through artifacts of practice mediated by language (i.e., peer observation, lesson plans, diaries, written reports, etc.). (Contreras, Arancibia & Cárcamo, 2009; Levine & Marcus, 2010).

However, few studies have related teachers' attitudes to their actual knowledge and the uses of such ICT in the improvement of collaborative work. For that reason the present study aims at examining the role of ICT in the promotion of Collaborative learning (CL) by investigating primary and secondary in-service teachers at whose schools regional ICT scores are high or very high according to the Spanish Administration. In particular we shall look at a threefold relationship: (1) what *teachers think* (teachers' perceptions) about the use of ICT to improve CL; (2) what *teachers know* about ICT (their knowledge about technological tools and techniques, and (3) what *teachers do*: which ICT resources and methodologies they employ in practice to promote collaborative relationships.

Additionally, we shall explore whether having previous collaborative experiences with colleagues will lead to a broader use of ICT in the teaching practices. Our hypothesis is that professional experience in the use of collaborative work strategies through technological resources might determine the methodology that teachers use.

Distinguishing which of the three above teachers' levels (think, know, do) is most prevalent (Lave & Wenger, 1991) in in-service teachers is important because it is widely assumed that using ICT for collaborative learning activates connectivist learning

METHODOLOGY

The study focused on teachers from 5th and 6th grades in Primary Education and 1st and 2nd grades of the Spanish secondary education levels (student ages between 10-14 years). The schools they worked in were accredited as highly qualified schools in the use of ICT by the Regional Government of Castile and Leon (Spain). Primary and secondary schools are classified in 5 levels of performance in the use of ICT (level 1= very low; level 2= low; level 3= average; level 4= high and level 5= very high). The schools with a high level (levels 4 and 5) have the following characteristics: Technological equipment and infrastructures with an average of 80%- 100% use coverage for educational and administrative processes; on-going innovation projects in the schools aimed at the integration of ICT and teachers' training courses in the use of ICT at the schools. Currently, the number of schools under this qualification is 148. Although we assumed that all primary and secondary schools might work collaboratively using ICT, we randomly chose from those high performing schools since they are representative institutions of best practices in technology-enhanced learning environments Keppell, Suddaby & Hard, 2015).

The data gathered for this research come from all the teachers in the population. The answers obtained determined the size of the sample: 185 in-service teachers.

Three variables were considered in García-Valcárcel and Mena's (2016) study: (1) Teachers' opinion on ICT as a way to promote (or not) Collaborative Learning (*What teachers think*); (2) Mastery of ICT tools and strategies for Collaborative Learning (*What teachers' know*); (3) ICT methodology and resources used by teachers for Collaborative Learning in the classroom (*What teachers do*).

The data collection instrument was a 34-item questionnaire aimed at assessing the influence of ICT in teachers' collaborative work. The questionnaire was divided in three parts (sub-questionnaires) according to the three variables: Teachers' perceptions about how ICT promote Collaborative Learning: TP-Q (10 items); Teachers' knowledge about ICT resources and strategies: TK-Q (12 items); and Teachers' use of resources, methodology and strategies: TU-Q (12 items) To study teachers' opinion on ICT as a way to promote (or not) Collaborative Learning' variable, a 10-item sub-questionnaire (TP-Q) was used. The TP-Q collected information about the benefits of ICT in processes of collaborative work (4 items) and the possible limitations related to the time invested and assessment (6 items). The items had the same 5-point Likert-type scale, allowing expression of the degree of agreement or disagreement within the statement of the item. The scale ranged from 1 (completely in disagreement) to 5 (completely in agreement).

For the first dimension a semi-structured interview was also used to expand on the information from the questionnaire. In the interview, reference was made to five major aspects: (1) Activities developed in collaboration based on a joint project; (2) resources and materials employed; (3) Collaborative experiences with ICT; (4) benefits of using ICT in collaborative networks; (5) drawbacks of using ICT in collaborative network. For the second dimension, the TK-Q was employed to evaluate the knowledge of the teachers about different methodological strategies and resources that allow collaborative work to be carried out using ICT. These scales were also answered on a five point Likert scale. Finally, and for the third dimension, information about the methodological strategies and resources used by the teachers in their teaching activities was collected via the TU-Q, in which different collaborative work strategies were presented so that the teachers could indicate which ones they used. The TU-Q items had a dichotomous format since the answers were not exclusive.

The procedure for collecting data was based on the following phases: Phase 1: Construction of the questionnaires, the interview protocols, and ethical consents.; Phase 2: Selection of the schools and teachers who participated in the study; Phase 3: Application of the questionnaires and interviews; Phase 4: Data analysis.

The questionnaires were built as *ad hoc* tools for the proposed objectives and were uploaded onto an on-line platform so that the teachers of the selected schools (with accreditations of 4 and 5 in ICT) could assess them during the 2012-2013 academic year. The data analysis was based on a mixed methodology (Johnson & Onwuegbuzie, 2004; Paterson & Pentland, 2008) in which both quantitative (i.e. questionnaires) and qualitative (i.e. interviews) analytical methods were employed. This combined approach was adopted because it was the one that was best adapted to the descriptive nature (non-experimental) of the study (Cohen & Manion, 1989).

For the questionnaires, descriptive statistical analyses were conducted for the three variables considered. Furthermore, an inferential analysis was conducted to estimate the differences in the use of ICT among teachers who had more or less experience with respect to collaborative work (TU-Q). Statistical methods such as Student's *t* test for comparing the means of two independent samples, *Phi* coefficient and *Chi*-square test were used to explore this relation.

Regarding the qualitative analysis, a content analysis of the 'benefits and limitations' that had emerged in the interviews was performed. Topical analysis (Grounded Theory Analysis; Corbin & Strauss, 1990) was used to analyse teachers' answers to the open-ended questions, which were divided into utterances (Crasborn, Henissen, Brouwer, Korthagen & Bergen, 2011) and then grouped in categories following an inductive process of a two- and three-level hierarchy.

RESULTS

Regarding the first objective: *(1) what teachers think*, they believed that technological tools facilitate the implementation of projects; they allow time and information management to be optimized; they facilitate the monitoring and follow-up of students' work and they allow this to be updated. All the means were above four on the scale, except for an item referring to the work of each student, which had a score of 3.76

As for the limitations, the teachers reported that the time required for this activity, both for the teachers (who should plan it and follow it up) and for the students, were critical factors.

With regard to the interviews, a total of 144 comments were collected about the positive aspects of working collaboratively and 61 about associated problems. The usual length of these comments ranged from one sentence to four. Each comment was assigned to a general topic. Table 1 details the analysis of the comments referring to positive aspects.

Generally speaking, the professional dimension –teaching work- (47 declarations; 32.6%) was declared to be a stronger motivator to enhance the use of ICT for teaching collaboration, especially the interchange of ideas (23; 15.9%) and co-working (24; 16.6%). On the other hand, it seems that collaborative learning is facilitated by the elements of simplicity (7.5%), versatility (8; 5.5%) and operational functioning (6; 4.1%). The technical dimension of the ICT, that is to say, how to use them in classroom was valued in 26 of the utterances (18.05%) whereas the personal and social dimension was less considered to be a

Table 1. Positive aspects of ICT use to promote collaborative work

Dimensions	Categories			f	%
1.Professional dimension (teaching work) (47; 32.6%)	1.1.Interaction	1.1.1.Relation with colleagues	Participation with more teachers	6	4.16
			Possibility of having more meetings.	3	2.08
		1.1.2.Exchange of ideas	Share information and experiences	15	10.41
			Know what colleagues are doing	2	1.38
			Continuous updating	6	4.16
		1.1.3.Mutual work	Performing joint actions	3	2.08
			Follow-up of tasks/projects	5	3.47
			Distance working (synchronous and asynchronous	3	2.08
			Efficacy of the elaboration of knowledge.	7	4.86
	1.2.Materials and methodologies	1.2.1.Materials	Creation of new materials for class	5	3.47
			Saving costs of printed documents and other costs.	2	1.38
			Creation of programs and tools.	5	3.47
		1.2.2.Methodology	Development of new methodologies.	2	1.38
			Better quality in giving classes	2	1.38
			New perspectives in teaching methods.	5	3.47
	1.3.Time and spaces management	1.3.1.Time	Immediacy of communication.	14	9.72
			Schedule flexibility	3	2.08
		1.3.2.Space	Absence of physical space (virtual meetings)	2	1.38
			Easy coordination of face-to-face meetings	2	1.38
2.Technical dimension (ICT use) (26; 18.05%)	2.1.ICT ease of use	2.1.1Convenience	In coordinating work.	3	2.08
			In sharing ideas	3	2.08
		2.1.2.Simplicity	ICT are accessible tools (available at work and at home)	9	6.25
			ICT are intuitive tools	2	1.38
	2.2.Efficiency of use	2.2.1.Versatility	Multitude of technical applications-	3	2.08
			The ICT used are valid for many levels.	5	3.47
		2.2.2.Operativity.	ICT are very valid tools for collaborative work	2	1.38
			ICT allow doubts to be solved efficiently.	4	2.77
3.Personal and social dimension. (21; 14.58%)	3.1.Motivation	3.1.1.Interest	Working collaboratively with ICT can be exciting	4	2.77
			ICT are attractive	1	0.69
		3.1.2.Disposition	More time dedicated to ICT (by teachers) than to other matters.	3	2.08
			Tendency to evaluate ICT positively.	2	1.38
	3.2.Equity	3.2.1.Assumption of democratic values	More respect for the diversity of opinions	2	1.38
			Tolerance towards the work of others	2	1.38
		3.2.2.Other perspectives	Evaluation of the knowledge of others	1	0.69
			Broad scope of outlooks	1	0.69
	3.3.Comittment	3.3.1.Involvement in collaborative work	Need to perform the tasks corresponding to each participant.	2	1.38
			Evaluation of work carried out by others.	1	0.69
		3.3.2.Involvement in work in general	Greater commitment to work	1	0.69
			Higher degree of professionalism.	1	0.69
Total				144	100

In García-Valcárcel and Mena (2016).

motivator for the use of ICT (21; 14.58%). In fact, the personal commitment (5; 3.47%) and equity (6; 4.1%) were the aspects less addressed within the last dimension.

With regard to limitations, the sample of teachers in this study perceived twice as few negative aspects in the use of ICT to motivate collaborative networks. However, as shown in Table 2, there were some topics that they found particularly challenging.

Table 3 shows that teachers feel that they do not always get involved in the use of ICT (8; 11.9%) and when they do it, they should invest more working hours (7; 11.4%) and this can elicit dysfunctional relationships among colleagues (i.e. jealousy) (7; 11.4%). However, the distribution of frequencies is relatively homogeneous among the five major categories. It is to be remarked that greater difficulties are found in organizing the materials and time for collaborative learning when using ICT. The teachers from the sample foresaw fewer difficulties with technical issues such as computer handling (3; 4.9%).

In synthesis, the variables analyzed show that the participant school teachers maintain a positive assessment of the use of ICT as efficient tools for supporting collaborative work and at the same time they point out certain limitations to these strategies, such as the time required and the evaluation of the results achieved by each student.

As for *(2) what teachers know* the results show that the knowledge of teachers is moderate and does not reach the mean value of the scale (Table 3). The best-known tools were those related to social markers (3.51 ± 1.28), online conceptual maps (3.26 ±1.35), wikis (3.35 ± 1.22) and e-learning platforms (3.29 ± 1.19). Video repositories and blogs were less known by teachers as tools to promote collaborative networking.

Table 2. Drawbacks of ICT for collaborative work

Categories		*f*	%
1. Lack of knowledge in using ICT	1.1. Low level of teacher training.	6	9.83
	1.2. Different levels of knowledge among colleagues.	4	6.55
	1.3. Lack of knowledge about ICT tools.	4	6.55
2. Problems in personal relations	2.1. Poor relations with other colleagues (i.e., jealousy)	7	11.47
	2.2. Appropriation of the work of others.	2	3.27
	2.3. Lack of human contact.	3	4.91
	2.4. Different perspectives.	1	1.63
3. Lack of interest	3.1. Lack of motivation.	3	4.91
	3.2. Non-involvement.	8	11.94
	3.3. Too much effort.	1	1.63
4.Technical problems	4.1. Handling of computers or other devices (i.e., IDB).	3	4.91
	4.2. Connectivity and network problems.	3	4.91
5.Organizational problems.	5.1. Preparation of teaching materials.	4	6.55
	5.2. Excessive teaching load (more work at home).	7	11.47
	5.3. Difficulties in reaching consensus about shared work schedule.	5	8.19
Total		61	100

In García-Valcárcel and Mena (2016).

Table 3. Knowledge of technological resources for collaborative work

Resources	Mean	SD.
1. e-learning platforms (i.e., Moodle or similar)	3.29	1.19
2. Blogs	2.88	1.07
3. Wikis	3.35	1.22
4. Social networks (i.e., Edmodo)	3.14	1.26
5. On-line office suites (i.e., GoogleDocs)	2.84	1.20
6. Shared files (i.e., Dropbox)	2.77	1.32
7. Shared web sites (i.e., GoogleSites)	3.08	1.22
8. On-line audiovisual presentations (i.e.,SlideShare)	2.79	1.22
9. Video repositories (i.e.,Youtube)	2.34	1.15
10. On-line conceptual maps (i.e., Cmaptools. Popplet)	3.26	1.35
11. Shared photo albums (i.e.,Picassa)	2.50	1.25
12. Social markers (i.e., Delicious)	3.51	1.28
Total	2.98	.86

Finally, with respect to *(3) what teachers do* (i.e., teaching resources and pedagogical strategies that teachers used to foster collaborative work) it may be seen that the search for resources on the Internet (85; 48.6%) and the preparation of a subject topic for class by the students (64; 34.6%) were the most frequent activities. Working towards solving teachers' problems or collaborating on common projects were activities used by around 25% of the teachers in the sample (see Table 4)

Table 4. Teachers' use of pedagogical strategies for collaborative work (n=185)

Methodological Strategies	*N*	%
1. Working on projects	45	24.3
2. Problem solving assignments	51	27.6
3. Preparation of a subject topic for class	64	34.6
4. Internet search for resources	85	45.9
5. Working on Webquest	23	12.4
6. Discussion in forums	15	8.1
7. Wikis created by classroom students and teachers	12	6.5
8. Group blogs	32	17.3
9. Compilation of glossaries	10	5.4
10. Work in social networks	12	6.5
11. Playing electronic games	22	11.9
12. No strategy used.	88	47.6

In García-Valcárcel and Mena (2016).

On calculating the items chosen by each subject, it is interesting to note that 48% of the teachers did not use any of the listed strategies or resources, whereas 44% of them used between 2 and 5, or even more (7%).

A further analysis of the data led us to differentiate two very different collective groups between the teachers studied. The first group would show a high degree of coherence between their conceptions about the potential of collaborative learning and the use of ICT for such purposes (formed by teachers who implement ICT collaborative work in practice), while the second group would comprise teachers who would show a clear degree of incoherence between their conceptions and their teaching practice. However, the teachers who expressed a good evaluation of the potential of ICT for collaborative learning did not use any resources or strategies for this purpose.

DISCUSSION

In this study three dimensions of in-service teachers' positioning towards ICT implementation in school collaborative networks were scrutinized: teachers' opinion (think), teachers' technical knowledge (know) and teachers' actual use in practice (do). These three aspects are often studied in Teacher Education research but not in an intertwined form within the same sample of study.

This work shows the discrepancies between what the teachers of this study *think*, *know* and *do* in collaboration by means of the use of ICT. Firstly, teaching and learning with ICT clearly improves collaboration according with the teachers' beliefs. In other words, collaborative working methods were valued highly by teachers to achieve professional competencies sin ICT tools were seen as great facilitators. Virtual spaces provide the opportunity to share resources and have updated and well-organized information. In particular they are permanently available. This helps in the development of projects in which different students or teachers are involved and can interact in a flexible way without spatial or temporal limitations. ICT are now becoming polyvalent tools that allow the development of a large number of teaching activities in which participation is becoming a basic element of the process of constructing knowledge: debates, discussions of cases or simulations in the sense suggested by Hilz & Turoff (2002). At the same time, they offer the possibility of continuous training spaces for teachers and of updating based on sharing the situations and problems that emerge in educational practices with their colleagues, giving rise to virtual communities for practice, as pointed out by Wenger (1998) and Levine and Marcus (2010).

Secondly, it was evidenced that teachers' daily practice brings some limitations in the use of ICT for collaborative learning, essentially two aspects can be questioned: namely, the time involved in the development of such activities and the difficulty in evaluating the involvement of each of the students in the work carried out and the repercussions of this on the acquisition of knowledge and skills. Regarding these issues, teachers seem to develop their curricular proposals under a certain social pressure (from both the school itself and from students' families) to fulfil the program established and, in many cases, must commit to completing the modules explained in text books that are assumed to be the official and basic curriculum. Accordingly, the time invested in the development of projects, in which the teacher must dedicate 'extra' time for planning and follow-up and the students must dedicate considerable time and effort to complete the tasks set them, is considered a key element that may inhibit teachers when attempting to set up a working methodology in less traditional classrooms, more focused on collaborative learning though ICT. Many authors have addressed this issue, such as García, Gros and Noguera (2010).

Other types of limitation observed have to do with the lack of familiarity with ICT tools, difficulties in peer relations, lack of interest and involvement in the development of innovative teaching projects, technical problems that may arise in on-line courses and in the use of certain applications, problems of organization, etc. These findings fairly coincide with those reported by Authos, 2014).

These limitations might explain why in educational contexts with sufficient technological resources –as the schools selected in the present study- present a very limited use of ICT along with teaching methodologies that have not been transformed in the direction of more collaborative and dynamic learning processes (An, Kim & Kim. 2008).

Thirdly, regarding the technical knowledge teachers shown of ICT, it is to be recognized that prior experience of teachers in collaborative learning activities mediated by ICT is strongly related to teaching practices. This should lead us to reflect on the strategies used in teacher training, which require not only a theoretical knowledge of the teaching methods that can be applied to achieve the aims of collaborative learning but also in-context experience with ICT tools: i.e., participation in virtual communities, networks and projects. Accordingly, we agree with the ideas reported previously by other colleagues (Ball and Forzani. 2010; Rigelman & Ruben. 2012; Scharff & Brown. 2004) regarding the interest in fostering communities of practice through the emerging technologies with a view to generating new experiences and perspectives about ICT as dynamic tools for new ways of learning in tune with attitudes of progress and research in education.

REFERENCES

An, H., Kim, S., & Kim, B. (2008). Teacher perspectives on online collaborative learning: Factors perceived as facilitating and impeding successful online group work. *Contemporary Issues in Technology & Teacher Education*, *8*(1), 65–83.

Bakkenes, I., Vermunt, J. D., & Wubbels, T. (2010). Teacher learning in the context of educational innovation: Learning activities and learning outcomes of experienced teachers. *Learning and Instruction*, *20*(6), 533–548. doi:10.1016/j.learninstruc.2009.09.001

Ball, D. L., & Forzani, F. M. (2010). What does it take to make a teacher? *Phi Delta Kappan*, *92*(2), 8–12. doi:10.1177/003172171009200203

Cohen, L., & Manion, L. (1989). *Research Methods in Education* (3rd ed.). London: Routledge.

Contreras, P., Arancibia, M., & Cárcamo, L. (2009). Hipermedios y cooperación: la Web 2.0 y su incidencia en los procesos educativos [Hypermedia and cooperation: Web 2.0. and its influence in educational processes]. In M. Segura & B. Onetto (Eds.), Diálogos Culturales II. Interfaces Viciadas, comunicación visual y otras mediaciones (pp. 177-221). San José de Rio Preto: Bluecom.

Corbin, J., & Strauss, A. (1990). Grounded theory research: Procedures, canons, and evaluative criteria. *Qualitative Sociology*, *13*(1), 3–21. doi:10.1007/BF00988593

Crasborn, F., Hennissen, P., Brouwer, N., Korthagen, F., & Bergen, T. (2011). Exploring a two-dimensional model of mentor teacher roles in mentoring dialogues. *Teaching and Teacher Education*, *27*(2), 320–331. doi:10.1016/j.tate.2010.08.014

Crook, C. (2012). The Digital Native in Context: Tensions Associated with Importing Web 2.0 Practices into the School Setting. *Oxford Review of Education, 38*(1), 63-80.

Dillenbourg, P. (2002). Over-scripting CSCL: The Risks of Blending Collaborative Learning with Instructional Design. In P.A. Kirschner (Ed.), Inaugural Address, three Worlds of CSCL. Can We Support CSCL? (pp. 61-91). Heerlen: Open Universiteit Nederland.

Dirkx, J. M., & Smith, R. O. (2004). Thinking out of a bowl of spaghetti: Learning to learning in online collaborative groups. In T. S. Roberts (Ed.), *Online collaborative learning: Theory and practice* (pp. 132–159). Hershey, PA: Information Science Publishing. doi:10.4018/978-1-59140-174-2.ch006

Dobber, M., Akkerman, S. F., Verloop, N., & Vermunt, J. D. (2012). Student teachers' collaborative research: Small-scale research projects during teacher education. *Teaching and Teacher Education, 28*(4), 609–617. doi:10.1016/j.tate.2012.01.009

Domingo Coscollola, M & Marquès, P. (2011). Classroom 2.0 Experiences and building on the use of ICT in teaching. *Comunicar: revista científica iberoamericana de comunicación y educación, 37,* 169-175.

Engel, A., & Onrubia, J. (2010). Patrones de organización grupal y fases de construcción del conocimiento en entornos virtuales de aprendizaje colaborativo [Group patterns and knowledge construction phases in virtual collaborative learning]. *Infancia y Aprendizaje: Journal for the Study of Education and Development, 33*(4), 515–528. doi:10.1174/021037010793139608

García, I., Gros, B. Y., & Noguera, I. (2010). La relación entre las prestaciones tecnológicas y el diseño de las actividades de aprendizaje para la construcción colaborativa del conocimiento [ICT and learning activities design relation for the construction of collaborative knwoledge]. *Cultura y Educación, 22*(4), 395–418. doi:10.1174/113564010793351867

García-Valcárcel, A., & Mena, J. (2016). Collaborative learning and ICT use: What in-service teachers think and do in their classrooms. *Journal of Information Technology Research, 9*(1), 1–17. doi:10.4018/JITR.2016010101

García-Valcárcel, A., & Mena, J. (2013). *Recursos tecnológicos para la enseñanza e innovación educativa* [Technological resources for teaching and innovation]. Madrid: Síntesis.

García-Valcárcel, A., & Mena, J. (2014). Las TIC en el aprendizaje colaborativo en el aula de Primaria y Secundaria [ICT for collaborative learning in Primary and Secondary Education]. *Comunicar, 42*, 65-74

Garrison, D. R. (2006). Online Collaboration Principles. *Journal of Asynchronous Learning Networks, 10*(1), 25–34.

Gorghiu, G., Lindfors, E., Gorghiu, L. M., & And Hämäläinen, T. (2011). Acting as tutors in the ECSUT on-line course - how to promote interaction in a computer supported collaborative learning environment? *Procedia, 3*, 579–583.

Gress, C., Fior, M., Hadwin, A., & Winne, P. (2010). Measurement and assessment in computer-supported collaborative learning. *Computers in Human Behavior, 26*(5), 806–814. doi:10.1016/j.chb.2007.05.012

Hernández, N., González, M., & Muñoz, P. (2014). La planificación del aprendizaje colaborativo en entornos virtuales [Collaborative learning design in virtual envoronments]. *Comunicar, 42*(21), 25–32.

Hiltz, S. R., & Turoff, M. (2002). What makes learning effective? *Communications of the ACM, 45*(4), 56–59. doi:10.1145/505248.505273 PMID:12238525

James, R. (2013). ICT's participatory potential in higher education collaborations: Reality or just talk. *British Journal of Educational Technology*, 1–14. doi:10.1111/bjet.12060

Jirotka, M., Procter, R., Rodden, T., & Bowker, G. (2006). Special issue: Collaboration in e-research. *Computer Supported Cooperative Work, 15*(4), 251–255. doi:10.1007/s10606-006-9028-x

Johnson, R. B., & Onwuegbuzie, A. J. (2004). Mixed Methods Research: A Research Paradigm Whose Time Has Come. *Educational Researcher, 33*(7), 14–26. doi:10.3102/0013189X033007014

Jones, C., Dirckinck-Holmfeld, L., & Lindstrom, B. (2006). A relational, indirect, meso-level approach to CSCL design in the next decade. *Computer-Supported Collaborative Learning, 1*(1), 35–56. doi:10.1007/s11412-006-6841-7

Keppell, M., Suddaby, G., & Hard, N. (2015). Assuring best practice in technology-enhanced learning environments. *Research in Learning* Technology, *23*, 1-DOI:.10.3402/rlt.v23.25728

Killion, J. (2015). High-quality collaboration benefits teachers and students. *The Learning Professional, 36*(5), 62-64.

Kirschner, P. A. (2002). *Three Worlds ofCSCL. Can We Support CSCL.* Open University of the Netherlands.

Kreijns, K., Kirschner, P. A., & Jochems, W. (2003). Identifying the pitfalls for social interactions in computer-supported collaborative learning environments: A review of the research. *Computers in Human Behavior, 19*(3), 335–353. doi:10.1016/S0747-5632(02)00057-2

Lave, J., & Wenger, E. (1991). *Situated Learning: Legitimate Peripheral Participation.* Cambridge, UK: Cambridge University Press. doi:10.1017/CBO9780511815355

Levine, T., & Marcus, A. S. (2010). How the structure and focus of teachers' collaborative activities facilitate and constrain teacher learning. *Teaching and Teacher Education, 26*(3), 389–398. doi:10.1016/j.tate.2009.03.001

Little, J. W. (1990). The persistence of privacy: Autonomy and initiative in teachers' professional relations. *Teachers College Record, 91*(4), 509–536.

Livingstone, S. (2012). Critical reflections on the benefits of ICT in education. *Oxford Review of Education, 38*(1), 9–24. doi:10.1080/03054985.2011.577938

McInnerney, J., & Robert, T. S. (2004). Collaborative or cooperative learning? In T. S. Roberts (Ed.), *Online collaborative learning: Theory and practice* (pp. 203–214). Hershey, PA: Information Science Publishing. doi:10.4018/978-1-59140-174-2.ch009

McLaughin, M. W., & Talbert, J. E. (2006). *Building school-based teacher learning communities: Professional strategies to improve student achievement.* New York: Teachers College Press.

Monereo Font, C. Y., & Durán, G. (2003). Entramados métodos de aprendizaje cooperativo y colaborativo [Coopertative and collaborative learning methods]. Barcelona España: Edebe.

Oakley, B. R., Felder, M., Brent, R., & Elhajj, I. (2004). Turning student groups into effective teams. *Journal of Student Centered Learning*, *2*(1), 9–34.

Palloff, R., & Pratt, K. (1999). *Building Learning Communities in Cyberspace: Effective Strategies for the Online Classroom*. San Francisco: Joseey-Bass.

Paterson, M., & Pentland, W. (2008). The Use of Hermeneutics in a Mixed Methods Design. *Qualitative Report*, *13*(1), 116–134.

Paulus, T. M., Woodside, M., & Ziegler, M. F. (2010). 'I tell you, it's a journey, isn't it?' Understanding collaborative meaning making in qualitative research. *Qualitative Inquiry*, *16*(10), 852–862. doi:10.1177/1077800410383124

Pérez, M. A., & Delgado, Á. (2012). De la competencia digital y audiovisual a la competencia mediática: Dimensiones e indicadores [From digital and audiovisual competence to media competence: dimensions and indicators]. *Comunicar*, *39*, 25–34. doi:10.3916/C39-2012-02-02

Pérez-Mateo, M., Romero, M., & Romey, T. (2014). La construcción colaborativa de proyectos como metodología para adquirir competencias digitales [Collaborative projects elaboration as methodology to acquire digital competences]. *Comunicar*, *42*(21), 15–24.

Pontecorvo, C. (2007). On the conditions for generative collaboration: Learning through collaborative research. *Integrative Psychological & Behavioral Science*, *41*(2), 178–186. doi:10.1007/s12124-007-9017-8 PMID:18193519

Rigelman, N. M., & Ruben, B. (2012). Creating foundations for collaboration in schools: Utilizing professional learning communities to support teacher candidate learning and visions of teaching. *Teaching and Teacher Education*, *28*(7), 979–989. doi:10.1016/j.tate.2012.05.004

Rubia, B. (2010). La implicación de las nuevas tecnologías en el aprendizaje colaborativo [ICT implications in collaborative learning]. *Tendencias Pedagógicas*, *16*, 89–106.

Scardamalia, M., & Bereiter, C. (1994). Computer support for knowledge-building communities. *Journal of the Learning Sciences*, *3*(3), 265–283. doi:10.1207/s15327809jls0303_3

Scharff, C., & Brown, H. (2004). Thinking through computing: The power of Learning Communities. *Computer Science Education*, *14*(4), 297–320. doi:10.1080/0899340042000303456

Schoor, C., & Bannert, M. (2011). Motivation in a computer-supported collaborative learning scenario and its impact on learning activities and knowledge acquisition. *Learning and Instruction*, *21*(4), 560–573. doi:10.1016/j.learninstruc.2010.11.002

Simon, B., Anderson, R., Hoyer, C., & Su, J. (2004). Preliminary Experiences with a Tablet PC Based System to Support Active Learning in Computer Science Courses. *ITiCSE 2004: Proceedings of the 9th Annual Conference on Innovation and Technology in Computer Science Education*, 213–217. doi:10.1145/1007996.1008053

Strijbos, J. W., & Fischer, F. (2007). Methodological challenges for collaborative learning research. *Learning and Instruction*, *17*(4), 389–393. doi:10.1016/j.learninstruc.2007.03.004

Tolmie, A., & Boyle, J. (2000). Factors influencing the success of computer-mediated communication (CMC) environments in university teaching: A review and case study. *Computers & Education*, *34*(2), 119–140. doi:10.1016/S0360-1315(00)00008-7

Vygotsky, L. S. (1978). *Mind in society: The development of higher psychological processes*. Cambridge, MA: Harvard University Press.

Wegerif, R. (2006). A dialogic understanding of the relationship between CSCL and teaching thinking skills. *Computer-Supported Collaborative Learning*, *1*(1), 143–157. doi:10.1007/s11412-006-6840-8

Wells, G. (2001). The case for dialogic inquiry. In G. Wells (Ed.), *Action talk and text: Learning and teaching through inquiry* (pp. 171–194). New York: Teachers College Press.

Wenger, E. (1998). *Communities of Practice: Learning, Meaning, and Identity*. Cambridge, UK: Cambridge University Press. doi:10.1017/CBO9780511803932

Wenger, E., Mcdermott, R. A., & Snyder, W. M. (Eds.). (2002). *Cultivating communities of practice: A guide to managing knowledge*. Boston: Harvard Business School Publishing.

Section 3
Data-Driven Intelligent Ecosystems

Chapter 9
Clinical Decision Support Systems Question and Answering

David José Murteira Mendes
Universidade de Évora, Portugal

Irene Pimenta Rodrigues
Universidade de Évora, Portugal

César Fonseca
Universidade de Évora, Portugal

ABSTRACT

A question answering system to help clinical practitioners in a cardiovascular healthcare environment to interface clinical decision support systems can be built by using an extended discourse representation structure, CIDERS, and an ontology framework, Ontology for General Clinical Practice. CIDERS is an extension of the well-known DRT (discourse representation theory) structures, intending to go beyond single text representation to embrace the general clinical history of a given patient represented in an ontology. The Ontology for General Clinical Practice improves the currently available state-of-the-art ontologies for medical science and for the cardiovascular specialty. The chapter shows the scientific and philosophical reasons of its present dual structure with a deeply expressive (SHOIN) terminological base (TBox) and a highly computable (EL++) assertions knowledge base (ABox). To be able to use the current reasoning techniques and methodologies, the authors made a thorough inventory of biomedical ontologies currently available in OWL2 format.

Este estudo é integrante do Programa de Cooperação INTERREG V - A Espanha - Portugal (POCTEP 0445_4IE_4_P), no âmbito da criação do Instituto Internacional de Investigação e Inovação do Envelhecimento (com financiamento de 1 346 288,05 €).

DOI: 10.4018/978-1-5225-4944-4.ch009

1. WORK IN PROGRESS AND MOTIVATION

We will present in this paper two sections that illustrate our present work status:

1. The steps that we consider are involved in the complex acquisition procedure of clinical concepts expressed in English from text in Portuguese.
2. A proposal of a Software architecture needed for an automated acquisition to articulate the steps presented before.

Our work final objective is to enrich/populate an ontology that shall allow us to devise AI (Artificial Intelligence) tools that reason about Clinical Practice. Given the reasons explained in (D. Mendes & Rodrigues, 2011b) we chose CPR (Computer-based Patient Record) Ontology as target for population. CPR is a W3C (World Wide Web Consortium) standard for representing clinical practice knowledge. In the mentioned paper, we itemize the different issues faced and explore some of the possible solutions in population/enrichment of the ontology. The major problem of personal jargon creation was not properly addressed, however, and so is fully discussed in the present article. We are in the process of demonstrating the possibility of information extraction from free-text clinical episode reports in an automated manner.

When developing a methodology for automatic population of a CPR Ontology we faced the particular problem of clinical concept recognition when dealing with Portuguese natural language text. After consulting with several MD (Medical Doctors) whose activity is the main subject of representing knowledge that way, we found that we can take into our advantage the fact that each one usually develops his/her own way of writing down their daily chores. What we have to develop is a way of maintaining an acquired controlled vocabulary. Assuming that this is a task involving NLP (Natural Language Processing) specifically for Portuguese, we presented in (D. Mendes, Rodrigues, Rodriguez-Solano, & Baeta, 2012) some techniques that can be easily applicable to different languages with the same set of constraints presented ahead.

2. EXPERIENCIES IN THE FIELD

2.1 The ULSNA Experiment

The ULSNA, E.P.E. (http://www.ulsna.min-saude.pt/) has as its principal object the provision of primary and secondary health care, rehabilitation, palliative and integrated continued care to the population and the means necessary to exercise the powers of the health authority in the geographic area affected by it. ULSNA is a healthcare providing regional system that includes two hospitals (José Maria Grande in Portalegre and Santa Luzia in Elvas) and the primary care centers in all the district counties. Universidade de Évora signed an agreement with ULSNA that enabled the usage of de-identified (according to safe-harbor principles as reviewed in (Meystre, Friedlin, South, Shen, & Samore, 2010) clinical data from the SAM system in use both in the Primary Healthcare units and in the Hospitals. Using the clinical data that is available for us we intend to take advantage of the tooling available to reach the objectives mentioned in section 1.

2.2 Inevitability of English Usage in Bioinformatics

In the Biomedical informatics domain the knowledge representation of choice has been evolving since the initial developments of the Gene Ontology project (http://www.geneontology.org/) in the beginning of the century into ontology based representations. With the advent of research about reasoning capabilities and its applicability in the Semantic Web a lot of services that expose that capabilities have appeared but all of them, that matter, exploring only English based terminology. Recently we have been testifying the appearance of a strong set of Web Services in the area of semantic annotation of texts in the Biomedical domain. These specialized, domain oriented services, are made available mainly by the U.S. NCBO (National Center for Biomedical Ontologies) through BioPortal (http://bioportal.bioontology.org/) or by the NLM itself through the UTS (Unified Terminology Services) interface. We can find among these, services that do all kind of text processing for an information extraction pipeline like POS (Part of Speech) tagging, NER (Named Entity Recognition), and for clinical concepts both CSD (Concept Sense Disambiguation) and SSE (Semantic Similarity Estimation) for instance. To be able to fully exploit these in our work we definitely need to translate very precisely from the Portuguese personal medical jargon to an English completely understandable by the annotating services.

3. HOW TO REPRESENT AN ACQUIRED CONTROLLED VOCABULARY

3.1 Translation Memory as a Jargon Repository

In the world of professional CAT, Computer Aided Translation, the use of personal terminologies has been subject to standardization using the TM (Wikipedia, 2011), translation memory, model. A TM is a database that stores so-called "segments", which can be acronyms, nicknames, words, phrases and paragraphs that have already been translated, in order to aid human translators through the CAT software. The source text and its corresponding translation in language pairs called translation units are stored in the TM in a standardized exchangeable form.

3.2 TMM[1] Tooling

From the huge listing of resources available currently in the internet we checked carefully those that can handle TM[2] based in the TMX Standard. TMX was originally developed and maintained by OSCAR (Open Standards for Container/Content Allowing Re-use) a special interest group of LISA (Localization Industry Standards Association). The format allows easier exchange of translation memory between tools and/or translators with little or no loss of critical data. TMM, Translation Memory Managers, that handle creation and maintenance of the TMX files can be desktop centric or centralized TM systems that store on a central server. They work together with desktop TMM and can increase TM match rates by 30-60% more than the TM leverage attained by desktop TMM alone. They can very fruitfully develop Machine Translation based TM that can then be exported and further refined by the desktop systems. We explored two systems that proved to be interoperable given the standardized formats that are handled: Google translator toolkit (https://translate.google.com/toolkit/) and mymemory translation services (http://mymemory.translated.net/) that will be detailed further with an appropriate example.

3.3 Structure and Extensibility

The rst step is the creation of a seminal TM for some sample documents by using MT[3]. Clearly all the acronyms and personal defined nicks were not matched but we can then download a base TM to be further refined. With the possibility of extending and refining the XML based les we proceeded with the aid of another 2 desktop tools. TM les have an incremental XML structure that can be very easily manipulated trough adequate XSLT transforms that can be manually applied or enqueued in an automated Work ow as we show in the section 5. However, some tools nurtured our possibilities and rendered our work even simpler. Those are the open source OmegaT (http://www.omegat.org/en/resources.html/) and its companion the Okapi framework (http://okapi.opentag.com/). For every translation job, OmegaT creates a set of project folders which contain or can contain specific files. With appropriate file positioning, as can be learned from the OmegaT documentation, one can manipulate the sources and translations. We use the le positioning capabilities of our machine introduced in section 5 to automate all TM management. The knowledge accumulation for our specific translation tasks are just the positioning of the subsequent developed TM in the specified folder. The sidekick okapi framework companion allow us to automate all those steps and integrate them in a local or even, web based, work ow. We have then a TM enrichment work ow that can incrementally match more and more concepts. TMs can even be orchestrated by service or specialty for example rendering a fewer amount of work left to be done individually.

4. AUTOMATED ACQUISITION FROM CLINICAL EPISODES TEXT

As reviewed by the authors in (Mendes, D., & Rodrigues, I. P., 2011a) the state-of-the-Art for acquisition from Clinical Text has enjoyed strong developments in recent years. In the mentioned paper, we present a proposal for automated acquisition from HL7 messaging but here we are delving into the more generic possibility of extracting from free text present in most interfaces used by clinicians. Going from clinical episodes free text that is usually presented in a human friendly format to one adequate for computer processing involves a fair amount of text processing to handle different situations because: (1) Reports aggregate information from different clinical episodes that are not uniquely identified or not even individually dated, (2) The clinician is only identified by his/her name if any identification is made at all, (3) The information conveyed in free text is intended only to be understandable by fellow practitioners or even by the clinician him/herself making use of pragmatic jargon normally plagued with acronyms and nicknames abundant in their specific community, (4) Text is profoundly intermixed with decorative elements for better legibility, normally in PDF or HTML, (5) The clinicians natural language is other than English without concepts defined in a foundational thesaurus like SNOMED CT or FMA for instance that don't exist in that particular language, (6) The time spanning of the processes depicted in natural language is difficult to represent formally.

4.1 The Adequate Annotation Workflow

A set of sequential steps must be used to go from the pure text to the extracted concept instance. Those steps work ow can be configured declaratively using the software architecture shown in section 5. The translation steps involved are: (1) Manual translation (that is indispensable for the translator tutoring) with the precise clinicians validation of their jargon adequately translated into English, (2) PDF to raw text, or to structured (XML), converting for adequate documents cleansing. The remote annotation steps are: (3) NER (Named Entity Recognition) of all the patient names, clinician names, addresses, symptoms, signs and prescriptions with all the acronyms, units, time-spanning and timesnapping involved with the usual short forms that a particular doctor usually uses. In our particular situation we maintain a local cache of the previously identified vocabulary to check exact concept matching, (4) WSD (Word Sense Disambiguation) where terms can be disambiguated without technical clinical expertise. Most of them however have to be disambiguated according to the previously acquired concepts in our controlled vocabulary, (5) EAV (Extraction of Attributes and Values) is the final pure, single language, task that has to be performed and in which we need the tooling that this paper refers to filter the concepts from the annotated text to extract concept instances, (6) Semantic annotation using the Web interface of any of the services that we introduce in 4.2, either manually using the interactive interfaces or automatically with the Web Services available.

Given the array of Web Services that can semantically annotate bio-medical concepts in English in 4.2, we choose to use an evolutionary approach for use of the BioPortal annotator. We first use the annotator after manual pre-processing for the TM tutoring and later a fully automatic workflow based in Web Services orchestration.

4.2 SAM Corpus

The number of registered clinical episodes from which we can extract our information induces any ontology or knowledge engineer in a very hard problem to solve just trying to figure out the granularity of the ontology. When trying to apply the principles of well-defined formal ontologies depicted in (Smith et al., 2007) and trying to avoid the errors mentioned in (Ceusters, Smith, Kumar, & Dhaen, 2004) we decided to get our hands wet with a simple approach to the representation of disease and diagnostic illustrated in (Scheuermann, Ceusters, & Smith, 2009). In our particular case we face a shortage of structure of the reports in SAM in order to extract the intended information in a systematic way. We illustrate below that no complete demographic information is available in any of the used reports or that the problems enumerated in the Problem List do not have any kind of severity or progress information associated with them. Reports that are produced by the SAM system and have interest for our work are:

- **List of Medical Problems:** This report provides a history line of clinical problems without any relation to any procedures taken whatsoever. One of the interesting contributions of automatic acquiring this information into our ontology is the possibility of generating a problem oriented timeline into which all the further discovered clinical procedures can be related to. It lacks, however, any kind of data mentioning intensity or gravity of the problem or its evolution along the period since it was declared active until the problem dismissal. So far, as what is of interest to our Ontological Realism approach, it can be seen as a possibility of reasoning by fact inferring and it renders our proposal more interesting because that possibility can be easily demonstrated. The pre-processing step of this type of documents has to take care of the fact that concepts extracted can be found in a feeder ontology like MSHPOR and thus have an ID from a standard vocabulary associated with it. From the problem we can then extract possibilities that will allow us to identify if there is an associated cpr:etiologic-agent, a cpr:pathological-disposition or a mere sign recording for instance. These medical problems are, of course, a cpr:hypothesizedproblem of a cpr:clinical-diagnosis but that is left to the clinician to judge accordingly and here it is of course based on his/her own reporting so it should be evident.

- **SOAP Report:** This report is the most informative of all available as it depicts a clinical encounter in a semi-structured way in a datable point of time. The Subjective, Objective, Assessment, Plan (SOAP) framework, used to structure progress notes to facilitate problem speci c, clinical decision making by physicians, is a well-known, canonical structure in the medical domain. The underlying structure of the SOAP report induces some very important assumptions to be true. We find sections that can be associated with Subjective, the symptoms section S where we extract directly into a cpr:symptom record, medications found here are those administered only during the patient visit. Objective, the objective section O that are sign records that we take as generator for cpr:clinical findings. Assessment, the analysis section A which are the clinical investigation acts whose outputs can be clinical artifacts to investigate things that can be consequence of any of physiological or pathological processes. finally Plan, the plan section P where the therapeutic acts can be extracted with all the timing, posology and prescriptions registered in that particular clinical encounter, medications here are prescribed for discharge(Mowery, Wiebe, Visweswaran, Harkema, & Chapman, 2012). We now take advantage of the fact that we have to translate from jargon to English to customize our centralized TMM like the Google translator toolkit or mymemory translation services enhanced with our own Translation Memories and Glossaries. Let us introduce some demonstrative examples taken from a sample document gently provided by Dr. Carlos Baeta and properly de-identified (Figure 1).

We will, in the process of using one of the services, create Translation Memories with the identified personal acronyms like:

AP (Antecedentes Pessoais) into Personal History, HTA (Hiper Tensão Arterial) into High Blood Pressure, FA (Fibrilhação Auricular) into Atrial Fibrilation and V. Mitral (Válvula Mitral) into Mitral Valve. Some of which are acronyms that can be given the suitable translated concept like: ECG (Electro Cardio Grama) into Electro Cardio Gram or those that are even English acronyms themselves like INR (International Normalized Ratio) into International Normalized Ratio and some which are not really needed because the conveyed information is irrespective of what language is in, like: CENTRO DE SAÚDE into HEALTH CENTER or SEDE into MAIN OFFICE.

Included in this sample are notorious some more complex problems that are not related to the translation itself but with some other problems like the time spanning of concepts like 1 comp/dia which is adequately translated to 1 tablet per day using the defined Translation Memory but has to be posteriorly well defined as a time delimited occurring process in the process of CPR ontology enrichment.

Services to annotate clinical concepts in free text - Apart from being able to provide our own Web Services for various tasks given the availability of downloading several types of terminologies like MeSH or SNOMED CT CORE and generally the UMLS Metathesaurus, currently there are a myriad of WS at reach that can be configured to be connected to our CP-ESB presented in section 5 as service providers. Among those we think that are worth mentioned here the BIOPortal, OntoCAT (http://www.ontocat.org/) and UTS.

All these provide Web Services that offer specific tasks for Biomedicine terminology. Carefully chosen endpoints provide features that range from simple term lookups to complete semantic concept acquisition. With the concepts identified in our, purpose directed, controlled vocabulary it's only a matter of direct translation with the aid of the mentioned TM to be able to semantically annotate.

Figure 1. SOAP-381

CENTRO DE SAÚDE
PONTE DE SOR
SEDE

Registo Clínico da Consulta

Paciente 381_SOAP
XXXXXXXX
Data Nasc XX/XX/XXXX (XX anos)
XXXXXXXXX
XXX XXXXXXX XXXXXX
XXXX XXXXX

ESPEC. 24/01/2011 15:32 Dr(a) Carlos Baeta

Dr(a) Carlos Baeta

S OAP
-F; 81 anos;
AP: prolapso da V. Mitral; Dislipidemia e HTA;
TA controlada
Assintomática

s O AP
TA-130/75mmHg
AP - N
AC - Tons arritmicos
Pulso arritmico

SO A P
ECG (26/2/2010) - FA

SOA P
Varfine, segundo INR
Bisoprolol - 5 mg/dia
Lasix - 1 comp/dia

-Prolapso da V. Mitral
-FA com resposta ventricular controlada
-HTA medicada e controlada.

Mantém medicação. Deverá manter dose de varfine para manter INR de 2 a 3;
Poderá ser enviada à consulta em caso de descompensação

	Nome Comercial	Qt
1	Temazepam [Normison], 20 mg, Cápsula mole, Blister - 30 unidade(s)	1
	Posol:	

Complex Clinical Concepts - The formal way that CPR was precisely defined and is expected to be populated urges us to handle some more complex problems like time-spanning, time-snapping of processes or the anatomical location of the collected phenomena using the suggested Ontological Realism. Our approach is to just extract the instances, populate according to the suggestions in (Scheuermann et al., 2009), and leave all the inferring possibilities to the OWL DL reasoners and interested clinicians. Although more comprehensively explored by us in (D. Mendes & Rodrigues, 2011a) we are collecting essentially cpr:medical-problems and to render evident the complexity of the population task and the CPR ontology as an illustrative example we can see how a therapeutical prescription instances a span:ProcessualEntity that cpr:occursDuring a given dateTime and we have to cleverly delimit it's beginning as cpr:startsNoEarlierThen and cpr:startsNoLaterThen and its conclusion between cpr:stopsNoEarlierThen and cpr:stopsNoLaterThen. A significant number of this kind of "smart thinking" commitments has to be done to embrace notions of snapping (like medication administration obviously) or anatomical placement where the FMA (Foundational Model of Anatomy) is used as a feeder/validator ontology. We use all the alignment of ontologies used to form the CPR as both feeding as well as validating models in the process of instance extraction from the texts, These are: FMA for anatomy, OWL Time for temporal relations, BioTop for physics and structural/systems biology, BFO for the ontology philosophy, SNOMED CT for the clinical documentation and reporting and nally the Relations Ontology that formalizes all the relations. Very important is the use of coding of all of the occurring episodes according to the SNOMED CT standard that will allow our representations to be very widely interoperable.

5. SOFTWARE ARCHITECTURE

We have presented the full conception of an architecture in (D. Mendes & Rodrigues, 2011a). We are proposing an extensible Enterprise Architecture based in an ESB, which we call CPESB (Clinical Practice - Enterprise Service Bus). This is a common rail where messaging can now using a subscription model that enables the communication to be detached from any two particular services but instead be available on-request by one and served by another in a loosely coupled way. The ESB can then intervene in message exchange and overwrite standard rules for service execution. The case of an intervention here is the ability to filter and redirect invocations to the appropriate NLP task processors depending on the source being labeled with the status of the load it carries. The REST philosophy is suggested in our proposal as the best way of implementing a Service Oriented Architecture that serves as the communication underlying structure of our system. REST end-points are available for the generality of our needs. The available Web Services can render responses in highly structured forms like JSON or in any of the standardized mime types that can be handled by the filtering and enqueuing capabilities of any configurable available ESB like those based in Apache ServiceMix (http://servicemix.apache.org/home.html) or Mule (http://www.mulesoft.org/) for instance. They can be configured to compose a complete pipeline very easily (Figure 2).

In the picture is shown the possibility of defining services (daemons) that monitor the presence of new reports and appropriately apply the needed transformations according to their status and content until a translated document is delivered to the adequate end-point for annotation. All of this can even be done with due care about scalability, availability and all problems associated with a modern state-of-the-art software architecture as presented in (D. Mendes & Rodrigues, 2011a).

Building over the suggested infrastructure the systems are rather composed as opposed to monolithicaly built and so manifest high capabilities of plug-and-play configuration allowing for interchangeable

Figure 2. TM Creation workflow

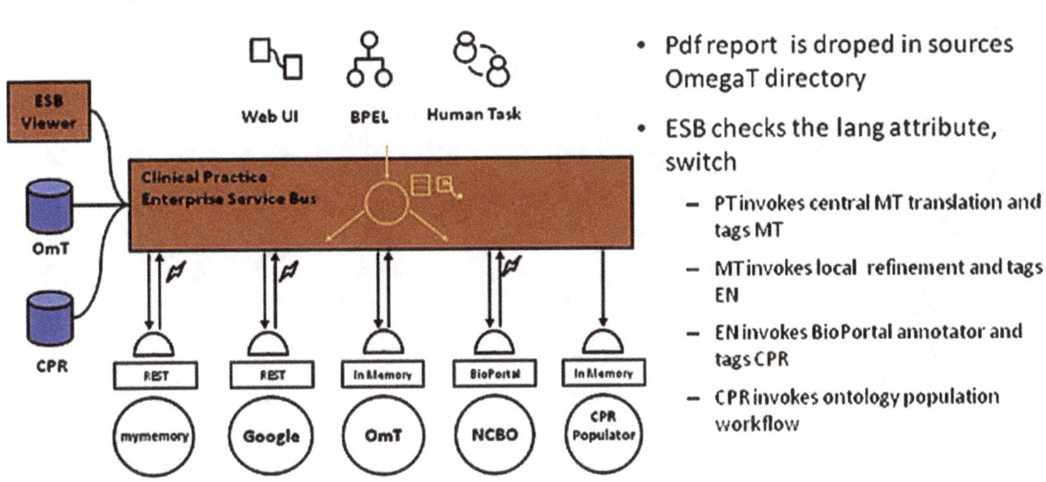

providers (as Web Services), Reference Ontologies (Feeders), and target ontologies. Having the foundations available with the right weapons provided one has to take a practical approach to the development of a target system using, in our case, the Java best-practices for pragmatic development that include a number of Patterns as in JEE (Java Enterprise Edition) or the pragmatic approaches developed in such successful projects as Spring (http://www.springsource.com/).

The flowchart that depicts graphically the acquisition from the source texts in Portuguese to the creation of the appropriate CPR instance is shown in Figure 3.

6. FUTURE WORK

In 2007 the work done under the W3C - Semantic Web Healthcare and Life Sciences Interest Group auspices rendered amazing contributions done with the most interesting technological landscape available at the time. Some realities that have since evolved and we could use for the present work are:

- The POMR Ontology that was started to be developed in 2006 was the target of the fruitful GRDDL e orts that appeared during 2007 in which base we are now working on. This efforts however were rdfs based for the OWL2 developments were still incipient. Only in November 2009 W3C standardized an OWL based ontology called the CPR that we use in our work.
- For a given HL7 CDA document the GRDDL gleaning presented in 2007 with it associate transform creates one cpr v0.5 instance with one cpr:patient.record with as many cpr:screening-acts

Figure 3. Acquisition flowchart

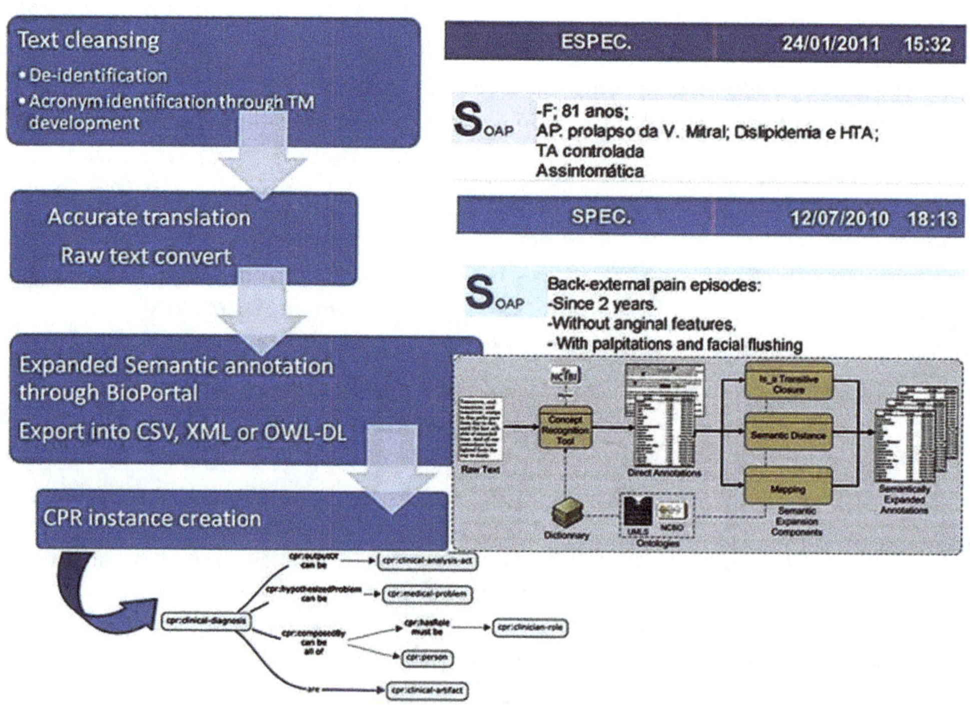

and associated cpr:clinical-descriptions as episodes are referred in the CDA source. This approach, though impressive technology, lacks the granularity that can be rendered now with further developments that CPR has gained in its latest versions.

Taking advantage of the developments done so far, we intend to build upon what was then achieved and propose work to be developed like:

6.1 Extend Our Proposal to Use GRDDL for Gleaning HL7 Documents Into CPR

The same architecture that we are proposing for gleaning from BioPortal's OBA output may be, as well, used to extract resources from CDA documents compliant with RIM V3. However it seems obvious that the complexity that the specification conforms could easily hinder the trees and not show the forest . Many recent proposals are being developed to apply restricted templates to make RIM tractable. The most notorious, that we intend to make reference to here, is the greenCDA project that already has delivered results. In fact, the greenCDA specification has been oficially balloted via HL7 and the version 1.0 is already available for download.

6.2 Using Only Refined Templates of Hl7 V3 Like Greencda Instead of the Full CDA

The idea was that the CDA was capable of representing any aspect of the medical record for any purpose. That sounds like a noble idea but in practice it creates a fixed overhead for even the simplest data exchange. The greenCDA project has developed a pragmatic methodology for creating simplified schemas that can be transformed directly to or from normative CDA. Referring to the excellent introduction by John Halamka the initiative can be brie y described as a: simple to use, XML construct that incorporates structured and unstructured clinical summary information. Simpler than CDA but full featured is exactly what Green CDA tries to do. It is a streamlined, human readable and computable lighter version of CDA that includes just the data and metadata necessary to do the job of representing a clinical summary. There are also tools available to convert CDA into Green CDA and additional tools are being developed to enable easy creation of Green CDA constructs by navigating the RIM, selecting attributes, and selecting associations to consolidate to make the XML atter. We feel that the development of XSLT for GRDDL processing from greenCDA documents is a very viable process of rendering a quasi complete picture of clinical practice extracting from HL7 messaging. We already have available the XML-Schema and transforms for back and forth CDA conversion and very easily will develop the OBA annotator output of greenCDA documents to CPR GRDDL transforms.

6.3 Restrict CPR to be OWL DL Well-Formed or Even Further

Unfortunately, CPR is not currently OWL DL syntax conformant. For the current reasoners to work effectively with OWL based ontologies they have to be OWL DL to be validatable in linear time. Although our clinical practice ontologies are restricted in their size most of the times, the reasoners check the validity a-prior and tend to refuse to work when some non-conformities are found. One interesting line of research may be the adequate restriction of CPR to be OWL DL conformant without lost of the needed expressiveness for clinical practice representation. We found even arguable that with adequate restrictions, like elimination of transitivity (D. Mendes & Rodrigues, 2011a) a Horn SHIQ ontology can be developed from CPR and then we reach the possibility of applying ConsequenceDriven Reasoners that have been shown recently to be very attractive for dealing with huge data-sets.

6.4 Pre-Curating of the Text Terminology for Clinical Concept Orientation

One of the main issues that were observable during the trial tests was that it is crucial to orientate the texts written in the SOAP reports for achieving a more direct clinical concept acquisition from SNOMED CT or CORE. When using the NCBO annotator even with only one level of expansion all the duplicate, obsolete and under-adequate terms were found and this rendered the need for a post annotation manual refinement task. This is incompatible with the full automation proposal that we intend to provide here. The solution then is to have local SNOMED CT validation based in AJAX completion to provide a-priory correct terminology for post annotation term adequacy.

7. RESULTS AND CONCLUSION

While the process of generating personal TMs has been developing simultaneously with other CPR population tasks the initial impressions look promising. The adequacy of the method and tools proposed here, however, are to be evaluated by the clinicians themselves. We are currently in the process of assessing the manual processes with a selected group that will after developing their own TMs, at their own pace, integrate their TMs in the complete ontology acquisition work ow to be able to assess the reasoning capabilities that can be made available. We are developing a process of quality assessment that will involve questionnaires that will be presented as soon as available and processed, expected during 2012. After this controlled group we expect that the generated TMs can be included in a regional corpus that will ease extremely the adherence of the rest of the community and try to expand to regional level which shall then be a straightforward step to achieve.

We demonstrate the articulation needed of different software tools, as well as individual medical knowledge, to extract the individual clinical concepts to develop personal TMs for automatic alignment of Portuguese texts to English ontologies, thesaurus and controlled vocabularies. The texts used are particular MD practice reports with instances collected automatically from their usage of the specific local SAM system. This proposed methodology and architecture is applicable to whatever EHR that produces clinical reports in whatever language in the world so it should be easily applicable globally.

REFERENCES

Ceusters, W., Smith, B., Kumar, A., & Dhaen, C. (2004). Mistakes in medical ontologies: Where do they come from and how can they be detected? *Studies in Health Technology and Informatics*. PMID:15853269

Mendes, D. (2011). A Semantic Web pragmatic approach to develop Clinical ontolologies, and thus Semantic Interoperability, based in HL7 v2.xml messaging. In *HCist 2011 - Proceedings of the International Workshop on Health and Social Care Information Systems and Technologies*. Paris: Springer-Verlag.

Mendes, D., & Rodrigues, I. P. (2011a). Automatic Ontology Population extracted from SAM Healthcare Texts in Portuguese. Jornadas de Informática da Universidade de Évora.

Mendes, D., & Rodrigues, I. P. (2011b). Well Formed Clinical Practice Ontology Selection. Jornadas de Informática da Universidade de Évora.

Mendes, D., Rodrigues, I. P., Rodriguez-Solano, C., & Baeta, C. (2012). Enrichment/Population of Customized CPR (Computer-based Patient Record) Ontology from Free-text Reports for CSI (Computer Semantic Interoperability). Procedia Technology, 5, 753–762). doi:10.1016/j.protcy.2012.09.083

Meystre, S. M., Friedlin, F. J., South, B. R., Shen, S., & Samore, M. H. (2010). Automatic de-identification of textual documents in the electronic health record: A review of recent research. *BMC Medical Research Methodology*, *10*(1), 70. doi:10.1186/1471-2288-10-70 PMID:20678228

Mowery, D., Wiebe, J., Visweswaran, S., Harkema, H., & Chapman, W. W. (2012). Building an automated SOAP classifier for emergency department reports. *Journal of Biomedical Informatics*, *45*(0), 71–81. doi:10.1016/j.jbi.2011.08.020 PMID:21925286

Scheuermann, R. H., Ceusters, W., & Smith, B. (2009). *Toward an Ontological Treatment of Disease and Diagnosis. In 2009* (pp. 116–120). San Francisco, CA: AMIA Summit on Translational Bioinformatics.

Smith, B., Ashburner, M., Rosse, C., Bard, J., Bug, W., Ceusters, W., & Lewis, S. et al. (2007). The OBO Foundry: Coordinated evolution of ontologies to support biomedical data integration. *Nature Biotechnology*, *25*(11), 1251–1255. doi:10.1038/nbt1346 PMID:17989687

Wikipedia. (2011). *Translation memory*. Retrieved from http://en.wikipedia.org/wiki/Translation_memory/

ENDNOTES

1 Translation Memory Manager.

2 Translation Memories.

3 Machine Translation.

Chapter 10
Towards an Intelligent System for the Territorial Planning: Agricultural Case

AMRI Benaouda
University of Salamanca, Spain

Francisco José García-Peñalvo
University of Salamanca, Spain

ABSTRACT

This chapter concerns the conceptualization of an intelligent system for the territorial planning, taking as an example the agriculture case as a tool in decision making. It is started by giving a comparison between the geographical information system (GIS) and the intelligent system (IS), demonstrating the limits of the GIS and the appeal to the artificial intelligence. Also, the chapter gives an overview of the application of decision support systems (DSSs), modeling and simulation applied in forest management, agriculture, ecology, and environment. Finally, the chapter proposes the methodology and the intelligent system proposed, setting up some indicators which help to aid decision making.

1. INTRODUCTION

Each year hundreds of books and Doctoral theses extend our knowledge of computers, or artificial intelligence.

Expert systems, artificial neural networks, fuzzy systems and evolutionary computation are major technologies used in intelligent systems. Hundreds of tools support these technologies, and thousands of scientific papers continue to push their boundaries.

Most of the literature on artificial intelligence is expressed in the jargon of computer science, and crowded with complex matrix algebra and differential equations (Negnevitsky, 2005).

This, of course, gives artificial intelligence an aura of respectability, and until recently, has kept non – computer scientists, at bay. But the situation has changed!

DOI: 10.4018/978-1-5225-4944-4.ch010

The personal computer has become indispensable in our everyday life. We use it as a typewriter and a calculator, a calendar and a communication system, an interactive database and a decision support system. But we want more we want our computers to act intelligently.

We see that intelligent systems are rapidly coming out of research laboratories, and we want to use them to our advantage.

An Intelligent System (IS) has the power, that allows the prediction of an output object for the given input object, often being able to model complex relationship between variables, here the performance depends on data: the learning group must follow propagation, however, the training phase can be time consuming, depending on the structure of the network (number of input variables and output, number of hidden Layers), the number of examples in the training sets, and the number of iterations.

Finally, Artificial Intelligence has recently been having remarkable success in fields such as natural Language processing, expert systems, and solving equations.

It is the right time for application of these techniques in simulation.

Its implementation will contribute to the organization, processing and analysis of knowledge of the subject.

2. TOOLS OF DECISION IN THE TERRITORIAL PLANNING

Until now, they have used the Geographical Information System (GIS) all over the world in the field of the territorial planning.

This is an impractical system with many defaults and deficiencies.

The GIS is well suited to differentiate and diversify the scale of the complex patterns of interactions in space, while the need of effectiveness of decision-making process requires the use of other systems for decision support.

The ability of GIS to store, process and analyze special data (either attribute geographic data) is clear. You can use the ability in real time to support the decision -making process, which are primarily systems data processing and, therefore, it is not well suited for addressing complex situations.

Creating scenarios, spatial analysis, and advanced modeling requires what is called procedural knowledge, while the so – called declarative knowledge is essential to represent the know-know, which incorporates methods and techniques of AI.

Other Limitations of the GIS:

- Policies incompatibilities
- Unresolved problem, issue certificates on the land already registered by the state cause serious conflicts.
- Very informative, firmly entrenched maps, with few references points.
- Limited – semiotics.
- More anthropological – Reading: Representation.
- Deficiencies.

The causes of the shortcomings of GIS for decision making can be classified into two classes:

1. General conceptual or methodological problems derived from the basic organization of GIS.
2. Technical problems related to the usual way of using these tools.
   a. Conceptual Problems:
      i. Initial design of GIS: Generalist and little specialization approach.
      ii. Fundamental concern: more "external" formal and geometric aspects of geographical problems.
      iii. No causal generators visible structures and "geometric" processes of geographical problems.
   b. Technical problems:
      i. "Hard" data (measured in precise terms, if possible quantitative: surveys, remote sensing, etc.).
      ii. Others considered "soft" date (derived from the opinions, preferences and judgments of people and are expressed in vaguer and qualitative terms).

3. ARTIFICIAL INTELLIGENCE (AI)

An Intelligent System (IS) must meet two objectives:

1. Being able to perceive the environment, and act rationally according to the perceived environment.
2. Requires reasoning (Logically), and acting on the conclusions drawn, as well as applying the rules drawing from interviews with experts.

3.1. Evolution of Modern Computational Intelligence

The history of the artificial intelligence passed from the era of great ideas and great expectations in the 1960s to the disillusionment and funding cutbacks in the early 1970s; from the development of the first expert system such as DENDRAL, MYCIN AND PROSPECTOR in the seventies to the maturity of expert system technology and its massive applications in different areas in the 1980s and 1990s; from a simple binary model of neurons proposed in the 1940s to a dramatic resurgence of the field of artificial neural networks in the 1980s; from the introduction of fuzzy set theory and its being ignored by the West in the 1960s to numerous 'fuzzy' consumer products offered by the Japanese in the 1980s and worldwide acceptance of 'soft' computing and computing with words in the 1990s.

Our brain stores the equivalent of over 10^{18}bits and can process information at the Equivalent of about 10^{15}bits by second.

By 2020, the brain could probably be replaced by a chip the size of a sugar cube-and perhaps by then there will be a computer that can play-even win-the Turing imitation game. (Negnevistsky, 2005).

For hardware, we really need to have human level speed and high memory machines. In terms of memory, things look seem are promising. For speed, we still have to wait. Human brain processing power ranges between 100 million MIPS to 100 billion MIPS.

(1 MIPS = 1 Million Instructions per second).

The other remaining problem is software. Once we get human intelligence level machines, software will be required.

In doing so, one has to understand how the human brain works.

Computational neuroscience and molecular nanotechnologies. Neuroscience is concerned with how the individual components of the human brain work. Up to date research reports good computational models of the primary visual cortex.

But simulating the whole brain requires enormous computing power. Molecular nanotechnologies work at Nano scale level which is 1 to 100 nanometers- from 1/1.000,000 to 1/10.000 of the thickness of an American dime. Many of the key structures of human nervous system exist in Nano scale.

The real challenge will be to use Nano machines to dissemble a frozen or a vitrified human brain. (Grosan and Abraham, 2011).

In parallel with getting the artificial intelligence, small steps have been taken in terms of algorithms and methodologies, which can be applied to solve simple or more challenging real-world problems.

Much of the current research focuses on the principle, theoretical aspects, and design methodology of algorithms gleaned from nature.

Examples are artificial neural networks are inspired by mammalian neural systems, evolutionary computation inspired by natural selection in biology, simulated annealing is inspired by thermodynamics principles and swarm intelligence is inspired by collective behavior of insects or micro-organisms etc.

3.2. The Necessity to Beneficiate From the Advancement of the AI.

Models implemented on computers are described by relation which can be equation, qualitative relationships, or logical rules.

Environmental systems have several attributes which make their formal representation, different from that of other system (particularly industrial systems).

Guariso and Werthner (1989) describe some of the distinctive features of environmental systems:

- **Dynamics:** Environmental systems evolve over time. Many environmental processes are better viewed as single events and require discrete event models.
- **Spatial Coverage:** Environmental systems involve physical processes which take place in a 2-or-3 dimensional space.

Modeling such processes commonly requires the solutions of sets of partial equations liking spatially referenced variables. Data are stored in spatial databases of within a geographical information system (GIS) to assist in spatial analysis.

- **Complexity:** environmental system is complex, usually involving interactions between physic-chemical and biological processes. Models of such systems require multi-disciplinary expertise, and to be successful require periodicity: Many environmental processes are periodic in time, which adds a degree of complexity to parameter calibration and validation. Data storage can also be affected by the periodicity of data.

In addition, the following important features of environmental systems can also be identified:

- **Heterogeneity and Scale:** The media in which environmental processes take place are not homogeneous and cannot easily be characterized by measurable parameters. Processes mi different media may have quite different characteristics time and space scales.
- **Paucity of Information:** Observational data on environmental systems, particularly on internal states, are typically only sufficient for the characterizations of simply parameterized models.

While all of these characteristics are encountered separately in other fields, all these are presented simultaneously in environmental modeling.

These ideal tools for environmental modeling and simulation need to be well suited to cope with these characteristics. The problems which faced the Ecological Decision Support Systems (EDSS):

To achieve model integration, we must make existing models fully accessible to the modeling community, and we must supply integration software tools and standards to make model integration possible.

Providing integration software and standards is particularly difficult because of:

1. The lack of common specification for model interfaces;
2. The difficulty in extracting a sub-model from an existing model;
3. The unavailability of machines independent code;
4. The frequency with which models are integrated in software packages and cannot be used independently.

3.3. The Disciplines of Artificial Intelligence

Artificial intelligence (AI) has no defined academic research object except the interest in the "mental" and knowledge representations. It is divided into many sub-disciplines focusing on distinct problems (such as vision, problem solving, language comprehension, learning, etc.). There is no unified research paradigm and some branches of AI have become multidisciplinary exchanges of fields which combines philosophers, psychologists, computer scientists and others interested in various issues of intelligence. Thus, the fundamental branches of AI are part of cognitive science. Inevitably, there are multitudes of ways to design basic and applied research. We will discuss five:

The "pure" Artificial Intelligence. The school that we can locate for example in some research groups at MIT aims to discover what intelligence is very general. The modeling of the human mind is not in the foreground. Some are more interested in the mechanisms that constitute intelligence in simple autonomous agents (Brooks, R., 1991), others seek to establish programs that solve some difficult and specialized tasks like winning chess.

The "hard" Artificial Intelligence. This is to model the human mind constructing a program which is, at some level of abstraction, the human mechanisms. Here is where are found the famous controversial theses as "cognition = computation" and thus "mind = Program" that we will discuss in detail later.

A "soft" Artificial Intelligence (weak AI). The purpose is similar to that of "hard" AI, it is to model the human mind. However, the program has a heuristic role model. The computer and its program are not put in direct contact with the brain and mind. There are inductive and deductive variations of this approach. The first seeks to build the object "understanding of intelligence" by building programs that can solve some very specific tasks and then generalize and empirically test a solution founded. The second is more proper: Programs and models to test some assumptions.

"Practice" Artificial Intelligence. This is especially (but not only) for expert systems, probably the branch of AI best known to the general public. An expert system must automate certain human intellectual tasks or assist a human decision maker ("decision support" or "advisory systems"). Our work is in this category.

Intelligent tutors ("intelligent tutoring systems") are another convenient branch as some branches of robotics. In all these disciplines, "practice" does not mean total absence of epistemological reflection; many theoretical impulses have no origin in practical work. Curiously the limits of AI have emerged in expert systems more than in other sectors.

A "modeling tool" Artificial Intelligence. Reference is made to researchers in other disciplines making use of some representation languages and inference to describe certain phenomena, and to formalize models or theories.

Necessarily, the boundaries between these fundamental guidelines of AI are not airtight. Any attempt to automate the resolution of an intelligent task must raise the question what intelligence or the "mental" and if the artificial resolution is really done in an intelligent manner. Conversely, even the most speculative artificial intelligence has to ask the question of operationalization of its concepts. Thus (Pylyshyn, 1984) noted that "... in cognitive science the gap between met theory and practice is extremely narrow."

The relationship between data model and artificial intelligence theory is based on two pillars: Most approaches adopted a strategy sometimes difficult to articulate where data models (program) and theory influence each other. (Deutsch, K.W., 1986). Faced with these guidelines, we can identify a more traditional line drawing of experimental psychology. It can be modeled the following ways: (1) joint of a verbal form theory, (2) formalization of at least one aspect of a program and (3) the test phase including experimentation with subjects. The experiments test the accuracy of the model and then the validity of the theory.

The "classic" approach in artificial intelligence is more difficult to describe. Like any paradigm, "schools" of AI know several research levels. There is often a goal of general research and "asymptotic research" to "understand natural language." At this level, we observe it more or less well articulated to the nature of the research topic. At an intermediate level, there are partial theory stage pre-modeling. For example, in the field of natural language understanding, we can cite the scripts theory and definition of their structure and their use in cognitive processes. At a "low level", we finally find the programs (modeling) that implement more or less a partial theory. Regarding our example, we can cite the series of software made at Yale to identify scripts in a limited set of written texts and answer questions that are asked in the software. When this kind of research is carried out in a computer science department, we often hear that the program is a theory. This is the concept of the theory sometimes used in science when we try to express the essence of a theory in a formal language. The creation of formal and computational theories received the epistemological priority in AI. However, in practice, this maxim does not always translate very well: The difference between the program and the verbal theory is large, sometimes the program is "unreadable" and the parties that represent a theoretical interest are confused with the uninteresting parts.

The AI is also a high vocation empirical engineering discipline. It takes long and multiple efforts to build a program that "works". The sources of inspiration are many: observation of subjects, introspection, speculation, theories and psychological "paradigms", etc. The nature of the data according to the research levels. (Minsky, M., 1979) uses the term epistemology applied to characterize the contribution of AI to philosophy. AI program is therefore not only a theory but an experience, an artifact that can be studied, reviewed and handled as a source of knowledge that must be confronted in return data and other theories. It is therefore both subject construction and source of reflection (a bit like a mathematical theory). Data model (quasi-formal theory) and verbal theory are constituted and reconstituted in a circular fashion. A scenario may be difficult to accept for a Popperian. But this traditionalist epistemology of social science has no revenue to provide for the study of complex phenomena.

Research in AI intelligence is primarily explorative, empirical and experimental. Its epistemological foundations are strongly discussed in literature, but for now no solution has been adopted by the community. Discipline is too recent and a very difficult task.

In the following sections, we will discuss in more detail some of epistemological keys issues that are of academic interest and / or practice. Despite the variations between the forms of AI (in the case of simple programs designed to behave intelligently, deal with a specific task, or autonomous agents are able to learn abilities and interaction with the environment and sensor motor and social), two big questions remain common: what does the term "intelligent" signify? And can we build intelligent artifacts? Cognitive science is nonetheless agreed on one point: we know what human beings are capable of:

"They can (1) discriminate, (2) Manipulate, (3) Identify and (4) Describe the objects, events and states of affairs in the world they live in, and they can also (5) Produce descriptions and (6) Respond to those descriptions of objects, events and states of affairs. Cognitive theory's burden is now to explain how Human Beings (or Any Other devices) do all this (Harnad., 1991).

AI and other cognitive sciences accept the difficulty of this research program, even if they have to live with some difficulties or some epistemological paradoxes. Approaches such as behaviorism or economic rationalism are only leaks ahead of problems of interest philosophy for 2,000 years. Other systemic sciences are facing exactly the same problems as the AI.

3.4. Representation and Computation

The dominant approach in cognitive science in recent years is cognitivism, the explanation of intelligent behavior by manipulating representations mental processes, that is to say encodings (implicit or explicit information.) There are two major variants of thought: symbolism and connectionism. Both apply content, "Representational content," handled by computations. The key role of representations in the mental phenomena is a paradigmatic postulate. The essence of the concept of "representation" is not modeled in systemic or functional terms according to Bikhard (1993).

We first turn to the "classic" Artificial Intelligence (symbolic) that is anchored in the basic principles of psychological information. In classical artificial intelligence, most models feature intelligent behavior with formal operations on symbolic structures. Hence the slogan:

Cognition = computation+ representation

It is a rationalist position (Winograd., and Flores, 1986; Pfeifer and Verscure, 1992) which consists in attributing to the intelligent agent the following behavior: (1) defines the situations in terms of identifiable objects with given properties, then it identifies the general rules applicable to that situation based on identified items and finally, (2) applies those rules to the situation to deduce actions. This model is popular in many disciplines, including cognitive psychology. It is useful to recall the contrast with the following step. The intelligent agent of classical AI is essentially a problem solver as it is represented in:

1. Complex interpretation situation.
2. Problem solving with rules (or other principles).
3. Complex reaction.

One may wonder how to justify such a model of information processing mentally and physically. At the mental level, the answer is obvious. The idea that one has a representation of the facts in the environment that is manipulated thereafter matches the naive image we have of ourselves. Now, how model the existence of these symbolic processes in the brain?

A frequently cited response was formulated and operationalized exactly by Newell and Simon in their thesis of "physical symbol system" (Newell and Simon, 1976). "Knowledge level" (Newel, 1981).

Act-based symbolic representations are explained by a multilevel model. There is a physical mechanism that manipulates physical symbols (brain). The performances are physically instantiated as symbols of physical structures. They therefore have a precise form, a form that can be manipulated through mechanical operations in some way. Here a strong analogy to the operation of the computer. This level one) biological, physical, or physiological) concern- according to these authors, -not- is a problem of biology (such phenomena yet "lower" like chemistry and physics). Just know that a symbolic physical system has the means to serve as a general support for intelligent action.

For the researcher in psychology or in AI, the lowest level of interest would be the symbolic level 2, also called syntactic or functional architecture. The AI should not be concerned with the mechanics of the computer or the brain; but, only through the interpretive mechanism can these symbolic structures have shape. The explanation of mental life begins at this level and continues at level 3, that of knowledge. (Newell, A., 1981), also called "intentional or semantic level."

We try examining more closely level 2 (symbolic), the "interface" between the description of the mind as physical system and the mind as a representational system: We give the floor to Pylyshyn (1978, 1980, 1981, 1984). This level is also called "functional architecture." The "cognition computing over symbols formula" of Pylyshyn is probably very radical, but it has the merit of being clear and demanding.

To summarize what we have said about computing, a computational process is one whose behavior is viewed as depending on the representational or semantic content of its state. This occurs by virtue of there being another level of structure - variously called the` symbol level ' or the syntax or logical form-which possesses the following two essential properties. One, the formal syntactic structure of particular occurrences (tokens) of symbolic expressions corresponds to real physical differences in the system, differences that affect the relevant features of the system's behavior. Two, the formal symbol structures mirror all relevant semantic distinctions to which the system is supposed to respond and continues to do so when certain semantically distinctions Interpreted rules are applied to them, transforming them into new symbol structures. For every apparent, functionally relevant semantic distinction there is a corresponding syntactic distinction. Thus, any semantic feature that conceivably can affect behavior must be syntactically encoded at the level of a formal symbol structure. By this means we arrange for the system's behavior to be describable as responding to the content of its representations - to what is being represented –in manner perfectly compatible with materialism (Pylyshyn, 1984).

This cognitive position of AI is heavily inspired by the computation theory and can summarize the concept of symbolic AI system by the following (Harnad, 1991).

1. A game token physical (such as markings on paper, events in a digital computer), which are
2. Manipulated on the basis of explicit rules that are

3. Also physical tokens or chips chains.
4. The handling of symbolic tokens by rules is based solely on the shape of the chips (and not on their meaning), it is purely syntactic, and consists to a
5. Combine and recombine in symbolic tokens as symbolic rules.
6. There are primitive and symbolic tokens and atomic.
7. Composite chains symbolic tokens.

The entire system and all its parts - atomic and composite chips, syntactic manipulation (current and potential) and the rules - are interpretable semantically: syntax can systematically assign a meaning (assign a link to objects or describe a mental state.

4. DECISION SUPPORT SYSTEMS

Decision support systems (DSSs) do not making decision, but aid decision making (e.g., Bayesian statistics, classification and regression tree analysis), optimization (e.g., Linear programming, simulated annealing), and a variety of simulation techniques and tools (Bunnell and BoyLand, 2003).

All of them are models because they simplify and abstract reality.

As following of (Bunnell and BoyLand, 2003) all DSSs share two broad features: simplification and translation. The simplification necessary to understand the model necessitates translation between reality and the model system. The translation is illustrated in its simplest form, a research hypothesis.

4.1. What Is a Decision Support System?

When we use models to guide a decision we can consider them decision support systems (DSS).

We begin with a question of nature. We translate this question into a research hypothesis. That translation process changes our initial question. The research hypothesis is tidier and easier to communicate, but some notions are omitted as we strive for example, testable statement.

Figure 1. Rephrasing our questions so they can be tested in the natural world involves several translations

Question

↓

Research hypothesis

↓

Statistical design

↓

Real world

4.2. Use of DSS

Data structures and algorithms are the comer stones of any model.

Algorithms capture processes and management options.

Data capture initial states and parameterize the algorithms.

For both data and algorithms, the process of creating a model is one of abstraction and simplification.

Elements of the complex, real world are translated (abstracted) into data, algorithms, and tautologies.

Only parts of the complex, real word are chosen (simplification), and these are restricted in their range and properties to those that are measurable and relevant.

We shift our focus from the very broad, real world to a very specific model. For this shift to be useful, the very specific must encompass those elements that success fully represents relevant features of the real world.

In the following steps, we give examples of the application of the decision support systems in many fields.

4.3. Cases in Territorial Planning

4.3.1. Some Issues in the Design of Agricultural Decision Support Systems

The development of crop and soil process models, and their incorporation into decision support systems for the practical management of agricultural production, has made a significant contribution to the focus and integration of research by agricultural scientists. (Cox, 1996).

However, such DSSs should not necessarily be used as vehicles for communicating with other groups, such as formers, with a view to effecting directed change in their behavior. Other, sometimes more effective and efficient (and often more enjoyable and elegant) models are available.

Interest in building agricultural DSSs has come primarily from the professional research community there has been pressure from industry to bring the technology to market as quickly as possible. The pressure has arisen from the increasing demands for accountability publicly – funded agricultural R & D programs, which need a tangible product to indicate that something has been done. Further pressure has come from the more towards commercialization of publicly – funded R & D, which is taken to mean the production of information packages which can be sold.

There are reasons for supposing that the expectation that farmers could use fully exploit DSS technology to improve their own practice is seriously flawed:

First, the design and presentation of DSSs are sufficiently complex to require considerable technical support before one can operate them effectively; they require frequently require monitoring data about the state of the system which it is unrealistic to expect farmers to collate; and they are based on an information platform (the computer) which few farmers currently use as a planning tool.

Figure 2. Structure of DSS

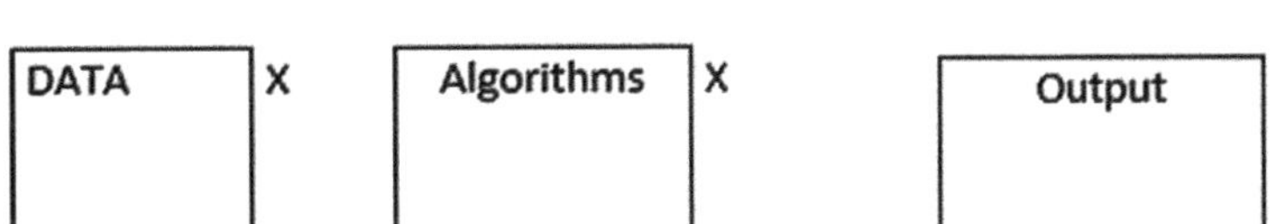

Second, most DSSs are justified on the basis that they provide information that is useful only for the exception; they are rarely applied to improving performance in routine situations where the answer is often clear. Thus, the potential gains from using the technology are marginalized: To those people who least need assistance (because of the extent of their current skills); and to situations where the balance between costs and benefits is problematical.

We can conclude the application of DSSs in agriculture saying that: The development of distinctive criteria for evaluating success in the construction and use of crop and soil process models in agricultural R & D, and for DSSs to support decision making in the management of agricultural production systems, are urgently needed if we are to capitalize on the gains that we have already made.

- Different research fields to find a common ground.
- Randomness: Many environmental processes are stochastic. In addition, the parameters of models representing such processes are usually uncertain, and it is common to know their rangers only approximately.

These attributes call for techniques such as statistical analysis of model equations.

4.3.2. DSS for Integrated Ecological Assessment

An ecological assessment is a science – based process and tool for determining the condition for an ecosystem relative to an identified condition or state (Mahan et al., 2015).

Ecological assessments, often based on monitoring data, are used for making or re-evaluating land management and/or regulatory decisions and reporting those decisions to the public.

These assessments have been adopted by many natural resource managers at governmental and non–governmental organizations.

Many researches established indicators for decision making.

Indicators are measures that describe the de condition of an ecosystem and/or its components (e.g., pH, biological indices, species diversity).

The objectives in many studies were as follow:

- Determining the scale of ecological assessment and dataset to use.
- Developing a general reference condition for the area chosen.
- Selecting ecological indicators to represent our ecosystem and associated thresholds.
- Using selected ecological indicators and thresholds both individually and in an integrated fashion to assess ecosystem condition.

Following these objectives (Mahan et al., 2015) developed and applied an indicator – based decision support system (DSS) model that incorporated geographic information system (GIS), fuzzy logic, expert judgment, best available data and ecological thresholds.

The authors cited above a dashboard which displays a variety of features that permits the reader (e.g., public, government stakeholders') to gain a rapid understanding of the assessment outcomes.

4.3.3. Decision Support System for the Sustainable Forest Management

Due to multi–faceted role of resource, sustainable forest management necessitated decision – making which recognizes and incorporates diverse ecological, economic and social processes; a multitude of variables; and conflicting objectives and constraints. Sustainable forest management also encompasses spatial as well as temporal dimensions (Montreal process, 1995; Varma, 2000).

1. **Definition of Sustainable Forest Management:** There is no universally accepted definition of sustainable forest management. It is of often regarded as a logical extension of Brandt land´s concept of sustainable development (Ferguson, 1996). While recognizing a responsibility for the current and the future generations (Toman and Ashton, 1996), it does not help in guiding or achieving sustainability at operational level or the treatment of trade-offs with other objectives.

Generally, the health and well–being of the people and the forest ecosystem together determine the progress of management practices towards the sustainability.

To be able to apply the concept of sustainable forest management as clearly and simply as possible, it has been necessary to describe it in terms of guiding principles, criteria and corresponding indicators.

In this regard, the results of international initiatives (ITTO, 1992, Helsinki Process, 1995 and Canadian Forest Service, 1995) are significant.

2. **Uncertainty in Decision Making:** Uncertainty plays an important role in any decision – making process. An estimate of uncertainty attached to field data can be obtained from the root mean square error of the measurement and sampling errors (Philip, 1994, Eastman, 1997).
3. **Selection of Criteria and Indicators:** In order to build on the results of earlier research ITTO, 1992, Helsinki Process, 1995 and Canadian Forest Service, 1995), the following criteria and indicators were chosen for this category.
    a. **State of Forest Resource:** The distribution and area of forest type, by native or exotic species, and by age – classes.
    b. **Conservation of Biological Diversity:** The distribution and area of the forest managed specially for the conservation and use of forest genetic resources, distribution and number of endemic species, and distribution and number of species according to protection – status (endangered, threatened, rare, and extinct).
    c. **Forest Health, Vitality and Integrity:** The distribution and area affected by disease, insects, fire, competition from exotic species and unregenerate part of the ecosystem.
    d. **Production of Wood and Other Forest Products:** The distribution and area of forest managed according to a management plan, distribution and area available for the production of wood and non – wood forest products, distribution and area dedicated to sustained production of wood and non-wood forest products.
    e. **Soil and Water Protection:** The distribution and area of forests managed specifically for the purpose of soil and water conservation, distribution and area subject to poor organic content, and compacted soil.
    f. **Socio-Economic Functions:** The distributions and area of forest managed primarily for leisure and tourism, for cultural values, for landscape and for meeting the needs of the forest – dependent population.

4. **Literature Review:** Forest resource planning is very complex problem mainly due to the multiplicity of wide-ranging criteria involved in the underlying decision making process (Diaz-Balteiro L., 2008). Thus, every decision made affects criteria of different nature like:
   a. Economic issue (e.g., timber, forage, livestock, hunting, etc.).
   b. Environmental issue (e.g., soil erosion, carbon sequestration, biodiversity conservation, etc.).
   c. Social issue (e.g., recreation activities, level of employment, population settlement, etc.).
   d. It's cultural and spiritual significance for people. (Kanowsky P.J., et al., 2009).

The scope and complexity of issues that need to be addressed is part of reason that decision-making in natural resource management is difficult. (Bruce, G.M., et al., 2012).

In many development countries, public and managers and planners consider several overarching issues regarding sustainability, including loss of native forests and grasslands, degradation of ecosystem services derived from them, effects of climate change, and managing in the face of ecosystem disturbances from fire, insects, disease, development pressures, changing demographics and settlement patterns.

4.3.4. Delivering Environmental Decision Support System: Software Tools and Techniques

Why Modeling and Simulation?

The practice of developing and using mathematical models of physical sciences with the advent of the computers.

The industrial sector has made wide use of the computer modeling, to the point where most of the research and development for new processes and products is dependent on computer simulations. The wide use of computer simulation by the industrial sector has strongly influenced the development of the commercially available modeling and simulations software (Rizolli and Young, 1997).

Computer modeling of environmental systems is also common, both a method for scientists to test hypotheses and so better understand the way such systems function, and also as a predictive tool for those who manage such system.

Modeling and simulation provide a rapid means of investigating the expected response of a system possible future changes, by under taking the necessary computations which are commonly complex and data intensive.

More and more, computers are also providing powerful ways of displaying model predictions to the user to allow better and quicker comprehension of the complex spatial and temporal dynamics which characterize environmental system.

4.3.5. Expert Help

Expert help can be provided by tool such as expert system shells, hyper textual help systems, multimedia presentation, and context – based interface which can assist the EDSS user.

For example, expert help can be very useful in accessing the knowledge base to help the user retrieve information on data preparation, model selection and result interpretation.

The main technology used in the implementation of expert help modules is expert systems, which are now a mature artificial intelligence technology. Tools used include logic programming, constraint programming and semantic networks.

Some examples of expert help in EDSS are given below:

- NEXSYS is an expert system for the assessment of nutrient generation rates in catchments (Young, W.J. et al; 1997). It was developed by CSIRO Land and water, Australia.
- FRAME (Caroli et al., 1994) developed at the politecnico di milano, is an expert system for model selection in the field of our pollution. It thus works in a concise problem domain, and this has been found to be critical for the success of such tool.
- FRAME uses an "expert debase" approach to match the user description of the problem with its model base. It can select on executable model or, if a pre-packaged model is not present, it can gather the necessary equations to build a suitable model for the situation understudy.
- Varma K.V. (2000) used the prolog language to build ecosystem models, in order to develop a modeling system able to reply to queries regarding the casual chain the led a particular result.

This work uses from the artificial intelligence field of qualitative modeling and model-based reasoning to achieve an expert help facility.

4.3.6. Some Remarks About DSS

Despite their popularity, the success of DSS development is uncertain and many computerized decision – support tools have failed when dealing with complex and unstructured problems (Mysiak et al., 2005).

Despite of high number of DSS developed, the risk of decision support systems failing to be up to the challenge of real-world problems is reported to be high (Bridge, 1997; Newman et al., 1999; OASIG 1996).

Moreover, judging whether a DSS has been successful or not is difficult, and sometimes cannot even be based on whether the DSS is used by the intended users. For example, in the case referred to by Newman et al. (1999) the DSS became obsolete because the end user (after having employed it for a short while) had become familiar with is logic and was able to apply it on his own.

5. AGRICULTURAL CASE

5.1. Methodological Aspect

The purpose of the study is to have a decision support tool. It will thus allow the decider to assert the management of the sector of agriculture through:

- The identification of priority points and the hierarchization of their action plans.
- The localization of investments and the apprehension of direct impact.
- The preparation of annual and multi-year plans.
- The evaluation of the impact of sectoral policies and the resolution of the development problems of the agricultural sector.

- The control of the evolution of the economic situations. The intelligent system is designed to be scalable (provided that its management and development is supported by the institution of agriculture). Thus, it will be a tool for controlling investment policy in strategic investment policies taking into account the situation of the agricultural sector and the orientations provided by specific planning instruments and studies. The evolutionary system granted to this decision help system will also allow: on one hand, to take charge of the priority areas of management expressed by the decision-maker.

On the other hand, to raise the level of effectiveness of the decision support system by improving the state of available information produced. This is done through the institution that manages the sector under study.

Thus, from a given situation, this decision-making system is developed by increasing the management capacity of the agriculture sector and by mobilizing more reliable and relevant data as time goes on measuring their production.

Thus, the methodology to follow in order to complete this study should focus on how information is mobilized in the form of a file, how is it processed, and finally, how is it organized.

Given the purpose of the study and the specificities of the sector in question, the results depend on surveys and consultation with the managers on two levels:

1. Inquiry and consultation phase with the agricultural services; this phase gives rise to the orientation of the study.
2. Survey phase of institutions producing information on the agriculture sector.

These surveys and consultations make it possible, on one hand, to identify the necessary data, and their origin (the structure that produces them, their periodicity, and their validity); on the other hand, based on the processing of these data, choose significant indicators of the situation of the agricultural sector.

It is therefore essential in this phase of the study to analyze the administrative organization chart of the institution that manages this sector as well as the activities and tasks of each department and office. This will make it possible to study the missions and the operation of each school of the institution that manages the sector and the supports received (or produced).

The methodology followed must lead to a diagnosis of the current state of information which will determine the degree of mastery and management skills. It is on this current state of information that the decision support system must act:

- Improve the data produced.
- Acting on its management capabilities by producing operational indicators. -To establish a system of relationship (a circuit of information) perennial and regular between the institution of agriculture and other institutions and structures that produce information about this institution. The diagram in Figure 3 explains the methodology study.

5.2. Proposed Model

It is divided into three parts: Database, Indicators, and results visualizations. It starts with:

Figure 3. Diagram: Methodological aspect

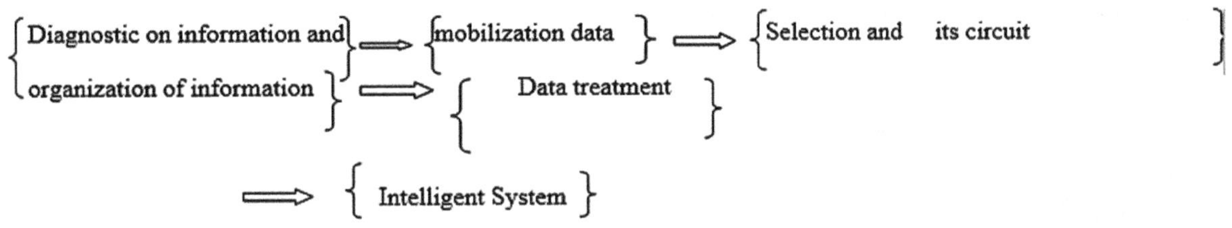

- Identification of the data necessary for the realization of the indicators (algorithms).
- The organization of these data as basic files. The set of files of the field of management of the treated agriculture constitutes a database stored on their computer support.
- Processing of these data to obtain a set of operational indicators.
- The visualization of the results and classification of the studied municipalities. Here, it is specified that the feeding of the decision aid system by data accounts for the evolution in time and space of the value of the selected indicator.

The diagram in Figure 4 shows in details the proposed model.

5.3. Identification of the Data Flow and Information Report

On the basis of the materials collected after institutions in the agricultural sector, the choice of the mediums containing the information corresponding to the installation of the decision-support system is made; and the supports that are missing are preset.

Then we proceed to the different treatments. Such processing should be in the form of files, which could contain the following information:

Figure 4. Diagram 2: Proposed model

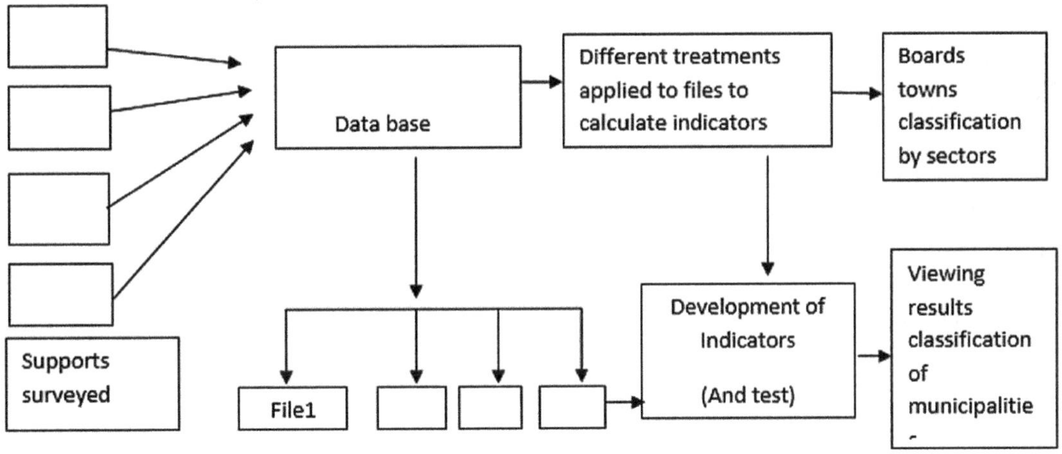

- General distribution of land;
- Distribution of land used by agriculture;
- Vegetable production;
- Animal production;
- Reorganization of livestock;
- Framing;
- Material Park.

Thus, the survey of institutions in the agricultural sector will have as objective the identification of the data produced, mobilized, or stored by the sector.

5.4. Proposed Treatment

On the basis of the concerns of the institution managing the agricultural sector, the various data processing operations are carried out. The aim of these trials is to propose certain indicators that will enable the institution to make sound planning and spatial planning decisions.

The steps of this treatment involve:

- A listing of the media harvested.
- A selection of the supports deemed useful for the proposal of the indicators.
- A proposal for files and materials deemed to be of interest to the institution in question and which the surveyed sector does not have.

5.5. Establishment of Indicators

The main objective of the study is to conceive and propose indicators that allow the management of the sector of the agriculture, necessary to the local management. It is about designing a data organization and data processing models to propose indicators. The surveys and consultation sessions that should be conducted with the agricultural services and the analysis of the information mobilized could lead to the development of the indicators. Thus the indicator can contain the following information:

- Code file.
- Wording.
- Entitled.
- Formula.
- Definition.
- Type of decision to be made.

Figure 5 shows a table that explains the indicators.

Figure 5. Table indicators

File code	Wording	Intitule	Formula	Definition	Complement of information to aide decision
	R.UAA	Rate of the useful agricultural area	UAA ------ X 100 TA	Ratio of utilized agricultural area and the total are.	Estimating the agricultural potencial
	R1	Rateof population occupied in agriculture on the total population	POA ------ % PT	Ration between the population and the total occupied population.	Giving the participation of agriculture in total employment.
	R2	Rate of population occupied in agriculture over the total occupied population.	POA ------ % TOP	Ratio of agricultural population and total occupied population.	Giving the agricultural sector in the support of the population.
	R3	Rate of population occupied in agriculture in the total labour force.	POA ------ % TLF	Ratio of agricultural population and the total labour force.	Giving the capacity of agricultural to the employment creation.
	Index of Anderson	POA by municipality ------ TOP by municipality IA=---------------------- POA province ------ TOP province	Measuring the degree of specialization of towns in the agriculture sector.		IA<1 Non-dominant agricultural activity. IA> 1 Town specialized in agriculture.

6. CONCLUSION

The overall objective of this study is to develop a decision-making tool with the aim of proposing a set of indicators structured, organized and elaborated on the basis of the information produced and stored at the level of the different structures of the zone concerned by the study.

The intelligent system results from a set of operations established on the basis of:

- Mobilization of relevant, comprehensive, reliable and renewable information.
- An information organization in the form of basic files, the set of which constitutes the database of the intelligent system, from:
  - A diagnosis on the information circuit of the priority management areas.
  - The collection of media from each department and office.

- ◦ The processing of data base files for the calculation of operational indicators for the assistance to the decision.
- ◦ The visualization of the indicators by using the computer tool and which gives an account of the evolution in time and space.

The evolution over time of the indicators and the spatial distribution of their value will make it possible to carry out several modes of classification to identify the situations of the municipalities, for example, and to measure the nature of the intensity of the intermunicipal disparities.

To be able to meet the objectives of the study: Identification and mobilization of the various criteria necessary for the design of the Intelligent system according to the data and needs expressed by the management of the municipalities, the choice of indicators must be closely linked to priority actions by local authorities; priorities which experience the situation and needs of the zone concerned by the study in the field of economic and social development.

REFERENCES

Bikhard, M. H. (1993). Representational Content in Humans and Machines. *Journal of Experimental & Theoretical Artificial Intelligence*, *5*(4), 285–333. doi:10.1080/09528139308953775

Bridge, M. C. (1997). The rise and fall of an executive information system: A case study. *Information Systems Journal*, *7*(4), 277–288. doi:10.1046/j.1365-2575.1997.00021.x

Brooks, R. A. (1991). Intelligence without reason. In *Proceedings of the 12th international joint conference on Artificial intelligence*. San Francisco, CA: Morgan Kaufmann Publishers.

Bruce, G. M. et al.. (2012). Recent advances in applying decision science to managing national forests. *Forest Ecology and Management*, *285*, 123–132. doi:10.1016/j.foreco.2012.08.024

Bunnel, F. L., & Boyland, M. (2003). Decision Support Systems: IT is a question not the model. *Journal for Nature Conservation*, *10*(4), 269–279. doi:10.1078/1617-1381-00027

Canadian Forest Service. (1995). *Criteria and Indicators for the Conservation and Sustainable Management of Temperate and Boreal Forests: Montreal Process*. Hull, Canada: Canadian Forest Service, Natural Resources Canada.

Caroli, . (1994). Knowledge based tool to support choice of the right air pollution model. In G. Guariso & B. Page (Eds.), *In Computer Support for Environmental Impact Assessment*. Amsterdam: North Holland.

Cox, P. G. (1996). Some Issues in the Design of Agricultural Decision Support System. *Agricultural Systems*, *52*(2-3), 355–381. doi:10.1016/0308-521X(96)00063-7

Deutsch, K. W. (1986). *The nerves of government*. New York, NY: The free press.

Diaz-Balteiro, L., & Romero, C. (2008). Marking forestry decisions with multiple criteria: A review and an assessment. *Forest Ecology and Management*, *255*(8-9), 3222–3241. doi:10.1016/j.foreco.2008.01.038

Eastman, J. R. (1997). *IDRISI for Windows*. Worcester, MA: User's Guide IDRISI Production.

Ferguson, I. S. (1996). *Sustainable Forest Management*. Melbourne: Oxford University Press.

Grosan, C., & Abraham, A. (2011). *Intelligent Systems: A modern Approach* (Vol. 17). Berlin: Springer. doi:10.1007/978-3-642-21004-4_1

Guariso, G., & Werthner, H. (1989). *Environmental Decision Support System*. New York: E. Horwood, Halsted Press.

Harnad, S. (1991). Other Bodies Other Minds: A Machine Incarnation of an Old Philosophical Problem. *Minds and Machines*, *1*(1), 43–54.

Helsinki Process. (1995). *European criteria and indicators for sustainable forest management adopted by the expert level follow-up meetings of Helsinki Conference in Geneva (24 June 1994) and in Antalya (23 June 1995)*. Author.

ITTO. (1992). *Criteria for the Management of Sustainable Tropical Forest Management*. Yokohama: International Tropical Organization.

Kanowski, P. J., & Williams, K. J. H. (2009). The reality of imagination: Integrating the material and cultural values of old forests. *Forest Ecology and Management*, *258*(4), 341–346. doi:10.1016/j.foreco.2009.01.011

Mahan, C. G., Young, J. A., Miller, B. J., & Saunders, M. C. (2015). Using Ecological Indicators and a Decision Support System for Integrated Ecological Assessment at Two National Park Units in the Mid-Atlantic Region, USA. *Environmental Management*, *55*(2), 508–522. doi:10.1007/s00267-014-0391-y PMID:25371192

Minsky, M. (1979). The Society Theory of Thinking. In P. Winston (Ed.), *Artificial Intelligence: An MIT Perspective* (Vol. 1). Cambridge, MA: MIT Press.

Mysiak, J., Giupponi, C., & Rosato, P. (2005). Towards the development of decision support system for water resource management. *Environmental Modelling & Software*, *20*(2), 203–214. doi:10.1016/j.envsoft.2003.12.019

Negnevitsky, M. (2005). *Artificial Intelligence. Intelligent Systems* (2nd ed.). Harlow, UK: Addison Wesley.

Newell, A. (1981). The Knowledge Level, Presidential Address, AAI80, Stanford University, 19 Aug 1980. *AI Magazine*, *2*(2).

Newell, A., & Simon, H. A. (1976). Computer Science as Empirical Enquiry: Symbols and Search. *Communications of the ACM*, *19*(3), 113–126. doi:10.1145/360018.360022

Newman, S., Lynch, T., & Plummer, A. A. (1999). Success and failure of decision support systems: Learning as we go. *Journal of Animal Science*, *77*(E-Suppl), 1–12. doi:10.2527/jas2000.77E-Suppl1e

OASIG. (1996). *The performance of information technology and the role of human and organizational factors*. Organizational Aspect of Information technology Special Interest Group.

Pfeifer, R., & Verscure, P. (1992). Beyond Rationalism: Symbols, Patterns, and Behaviour. *Connection Science*, *4*(3-4), 313–325. doi:10.1080/09540099208946621

Philips, M. S. (1994). *Measuring Trees and Forests* (2nd ed.). Wallingford, UK: CAB International.

Pylyshyn, Z. W. (1978). Computational Model and Empirical Constraints. *Behavioral and Brain Sciences, 1*(01), 93–127. doi:10.1017/S0140525X00059793

Pylyshyn, Z. W. (1980). Computation and Cognition: Issues in the Foundation of Cognitive Science. *Behavioral and Brain Sciences, 3*(01), 111–169. doi:10.1017/S0140525X00002053

Pylyshyn, Z. W. (1981). Complexity and the Study of Artificial and Human Intelligence. In J. Haugeland (Ed.), *Mind Design*. Cambridge, MA: MIT Press.

Pylyshyn, Z. W. (1984). *Computation and Cognition, Towards a Foundation for Cognitive Science*. Cambridge, MA: MIT Press.

Rizolli, A. E., & Young, W. J. (1997). Article. *Environmental Modelling & Software, 12*(2-3), 237–249. doi:10.1016/S1364-8152(97)00016-9

Toman, M. A., & Ashton, P. M. S. (1996). Sustainable forest ecosystems and management: A review article. *Forest Science, 42*(3), 366–377.

Varma, K. V., Ferguson, I., & Wild, I. (2000). Decision support system for the sustainable forest management. *Forest Ecology and Management, 128*(1-2), 49–55. doi:10.1016/S0378-1127(99)00271-6

Winograd, T., & Flores, F. (1986). *Understanding Computers and Cognition*. Norwood, NJ: Ablex Publishing Company.

Young, W. J. (1997). An Expert System for estimating diffuse source nutrient export rates. *AI Applications, 11*(2), 11–18.

Chapter 11
Enhanced Bio-Inspired Algorithms for Detecting and Filtering Spam

Hadj Ahmed Bouarara
University of Dr. Tahar Molay, Algeria

ABSTRACT

The internet era promotes electronic commerce and facilitates access to many services. In today's digital society, the explosion in communication has revolutionized the field of electronic communication. Unfortunately, this technology has become incontestably the original source of malicious activities, especially the plague called undesirables email (SPAM) that has grown tremendously in the last few years. This chapter unveils fresh bio-inspired techniques (artificial social cockroaches [ASC], artificial haemostasis system [AHS], and artificial heart lungs system [AHLS]) and their application for SPAM detection. For the experimentation, the authors used the benchmark SMS Spam corpus V.0.1 and the validation measures (recall, precision, f-measure, entropy, accuracy, and error). They optimize the sensitive parameters of each algorithm (text representation technique, distance measure, weightings, and threshold). The results are positive compared to the result of artificial social bees and machine-learning algorithms (decision tree C4.5 and K-means).

1. INTRODUCTION

In recent years, engineers and decision makers are confronted daily to NP-difficult problems where the classical techniques are unable to find effective solutions. They affect generally all sectors (the design of mechanical systems, image processing, information retrieval, clustering ... ... etc.).

The current scientific world was considerably built up with the inaugural appearance of novel concepts and paradigms. Our nature aged more than 5 billion years represents the largest reserve of solutions and ideas for many kinds of problems. The human being started to explore them recently (initiated from the most minor thing as bacteria to the complex systems collection of human body).

DOI: 10.4018/978-1-5225-4944-4.ch011

Nowadays, for each encountered problem we must observe the nature; it may already have the same problem where it had found solutions, long years ago. In the digital community, the world celebrates the birth of new interesting paradigms known under the name of bio-inspired techniques. They have demonstrated theirs strength face to different challenges. The foremost part of our work is the modelling of three fresh bio-inspired algorithms:

1. Artificial social cockroaches (ASC) inspired from the lifestyle of cockroaches and their movement as a decentralised system without the conductor. They use the interaction between them and with their environment in order to be grouped under the most secure and attractive shelter for hiding.
2. Artificial haemostasis system AHS inspired from the mechanism of stopping the loss of blood (external haemorrhagic).
3. Artificial heart-lungs system AHLS mimicked from the functioning of blood oxygenation. Recently, the e-mail service has become enormously used, and the principal vector of communication in our digital society despite the emergence of social webs and the web 2.0 tools. Moreover, it permits to users with a mailbox (BAL) and address mail to exchange messages (picture, files, and text documents) from anywhere in the world via internet. Regrettably, among all the messages received by an individual in his mail box, we recognize two cases:
    a. **Regular (HAM):** The email (Ham message) sent by friends or by websites subscribed in.
    b. **Irregular (SPAM):** The unsolicited emails (junk e-mail) sent in bulk by malicious people (spammers).

According to the most recent report of the Radicati Group (2013), who supplies quantitative and qualitative researches with details on the e-mail, the security, and the social networks, has exhibited that 70-80% of email traffic is composed of spam. It is a rigorous problem in the electronic life, which presents the main challenge for Mail server administrators, and responsible of information organizations.

For that matter, several spam detection systems have seen the light based on learning machine algorithms and probabilistic techniques including Bayesian classification, artificial neural networks and text compression. It represents a supervised classification task, which has witnessed a burning interest from companies and particles. Merely, the spammer techniques were dramatically evolved where the conventional systems are inefficacious face to several limits:

- **Choice of Parameters:** The majority of spam detection techniques are based on the parameters (distance measure and messages pre-processing techniques). A poor choice of these parameters may cause a degradation of detection performance.
- **False Positive Problem:** The classification of spam messages as ham.
- **False Negative Problem:** The classification of ham messages as spam.
- **The Messages Language:** The language of messages may be written in (English, French, Chinese ... ..etc.).
- **Ambiguity of Natural Language:** It is a major problem caused by the difference in the discourse of users. This is reflected by the difference of vocabulary used for expression of the contents of emails.
- **Representation of Spam Detection Results:** The majority of current spam detection systems are based on a primary visualization and interaction model. They present the spam message as an ordered list which makes the visualisation of all spam messages difficult.

- Execution time.
- **Multiplicity of Messages:** The content of emails can occur from different disciplines (Marketing, biological, multimedia, sports, computer science ...etc.).

The nuisance brought by the spam is not limited only on the influx of undesired mails or loss of legitimate mails; simply we can identify different cases of spam such as Scam, FUD, Hoax, the spam telephony (Spim) and Phishing.....Etc. The second component of our work is designed for the application of the algorithms (ASC, AHS and AHLS) for filtering and detection of SPAM.

In order to better understand the spam detection results given by our system, and to extract valuable information that assists us to combat this phenomenon. The third part of our work is designed for the construction of a visual result mining tool. In this paper, we are confronted to different questions

- Why do we use the haemostasis system, the heart lungs system and the lifestyle of social cockroaches for spam detection?
- Do the algorithms (ASC, AHS and AHLs) assure more security and protection against the SPAM problem compared to the classical techniques?
- What is the ideal configuration of each algorithm (ASC, AHS and AHLs)?
- Does the development of new visualization tool improve the quality of services to meet the needs of modern users?

2. REVIEW OF LITERATURE

This segment will be composed of two parts: Firstly, the work published around the bio-inspired techniques. Secondly, the different spam detection system existed in literature.

2.1. Bio-Inspired Techniques Existed in Literature

The computer revolution had changed human society: communication, transportation, industrial production, administration, accounting and technological advances …etc. Still, some problems qualify as “hard” cannot be processed with traditional hardware and software. Face different challenges, several bio-inspired techniques have been emerged inspired from natural phenomenon. They are classified into two large families as presents the next figure 1:

2.1.1. Evolutionary Algorithms (EA)

The EA represents a family of algorithms founded on the theory of evolution and the modern genetic. This family of algorithms is computationally expensive which do not guarantee the optimality of the results. We can recover:

- Genetic algorithms (GAs) appeared in the United States by john Holland in (Holland, 1973). It accepts as input a set of initial solutions that will evolve from generation to another using the operators (selection, mutation and crossover) based on a fitness function to find the best results.

Figure 1. Bio-inspire techniques existed in literature

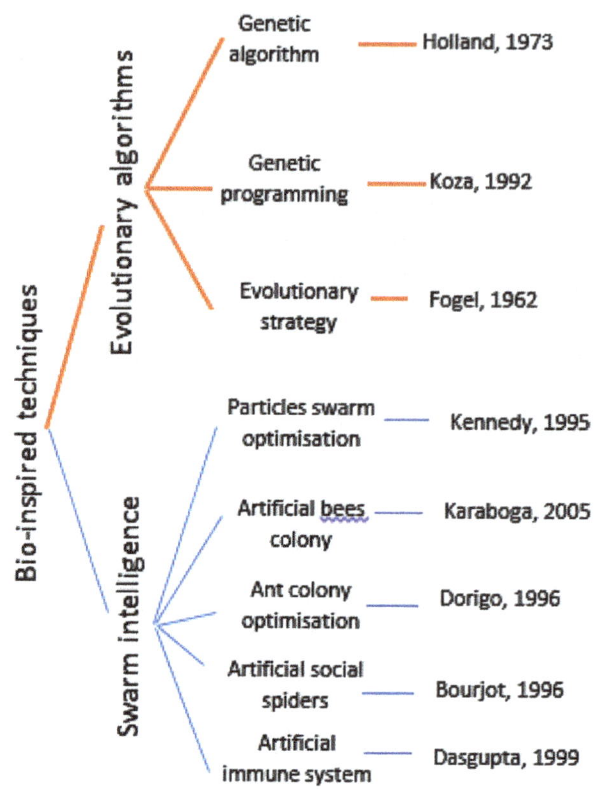

- Genetic programming (GP) proposed by John Koza in (Koza, 1992. Its challenge is to instruct the machine to execute tasks without having been specifically programmed for that.
- The evolutionary strategy (ES) was born in Germany in the 60s. The first work was presented in (Fogel, 1962) to solve numerical optimization problems.

2.1.2. Collective Intelligence Family (CI)

The CI family based on the work of a team where each member performs a specific work for the general interest of the group. Various techniques were elaborated following the functioning of this family:

1. **Particle Swarm Optimization (PSO):** The PSO was founded the first time in (Kennedy, 1995). It is based on a statistical population of points initially spread over a research arena that can be utilized to get an approximate answer to the problems extremely difficult.
2. **Ant Colony Optimisation (ACO):** The ACO is an algorithm proposed the first time by Dorigo & DI Caro in 1996 (Dorigo, 1996) to determine the shortest route between two locations. It was inspired from the behavior of ants.
3. **Bees' Colony Algorithm (BCA):** Like the same principle of ACO, the bees are social insects. They are driven to survive in highly organized colony, consisting of workers, males and a single

queen where each one has a specific task to perform. By observing this biological behaviour an approach called bees colony algorithm (BCA) was nominated in (Karaboga, 2007).

4. **Artificial Fishes Swarm Algorithm (AFSO):** The AFSO was copied from the collective life of fish invented by Li and al in (Li, 2002). It is based on a set of very simple rules constituted from the natural schooling behaviour of fishes and the relationship between them. The AFSO follows three main features: searching, swarming and following.
5. We can distinguish other methods as an artificial immune system (AIS) proposed by Dasgupta in (Dasgupta, 1999) based on the defensive behaviour of the human being against the alien invader.
6. **Artificial Social Spider (ASS):** The ASS is a method presented in (Bourjot, 2003) using the property of weaving and back to avoid isolation using a set of rules for the movement and a weight associated for each direction.

Cockroach Recent Works

The social life of cockroaches has been examined by researchers in (Halloy, 2007) for the purpose of coordination between multi-robot systems with the integration of robots into a group of insects, in order to control their final choice. The idea introduced in (Lihoreau, 2009) about the ability of insects to recognize their parents that represents a great importance in the evolution of the social life of cockroaches. The authors in (Moore, 1988) have attempted to study the influence of social experience on the behaviour of male cockroaches. A very significant work has been presented in (Jeanson, 2005) led by a group of researchers trying to study the collective behaviour of cockroaches and their evolution mode.

Concerning the use of social cockroach in the computer world in (Havens, 2008) a new adjustment of the particle swarm optimization (PSO) called roach infestation optimization (RIO), in this report the authors demonstrate the relationship between these two approaches and their ability to find the global optimum by a combination of different functions. Not forgetting the algorithm suggested in (Cheng, 2011) to work out the TSP problem using the daily life of cockroaches, this proposal is totally dissimilar from our suggested model.

2.2. Spam Detection Systems Existed in Literature

It is only in the late 90's that the problems of filtering and detection of spam drew the attention of the researchers. This section detailed the pertinent spam detection techniques existed in literature as presents the next figure

2.2.1. Classical Techniques

There are two large families of classical techniques to deal with spammers as detailed in the next

2.2.1.1. Digital Signature Family

This kind of techniques usually based on a set of digital signature. They need to have a detailed analysis of mails that circulate in the network. Among the methods according to this principle, we can distinguish (DNS registration in (Chirita, 2005), blacklists in (Grier, 2010) and Grey list in (González-Talaván, 2006).

2.2.1.2. Transmitter Analysis

These techniques are based on the user profile and the analysis of the spammers' behaviour such as the methods URLs analysis in (Benczur, 2005) and Teergrubing in (Jindal, 2007).

2.2.2. Machine Learning Family for Spam Detection

Among the techniques found in the literature following the principle of this family, we can find:

2.2.2.1. Naive Bayes NB

The NB is a probabilistic model based on Bayes theorem and belonging to the linear algorithms group that has been used by Sahami in (Sahami, 1998) to improve the anti-spam solution using the differential miss-classification concept.

2.2.2.2. K-Nearest Neighbours (KNN)

The KNN is a very simple algorithm that allows an instance-based classification based on a measure of similarity and parameter K. Firte et al in (Firte, 2010) used this principle for the construction of an online application that can detect undesirable mail.

2.2.2.3. Support Vectors Machine (SVM)

SVM is a classification technique that aims to find a function hyper plane based on the supports vector calculated by Lagrange in method (Suykens, 1999) to maximize the margin between classes. Ducker and his collaborators in (Ducker 1999) used this principle for the categorization of electronic messages (spam and ham). The quality of results of their experimentation is better, which were compared with three methods rip, and boosting Roccio (+ svm decesion tree).

2.2.2.4. K-Means

K Means is an unsupervised algorithm based on the initialization of the parameter K. Sasaki and al in (Sasaki, 2005) thought to apply this principle with K = 2 (spam and ham) for spam detection problem.

2.2.2.5. Decision Tree c4.5

After a new form of spamming based on image has appeared. In (Krasser, 2007), the researchers had applied the C4.5 to generate decision rules that will make a choice between the spam image and ham image.

2.2.3. Bio-Inspired Family for Spam Detection:

This section is dedicated to the different work done around the spam detection using the bio-inspired techniques.

2.2.3.1. Genetic Algorithms GAs:

A very interesting paper was presented in (Muhammad, 2011) as a recapitulation of the application of three bio-inspired algorithms (artificial immune system AIS, genetic algorithms and GAs Artificial neural network ANN) for the detection of spam.

2.2.3.2. Particle Swarm Optimization (PSO) for Spam Detection

Lai et al (Lai, 2007) had the idea of examining how particle swarm optimization algorithm can help select relevant features for discriminating the classification of solicited and unsolicited messages.

2.2.3.3. Ant Colony Optimization ACO for Spam Detection

Taweessiwate et al in (Taweesiriwate, 2012) had the idea to reformulate the principle of ACO for the problem of host spam detection using the webspam-UK2006 dataset for his experimentation.

2.2.3.4. Social Bees

Hamou et al in (Hamou, 2013) have formulated a new method inspired from the social life of bees, and their way to ensure the proper functioning of the hive using the filtering principle. Each filter is by ensuring bee workers (supervisor, cleaner and collector). A message is identified as spam if it succeeds spent three filtering

2.2.3.5. Artificial Immune System

To better counter the phenomenon of spamming the researchers considered spam as pathogens that are combated by humans where a spam detection method was introduced by Oda et al in (Oda, 2005) called artificial immune system AIS to effectively protect the email users. This algorithm has been tested on the Spam Assassin corpus and gave satisfactory results.

2.3. Text Pre-Processing

This is a necessary step in the process of each algorithm that deals with unstructured data. It unifies the representation of emails (messages written in natural language) to renders them interpretable by the machine by following the next steps:

1. Elimination of special characters and numbers.
2. Division of messages to a list of terms using the techniques bag of word and n-gram character (Bouarara, 2014).
3. Calculate the importance of each term in the message using the weightings TF-IDF and TFC (hamou, 2013).

At the end each message will be transformed into a vector and the dataset to a matrix (messages * terms). The lines represent the messages vector and the columns represent the attributes of each vector.

Figure 2. Different approach for filtering of spam

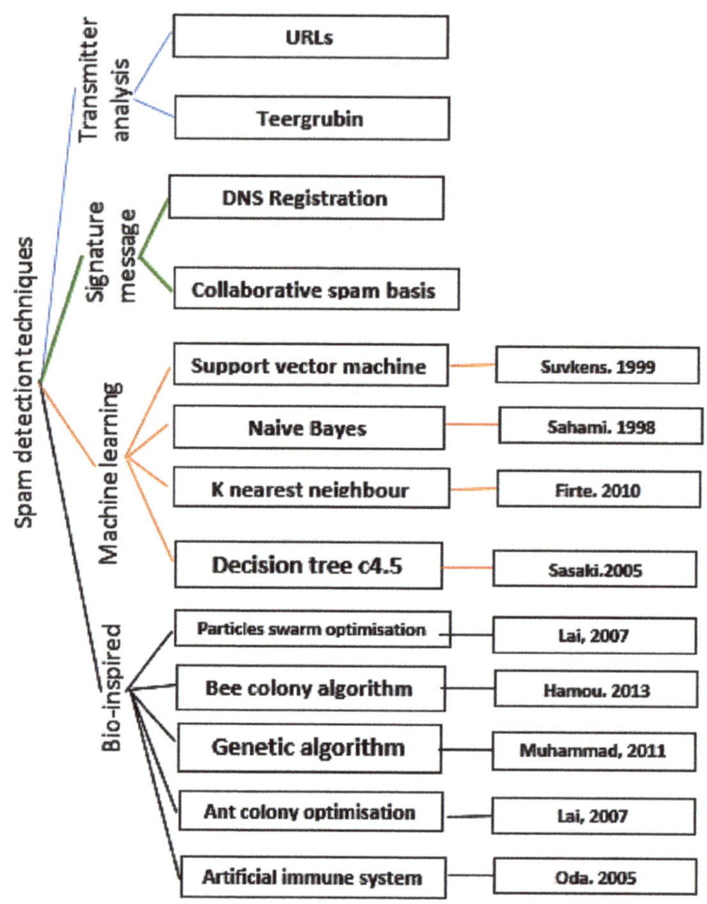

3. RESEARCH APPROACHES

We have chosen of the modelling of haemostasis system, the lifestyle of social cockroaches and the heart lungs system in order to filtering SPAM after the overlap which is explained in details in the accompanying table because they are based on the notions of filtering and classification.

3.1. Artificial Social Cockroaches (ASC)

The principle of artificial social cockroaches (ASC) is rather simple. Merely, we must explain the source of inspiration of this technique.

3.1.1. Experience

We put cockroaches in a basin where there is the light everywhere. We built two artificial shelters (Figure 3a) using two red circles because cockroaches do not discover the red colour (Figure 3b). This experience demonstrates the movement of cockroaches:

Table 1. Summary of the different algorithms (ASC, AHS and AHLS)

		Biological System	Artificial System
ASC	input	Cockroaches	Messages
	principal	- Cockroaches of the same family are hidden under the same shelter. - Cockroaches of different family are hidden in different shelters.	- Messages similar are classified in the same class - Dissimilar messages are classified in different classes
	Output	- Two shelters - the cockroaches hidden in each shelter	- Two classes (spam and ham). - Messages classified in each class.
AHLS	input	- Blood - Air	- Messages - External information (Learning basis)
	principal	- The heart and lungs work together to ensure the supply of body cells with oxygen and remove carbon dioxide CO2. - During the circulation the heart pumps the blood oxygen poor to the lungs. - Blood Poor with oxygen going to be deoxygenated by breathing out the carbon dioxide CO2 out of the body through the mouth. - The O2 will comes back via the pulmonary circulation to the left heart	- The artificial heart and the artificial lungs will work together to detect and eliminate spam messages. - The messages probably spam will be selected by the naive Bayes algorithm to move to the step of filtering lungs. - The messages probably spam will pass through two filters (left and right) in order to have a final detection of messages and eliminate them. - The ham messages will be returned to the inbox.
	Output	- Co2 - O2	- Spam message - Ham message
AHS	input	Platelets circulating in the plasma	Email received by user
	principal	Among all platelets qui circulating in the plasma only the platelets nearest to the site of the injury will be selected for stopping haemorrhagic.	Among all messages received by user only the messages similar to the spam message learning basis will be selected to be eliminated.
	Output	- Selected platelets - No selected platelets	- Spam - Ham

Figure 3. Description of the experience

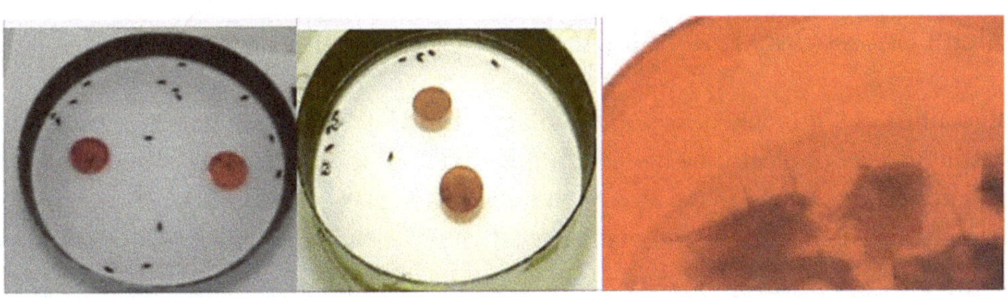

a- Choice of secure shelter b- grouping of cockroaches

- **Exploration Phase:** The cockroaches walk randomly in all direction in order to pick out the most attractive shelter.
- If it finds a more attractive shelter, it joins to it, and sends pheromones as odour to its congeners.
- Each cockroach joins the most attractive shelter that contains more darkness (less brightness) and pheromone (the smell of its congeners).
- Iteratively, if a cockroach does not feel good, it keeps changing its current shelter.

From the experience, we remark that:

- The cockroach chooses always the most secure and attractive shelter.
- The cockroaches from the same colony are grouped under the same shelter.

3.1.2. Biological Model

The cockroach is a gregarious insect, which thinks better with a group than solo. It is characterised by the societal phenomenon of searching the most attractive and secure place (shelter) for concealing. The cockroaches' colony is flexible, robust, decentralized and self-organized. In our work, we are interested by the cockroach that lives in apartments and it is never isolated. The cockroach has three essential features:

1. **Environment:** Cockroaches prefer warm and moist microclimate. Their nests and eggs are installed in dry locations.
2. **Aggregation Rules:** The movement of cockroaches is based on the interaction between them and with their environment. It is guided by three aggregation rules:
    - Shelter darkness
        - Cockroaches are attracted by the darker place such as corners and shelter.
    - Congeners Attraction:
        - Each cockroach follows the shelter where it finds more of its congeners (cockroaches from the same family).
    - Security Quality SQ:
        - The cockroaches positioned in the centre of the shelter have an elevate security quality. Differently, the cockroaches positioned at the edge of the shelter have a bass security quality.
3. **Communication:** The cockroaches perceive communicate between them using their antennas by sending an aggregation pheromone (odour).

3.1.3. Artificial Social Cockroaches (ASC) for Spam Detection

The ASC algorithm was mimicked from the biological model, and the experience previously considered. It has as input a set of electronic messages (cockroaches) that will be classified (hidden) in a class (shelter) spam or ham according to the attraction rate of each class. The process of detecting spam message is realised using three aggregation operators (shelter darkness, congener's attraction and security quality) as presents the Figure 4.

3.1.3.1. Artificial Cockroaches

For the problem of spam detection the artificial cockroaches represent emails.

3.1.3.2. Artificial Cockroaches Representation (ACR)

To make emails manipulated by the machine, they must be presented in digital format. The digitalisation of emails is realised following the text pre-processing process previously detailed.

Figure 4. Spam detection system based on artificial social cockroaches ASC

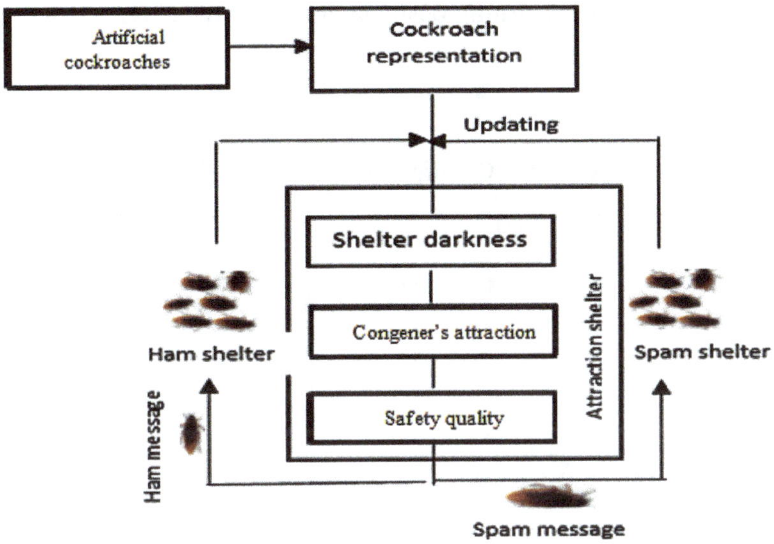

3.1.3.3. Artificial Communication

Two emails are from the same colony (similar) if the distance between them is less than a threshold fixed in advance.

3.1.3.4. Detection (Shelter Attraction)

The electronic message will be classified in the most attractive class (shelter). The attraction rate of each class for each email M_N is based on three aggregation operators (shelter Darkness (SD), congeners' attraction (CA) and safety quality (SQ)) and the objective function SA.

$$SA\ (M_N, S_i) = \alpha * SD\ (M_N,\ S_i) + \beta * CA\ (M_N,\ S_i) + ϰ * SQ(M_N,\ S_i)$$

Where:

M_N : New email number I (to be classified) received by user.
S_i : Class number i.
α, β, ϰ: Parameters to adjust the importance of each operator.

1. Shelter Darkness (SD) Operator

This procedure is based on the messages learning basis belonging to the class LB (Si) divided by the total number of messages learning basis.

$$SD\ (S_i) = \frac{LB(MS_i)}{\#MLB}$$

$\#\text{MLB}$: The total number of messages learning basis.
$\text{LB}\left(MS_i\right)$: The number of messages learning basis belonging to the class S_i (shelter).
I : can be spam or ham.

Example:
In our work, we use the dataset spam v.0.1 previously detailed. The darkness of each class is:

$$\#\text{MLB} = 300$$

$$\text{LB}\left(MS_{spam}\right) = 100$$

$$\text{LB}\left(MS_{ham}\right) = 200$$

$$\text{SD}\ (S_{ham}) = \frac{200}{300} = 0.66$$

$$\text{SD}\ (S_{spam}) = \frac{100}{300} = 0.33$$

2. Congeners' Attraction (CA)

A message is attracted to the class where there are more of similar messages.

$$\text{CA}\ (M_n,\ S_i) = \text{PC}\ (M_n,\ S_i)\ \text{CFN}\ (M_n, S_i) \left(\frac{\sum_{T=1}^{L} distance(M_n\ ,\ \ M_T S_i)}{\text{L}} \right)$$

M_n : Message to be classified.
L: the total number of messages in the class i.
$M_T S_i$: Message number T belongs to the class Si.
Distance (M_n , $M_T S_i$): the distance between the message to be classified M_n and the messages $M_T\ of\ the\ class\ S_i$.

- **Congener's Probability (CP):** CP depends on the number of messages belongings to each class. It is used in order to not promote a class over another.

$$\text{PC}\ (C_n,\ Si) = \frac{1}{\#\, D_T S_i}$$

$\# D_T S_i$: Number of messages of the class i.

- **Congener's Family Number (CFN):** The number of messages of the class S_i, which are most similar (from the same family) with the message to be classified M_N based on:
    - A distance (M_N, $M_T S_i$) between each new cockroach M_N and every cockroach M_T of the leaning basis of the class S_i.
    - A threshold fixed in advance.

If the distance between M_N and M_T <threshold then M_N and M_T are similar (from the same family). Else M_n and M_T are not from the same family.

3. Security Quality (QS) Operator

A message has a maximal safety quality if it is near to the centroid of the shelter. The security quality SQ(M_n , S_i) represents the distance between the message to be classified M_n and the centroid of each class.

$$\text{SQ}\,(M_n, S_i) = \frac{1}{\text{distance}\left(M_n, CS_i\right)}$$

CS_i : Centroid of the shelter S_i

Updating step: Iteratively the aggregation operators of each message will be updated and the messages will be reclassified.

4. Stopping Criterion

The stopping criterion of ASC algorithm is either the number of iterations fixed in advance or if the messages in each class remains the same for iteration i and iteration i-1.

Procedure: The following pseudocode summarizes the algorithm artificial social cockroaches for spam detection (Algorithm 1).

3.2. Artificial Haemostasis System (AHS)

The AHS algorithm was mimicked from the phenomenon called haemostasis system, which protects human corps against the loss of blood provoked by the external haemorrhagic.

As mentioned in Figure 5, the external haemorrhagic is a blood flow outside of the circulatory system caused by an injury.

Algorithm 1. Artificial social cockroaches for spam detection

Input
Threshold, Distance measure
Weightings adjustment κ, β, α.
SMS spam V.0.1 dataset
While not stopping criterion do
For each no-secure cockroach C_n (test basis messages) do
For each shelter (S_{spam} and S_{ham}) do
Calculate
/*Shelter darkness*/

$$SD(S_i) = \frac{LB(CS_i)}{\#CLB}$$

/*Congeners attraction*/

$$CA(M_N, S_i) = PC(M_N, S_i)\ CFN(M_N, S_i)\left(\frac{\sum_{T=1}^{L} distance(M_n, M_T S_i)}{L}\right)$$

/*Quality safety*/

$$SQ(M_N, S_i) = \frac{1}{distance(M_N, CS_i)}$$

/* the shelter attraction AS */
$SA(M_n, S_i) = \alpha * SD(M_n, S_i) + \beta * CA(C_i, S_i) + \kappa * SQ(M_n, S_i)$
Fin
If $SA(M_n, S_{spam}) > SA(M_n, S_{ham})$ then
$M_n \leftarrow$ spam Else $M_n \leftarrow$ ham
Fin
Updating the aggregation operators
Fin
Spam class and ham class
Messages of each class.

Figure 5. External haemorrhage

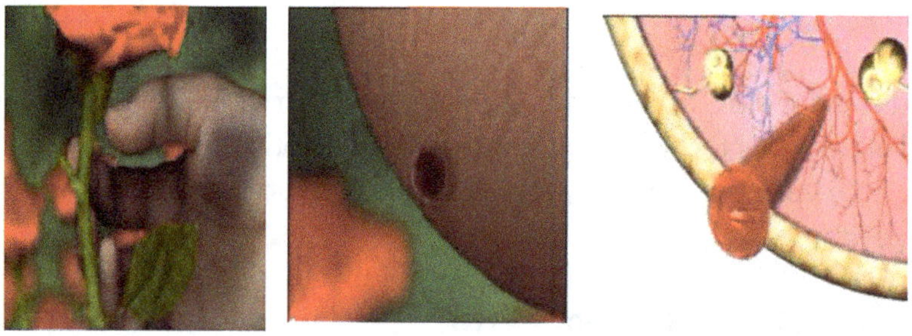

a) Injury b) rupture of blood vessel

3.2.1 Biological Haemostasis System BHS

The BHS has been evolved over millions of years. It encompasses all the necessary ingredients that assert to the bleeding stops with multilevel principal both in parallel and sequential fashion. Among all platelets existed in the plasma only a set of them are selected by the haemostasis process to stop haemorrhagic.

Figure 6 exhibited that the biological haemostasis process is consisted of a pre-haemostasis and three complimentary steps:

3.2.1.1. Vascular Phase VP (Pre-haemostasis)

The VP lowers the calibre of the blood flow.

3.2.1.2. Primary Haemostasis (PH)

The PH is activated to stop the haemorrhagic using two elements (platelets and fibrinogen). It selects the nearest platelets to the location of haemorrhagic to use in forming a white thrombus (platelet clot).

3.2.1.3. Secondary Haemostasis (SH)

The platelet clot must be consolidated because it is flimsy. The SH covers the damaged walled of the blood vessel in two steps:

1. **Fibrin-formation:** To construct the fibrin filter
2. **Combination:** Combine platelets clot and fibrin filter to construct haemostasis plug.

3.2.1.4. Fibrinolysis

The fibrinolysis scrubs the wastes by destroying the polymers of fibrin. It allows the reversal of the blood vessel.

Figure 6. Biological haemostasis system

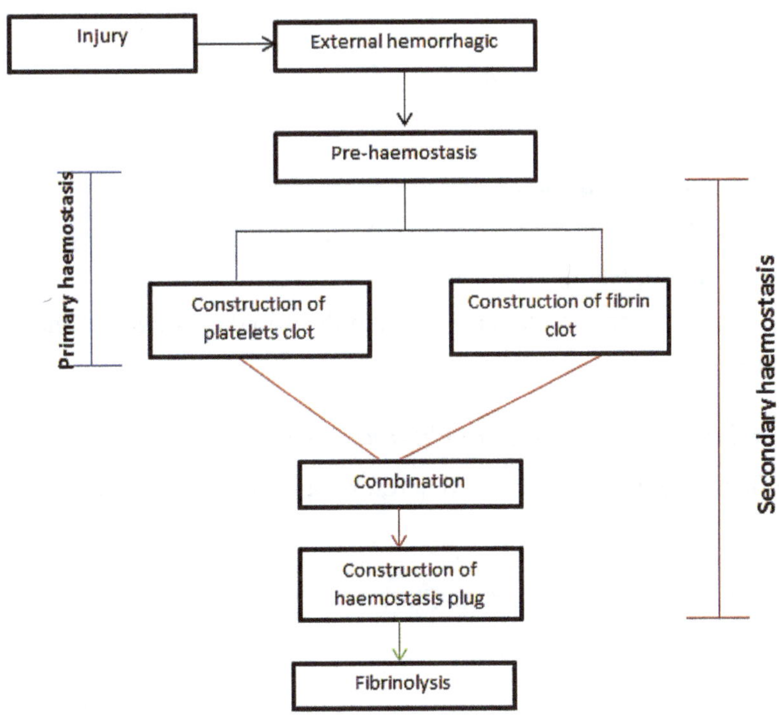

Figure 7. Secondary haemostasis

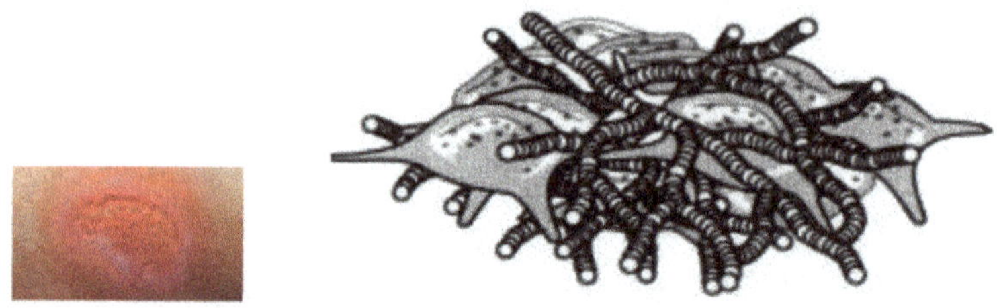

a) Haemostasis plug b) Fibrin

3.2.2. Artificial Haemostasis System for Spam Detection

Among all messages received by user email (platelets circulating in plasma) only messages that are similar to the spam messages of learning basis will be classified as spam and other messages are left in the inbox.

Figure 8 exhibited the AHS process based on the pre-processing of data (pre-haemostasis) and three main steps (primary detection, secondary detection and final detection).

3.2.2.1. Pre-Processing (Pre-Haemostasis)

This step is used to ensure the vectoring of the messages as detailed previously.

3.2.2.2. Primary Detection (Primary Haemostasis)

It constructs the first filter (platelets clot) by calculating and normalising the average distance between the new message M_i and the messages of the learning basis M_N.

$$\textbf{First filter } (\boldsymbol{M_i}) = \frac{\sum_{N=1}^{N} distance1\ (M_i, M_N)}{N}$$

N: The number of learning basis messages.
$\boldsymbol{M_i}$**:** Message to be classified.
M_N : Messages of the learning basis.
Distance 1: The distance between the message M_i and the Nime messages of the learning basis using the distance measure 1.

3.2.2.3. Secondary Detection (Secondary Haemostasis)

The solution of the first filter is not effective. Wherefore, it is consolidated by the secondary detection. This step is achieved in two steps:

1. Construction of the Second Filter (Fibrin-Clot)

Figure 8. Spam detection based on Artificial haemostasis system classifier

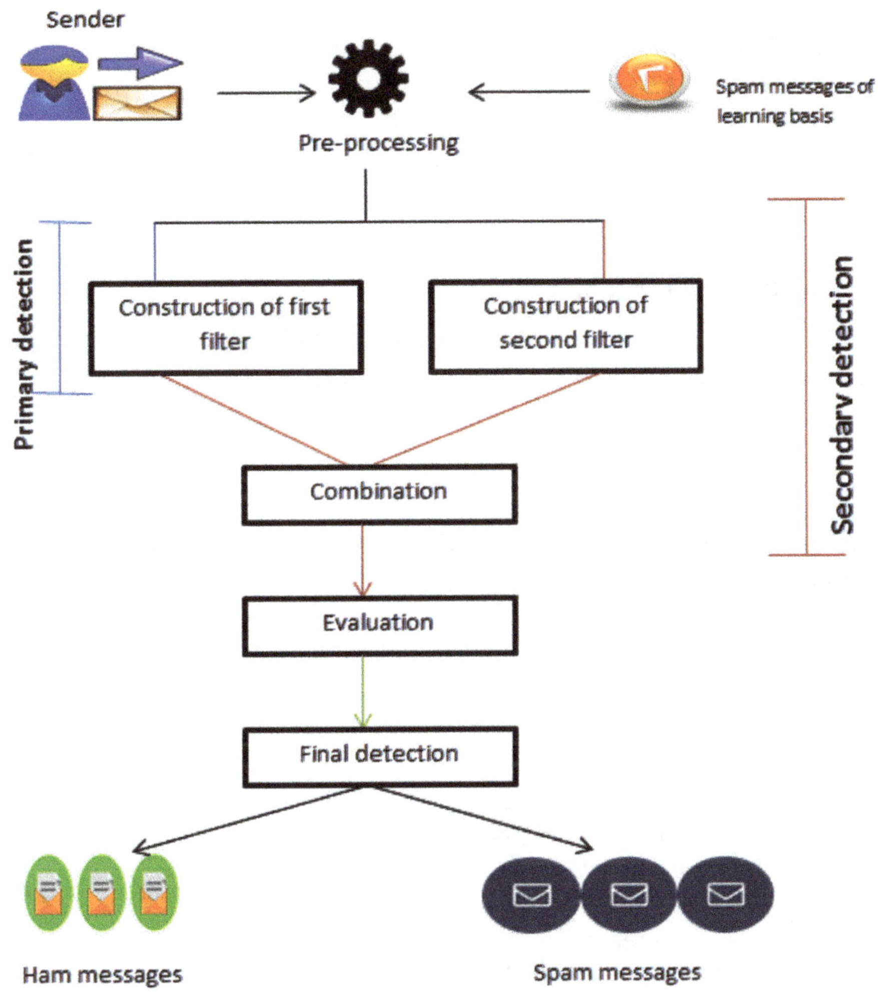

The second filter (fibrin clot) represents the average distance between the message to be classified M_i and the spam messages of the learning basis M_N using distance measure 2.

$$\textbf{Second filter}\ (M_i) = \frac{\sum_{N=1}^{N} distance2\ (M_i, M_N)}{N}$$

N: Number of spam messages learning basis.

Distance 2 (M_i, M_N): The distance between the message M_i and the Nime messages of the learning basis using the distance measure 2.

NB: The distance measure used for the construction of the first filter must be different with the distance measure used for the construction of the second filter.

2. Evaluation (Combination)

This procedure combines the results provided by the two previous filters to calculate the objective function for each message HP (M_i).

$$\mathbf{HP}\ (M_i) = \frac{(\text{first filter}\left(M_i\right) + \left(\text{second filter}\left(M_i\right)\right)}{2}$$

3.2.2.4. Final Detection (Fibrinolysis)

This procedure uses the objective function of each message to classify it. It is based on a threshold. If HP (M_i) <= threshold then the message M_i is classified as spam else the message M_i is classified as ham.

3.2.2.5. Procedure

The next algorithm provides a pseudo code listing of the artificial haemostasis system (AHS) for SPAM detection (Algorithm 2).

Algorithm 2. Artificial haemostasis system for spam detection

```
Input
Learning basis: messages pre-classified as spam
M_N: Message number N of the learning basis
M_i: Message number I to be classified
Choice of parameters: Threshold, two distance
measures.
Begin
/* pre-haemostasis */
Transformed of each messages to vectors.
For each message M_i do calculate
```

$$\textbf{First filter } (M_i) = \frac{\sum_{N=1}^{N} distance\ 1\ (M_i, M_N)}{N}$$

$$\textbf{Second filter } (M_i) = \frac{\sum_{N=1}^{N} distance\ 2\ (M_i, M_N)}{N}$$

```
/* evaluation */
```

$$\textbf{HP } (M_i) = \frac{(\text{first filter } (M_i)) + (\text{second filter } (M_i))}{2}$$

```
  If HP (M_i) <= threshold then
    The Message M_i <———— spam
     Else the message M_i <———— ham
End for
```

Return: list of spam messages.

3.3. Artificial Heart-Lungs System (AHLS)

The AHLS algorithm imitates the phenomenon of blood oxygenation ensuring by the work of heart and lungs system.

Before detailing the heart lungs system we must give you definition concerning the principal actors of this system:

1. **Lungs Definition:** The lungs consists the organ of respiration, which ensure the pulmonary circulation. The left lung is composed of two globes and right lung of three globes.
2. **Heart Definition:** The Heart is positioned between the lungs. It consists of two parts (right heart and left heart) that function as a tank to pump the blood.

The process of blood oxygenation is summarized in the next steps:

3.3.1. The Air Inhalation

The trachea brings air rich in oxygen coming from the upper aerial pathways to the alveoli bag (which is going to be filled with air like a balloon).

3.3.2. The General Circulation

The general circulation allows the filtering of blood by eliminating the carbon dioxide CO2 sent by the cells and replace it by the oxygen O2 brings via the air inhalation. This circulation is carried out by the intersection of two circulations:

3.3.2.1. The Systematic Circulation (High Traffic)

It corresponds to the circulation of oxygenated blood from the heart to all the organs of the body. It also includes the return of deoxygenated blood from the organs to the heart.

Figure 9 shows the steps of systematic circulation. The oxygenated blood is propelled by the left ventricle into the systemic circulation (next figure). This blood is firstly transported to the aorta along the path (2) and (3) until the gas exchange sites. Once, the exchanges between the blood and the cells were finished. The deoxygenated blood (blue) leaves the exchange sites following the path of the vena cava (the path 5 and 6) until reaching the right atrium of the heart

3.3.2.2. The Pulmonary Circulation (Small Traffic)

It corresponds to the circulation of blood poor of oxygen from the heart to the lungs to be oxygenated again. It also includes the return of oxygenated blood from the lungs to the heart.

Figure 10 demonstrates the process of pulmonary circulation. Once in the right atrium, the deoxygenated blood will be propelled into the lungs through the right ventricle (1), following the path (2) and (3) to reach the wall alveolar- capillary (filter) which is between (4) and (5). In this filter the blood gets rid of its carbon dioxide and it is re-oxygenated. The oxygenated blood reaches left atrium of the heart through the pulmonary veins (6) to be powered by the left ventricle into the systemic circulation, and so on for each inhalation. This step ensures the gas exchange between the alveolar air and the blood.

Figure 9. Systematic circulatory

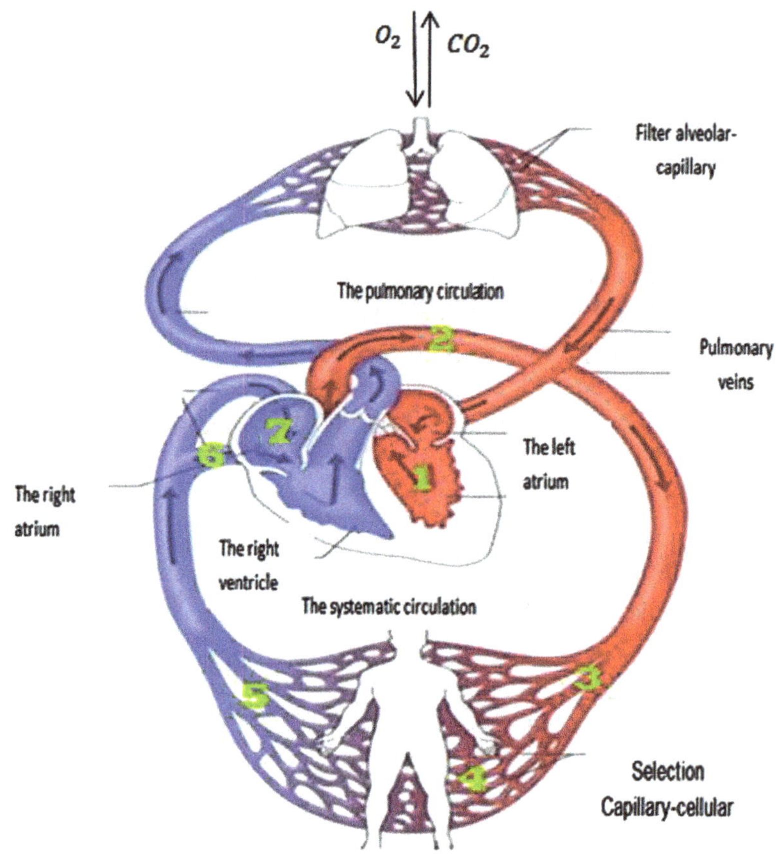

3.3.3. Expiration of Carbon Dioxide (CO2)

Carbon dioxide CO2 contained in the blood becomes gaseous and passes into the sac alveolar. The expiration propelled it outside of the body.

3.3.4. Artificial Heart Lungs System for Spam Detection

The AHLs algorithm was inspired from the phenomenon previously detailed. It is composed of four steps (artificial inhalation, selection heart (systematic circulation), Filtering lungs (pulmonary circulation), and artificial expiration) with a main role is the filtering of electronic messages (blood) received by the mail user to classify them as spam (Co2) or ham (O2).

3.3.4.1. Artificial Inhalation

The air (external information) provides all ham messages of learning basis (oxygen O2).

1. **Lungs Learning Basis (Pulmonary Alveolar Sac):** This basis will be filled by all ham messages (o2).

Figure 10. Pulmonary circulatory

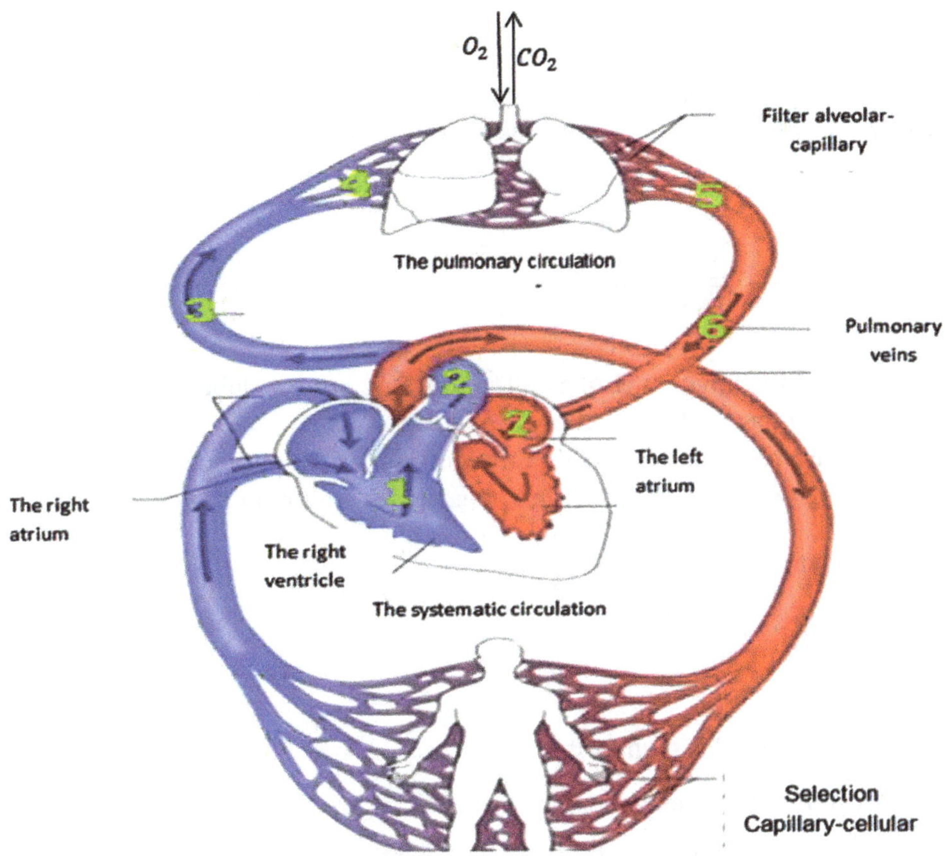

2. **Heart Learning Basis:** This learning basis is composed of a set of ham learning basis message and all spam learning basis messages.

3.3.4.2. The Largest Circulation (Classification)

This step based on the information provided by the inhalation phase that ensures the classification of email received by user (coming from the human body cells) following the next procedures:

1. **Heart Selection (HS):** The HS uses the heart learning basis to calculate the probability of each email received to be spam (Co2) and the probability to be ham (O2) using naive Bayes algorithm. All the messages probably spam (Co2) than ham (o2) will move to the next step (filtering lungs step). We have chosen naïve Bayes because in the natural system only the blood poor in oxygen must be transported to the pulmonary circulation.
2. **Filtering Lungs:** Each email probably spam will pass through two filters (left filter and right filter) using the lungs learning basis.
    a. **Left Filter:** This filter (left lung) calculates and normalize the average distance between each email probably spam and messages of lungs learning basis. Using a distance measure X.

$$\text{Left Filter LF } (MP_{spam_i}) = \frac{\sum_{N=1}^{N} distance1\left(MP_{spam_i}, M_N\right)}{\text{N}}$$

b. **Right Filter:** This filter (right lung) calculates and normalize the average distance between each email probably spam and messages of lungs learning basis. Using a distance measure Y.

$$\text{Right Filter RF } (MP_{spam_i}) = \frac{\sum_{N=1}^{N} distance2\left(MP_{spam_i}, M_N\right)}{\text{N}}$$

c. **Adjustments Parameters (Lobes):** The left lung has two lobes so it has a weighting = 2. The right lung has 3 lobes so it has a weighting = 3.
d. **Objective Function (Haematosis Function (HF)):** This function is calculated using the results of the two previous filters (left lung and right lung) that allows the final evaluation of each email.

$$\text{HF } (MP_{spam_i}) = \frac{\left(2^*\text{LF}\left(MP_{spam_i}\right) + 3^*\text{RF}\left(MP_{spam_i}\right)\right)}{5}$$

3.3.4.3. Artificial Expiration (Evaluation)

The final decision is based on the haemostasis function of each message MP_{spam_i}. Where If HF (MP_{spam_i}) < threshold then MP_{spam_i} is classified as ham and will be returned by the pulmonary circulation to the left heart Else MP_{spam_i} is classified as spam (co2 powered out of the body).

4. EXPERIMENTATION

Before starting our experiment we must define the data set and the evaluation metrics used in the realization of different tests.

4.1. Dataset (Spam V.0.1)

It is a public collection of messages pre-tagged by an expert. It is built for the research in the field of spam detection (Gómez Hidalgo, 2006). For our work we have decomposed this dataset in two subsets: learning basis and test basis as depicted in Table 2.

4.2. Validation Metrics

Evaluation measures are numerous, and most of them are based on recall and precision (between 0 and 1). To evaluate our algorithm we will use the following metrics:

Algorithm 3. Lungs heart classifier

Input
Spam learning basis, ham learning basis, Test basis
/* artificial inspiration */
Lungs learning basis $\longleftarrow$ Ham learning basis
Heart learning basis $\longleftarrow$ (spam learning basis (left part), ham learning basis (right part))
/* systematic circulation (heart selection)*/
For each new message M_i do
Calculate the probability of message M_i to be spam or ham (using naïve Bayes algorithm)
$P(ham/M_i) = P(W_1/ham)*P(W_2/ham)*..........P(W_N/ham)*P(ham)$
$P(spam/M_i) = P(W_1/spam)*P(W_2/spam)*..........P(W_N/spam)*P(spam)$
If $P(ham/M_i) < P(spam/M_i)$ then $M_i \longleftarrow MP_{spam\ i}$
Else $M_i \longleftarrow MP_{ham}$
/* pulmonary circulation (lungs filtering)*/
For each $MP_{spam\ i}$ do calculate
The left filter (LF) and right filter (RF)
Objective function
$HF(MP_{spam\ i}) = \frac{(2*LF(MP_{spam\ i}, M_N) + 3*RF(MP_{spam\ i}, M_N))}{5}$
/ * artificial expiration (evaluation of the Messages selected by the heart) */
If HF $(MP_{spam\ i})$ >threshold then $MP_{spam\ i} \longleftarrow$ spam
Else $MP_{spam\ i} \longleftarrow$ ham.
End for
End for
Return:
Spam messages
Ham messages

Table 2. Data set used in our work

	Decomposition	Message Number
Learning basis	Spam learning basis	100
	Ham learning basis	200
Test basis	Spam test basis	222
	Spam test basis	802

- Contingency matrix for each class that provides four essentials information (True Positive (TP), False Positive (FP) False Negative (FN) and True Negative (TN))
- **Recall:** Calculate the ability of the system to detect the message correctly spam.

$$R = \frac{TP_i}{TP_i + FN_i}$$

- **Precision:** Calculate the ability of the system to return only the message correctly classified

$$P= \frac{TP_i}{TP_i + FP_i}$$

- F-measure is a measure that combined the recall and precision:

$$F= \frac{2*R*P}{(R+P)}$$

- **Entropy:** Calculate the loss of information E=-p(log(p)).
- **Accuracy:** The success rate is the percentage of messages correctly classified by the system

$$Accuracy= \frac{TP_i + TN_i}{TP_i + TN_i + FN_{i+} FP_i} *100.$$

- **Accuracy:** The error rate is the percentage of messages incorrectly classified by the system.

4.3. Results

We will conduct a series of tests by applying the different algorithms (ASC, AHLS and AHS) on the dataset SMS spam V.0.1 by varying the following parameters:

- Text representation method 2, 3, 4, 5 gram character and bag of words.
- Distances measure: Cosine, Chebyshev, Euclidian, Manhattan
- Threshold [0.1 - 0.9].
- The adjustments parameters for each algorithm (ASC and AHLS).
- The weighting techniques (TFC and TF-IDF).

For each test we fix four parameters and varying the others. The most significant results are grouped into the following tables. Each table is accompanying by figures concerning the best results in term of distance measure used and text representation techniques used.

4.3.1. Analysis

To determine the ideal configuration of each algorithm (ASC, AHS and AHLs) for spam detection. We will analyse the influence of each parameter:

1. **Influence of Text Representation Method:** The results of every bio-inspired method change with the variation of the text representation method. By analysing Tables 3, 4 and 5 and the accompanying figures we can extract that the N-gram characters allows to the algorithms (ASC, AHS and

Table 3. Spam detection results using artificial heart lungs system (AHLS) algorithm

Distance measure	Text representation / Validation measures	2-gram	3-gram	4-gram	5-gram	Bag of word
Cosine +Euclidean	Recall	0.8939	0.9885	0.9842	0.9285	0.9625
	Precision	0.9839	0.9407	0.9204	0.9724	0.9372
	f-measure	0.9368	0.964	0.95126	0.95	0.9497
	Entropy	0.0159	0.057	0.0811	0.0272	0.0607
	Accuracy%	90.20 %	94.08 %	91.89 %	92.06 %	91.72 %
	Error %	9.79 %	5.91 %	7.63 %	7.93 %	8.27 %
Euclidean +Chebychev	Recall	0.7817	0.839	0.9054	0.8487	0.7068
	Precision	0.9791	0.963	0.9642	0.9758	0.984
	f-measure	0.8693	0.8967	0.9339	0.9078	0.8227
	Entropy	0.02	0.0363	0.0351	0.0239	0.0158
	Accuracy%	80.91 %	84.29 %	89.69 %	86.07 %	75.25 %
	Error %	19.08 %	15.7 %	10.3 %	13.92 %	24.74 %
Manhattan +Euclidean	Recall	0.5062	0.6573	0.7268	0.6691	0.6122
	Precision	0.9878	0.9937	0.9899	0.9769	0.9966
	f-measure	0.6694	0.7912	0.8382	0.7942	0.7585
	Entropy	0.012	0.0062	0.01	0.0228	0.0033
	Accuracy%	59.37 %	71.72 %	77.44 %	72.06 %	75.25 %
	Error %	40.62 %	28.27 %	22.55 %	27.93 %	24.74 %
Cosine +Manhattan	Recall	0.9459	0.9979	0.9968	0.9474	0.9854
	Precision	0.977	0.9031	0.8603	0.9685	0.8994
	f-measure	0.9614	0.9481	0.9236	0.958	0.9404
	Entropy	0.0227	0.002	0.1294	0.0307	0.095
	Accuracy %	93.83 %	91.13 %	86.73 %	93.24 %	90.97 %
	Error %	6.16 %	8.86 %	13.2 %	6.75 %	9.03 %

Figure 11. The best results of AHLS algorithm with different combination of distances

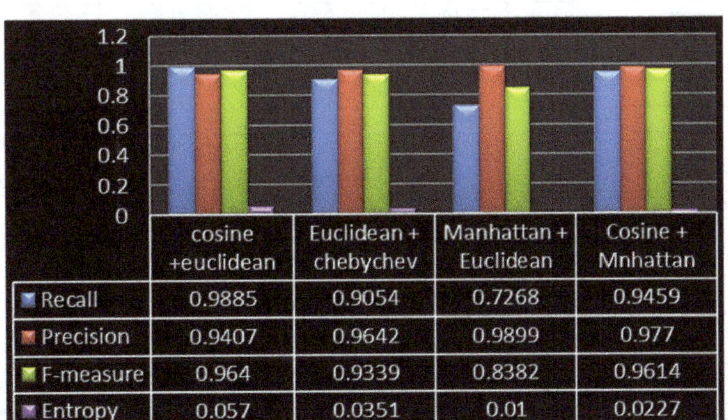

AHLs) to have a better performance compared to the bag of words technique. Generally, the N-gram technique is ideal for the representation of the corpus SMS spam V.0.1 because it provides several advantages:

a. It can lift the ambiguity linked to the word (ambiguity lexical, ambiguity syntactic and ambiguity semantic).
b. It offers more meaning to the text.

Figure 12. The best results of AHLS algorithm with different text representation technique

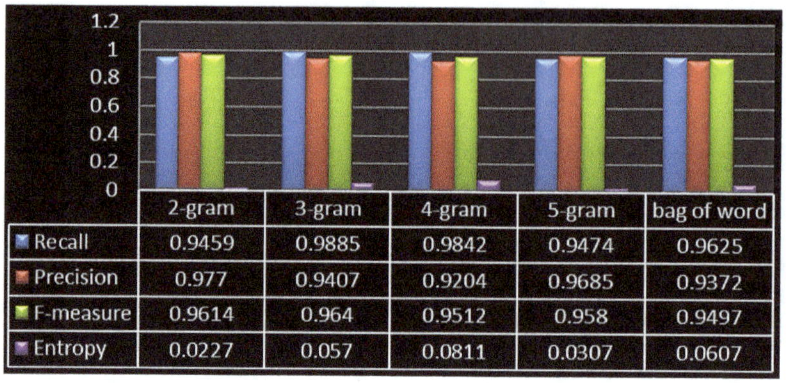

Table 4. Spam detection results using artificial social cockroaches (ASC) algorithm

Distance measure	Text representation / Validation measure	2-gram	3-gram	4-gram	5-gram	Bag of word
Euclidean	Recall	0.761	0.7927	0.7342	0.7117	0.7432
	Precision	0.6601	0.7489	0.6059	0.622	0.66
	f-measure	0.7042	0.7712	0.6684	0.6654	0.7
	Entropy	0.274	0.2165	0.3035	0.2953	0.2742
	Accuracy%	85.35 %	89.74 %	83.88 %	84.37 %	86.13 %
	Error %	14.65 %	10.26 %	16.12 %	15.63 %	13.87 %
Chebychev	Recall	0.7567	0.666	0.6801	0.6081	0.6711
	Precision	0.4827	0.5522	0.576	0.4179	0.5775
	f-measure	0.5934	0.6074	0.6264	0.503	0.625
	Entropy	0.3515	0.3279	0.3177	0.3646	0.317
	Accuracy%	77.14 %	81.05 %	82.22 %	73.14 %	82 %
	Error %	22.86 %	18.95 %	17.78 %	26.86 %	18 %
Cosine	Recall	0.6216	0.7792	0.8454	0.7072	0.6036
	Precision	0.4457	0.7456	0.8379	0.7584	0.563
	f-measure	0.5226	0.7682	0.8432	0.7393	0.585
	Entropy	0.3601	0.2188	0.1419	0.2097	0.3234
	Accuracy%	80.95 %	89.45 %	93.16 %	88.76 %	81.25 %
	Error %	19.05 %	10.55 %	6.84 %	11.24 %	18.75 %

Figure 13. The best results of ASC algorithm with different combination of distances

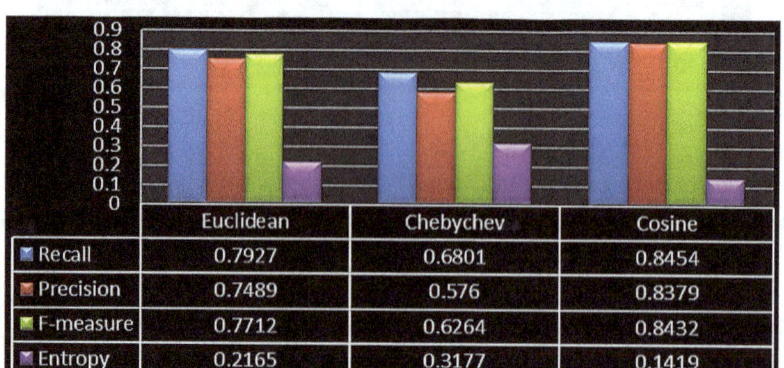

Figure 14. The best results of ASC algorithm with different text representation technique

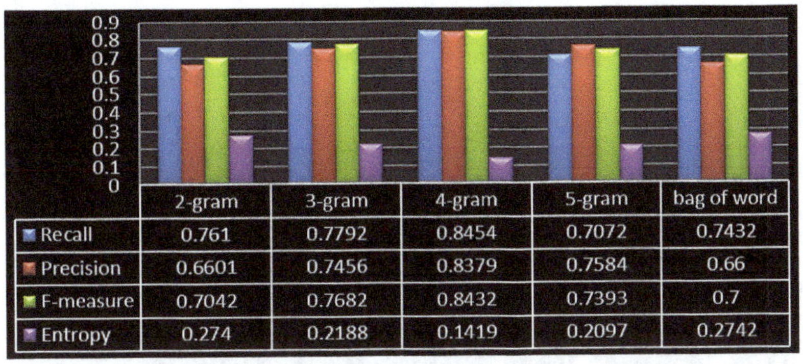

Table 5. Spam detection results using artificial haemostasis system (AHS)

Distance measures	Text representation / Validation measure	2-gram	3-gram	4-gram	5-gram	Bag of word
Cosine + Euclidean	Recall	0.9241	0.9937	0.9926	0.9401	0.975
	Precision	0.9801	0.9184	0.8831	0.968	0.9106
	f-measure	0.9513	0.9546	0.9347	0.9541	0.9417
	Entropy	0.0197	0.0781	0.1097	0.0734	0.0852
	Accuracy%	92.31 %	94.31 %	88.85 %	92.65 %	90.2 %
	Error %	7.68 %	5.68 %	11.14 %	8.35 %	9.79 %
Manhattan + Euclidean	Recall	0.8617	0.8587	0.9159	0.886	0.7328
	Precision	0.981	0.9605	0.9561	0.969	0.9791
	f-measure	0.9175	0.90679	0.935	0.9259	0.8381
	Entropy	0.0188	0.0387	0.0429	0.0305	0.0206
	Accuracy%	87.41 %	85.64 %	89.86 %	88.52 %	77.02 %
	Error %	12.58 %	3.87 %	10.13 %	11.47 %	22.97 %
Chebychev +Manhattan	Recall	0.449	0.6469	0.73	0.6313	0.5997
	Precision	0.984	0.9936	0.9872	0.9724	0.9965
	f-measure	0.6167	0.7836	0.8393	0.7656	0.7488
	Entropy	0.0158	0.0063	0.01279	0.0273	0.0034
	Accuracy%	54.64 %	70.94 %	77.53 %	68.8 %	67.31 %
	Error %	45.35 %	29.05 %	22.46 %	31.2 %	32.68 %
Cosine + Chebychev	Recall	0.9158	0.9273	0.9443	0.9117	0.7962
	Precision	0.9778	0.9479	0.939	0.9709	0.9795
	f-measure	0.9457	0.9375	0.9418	0.9404	0.8784
	Entropy	0.0219	0.0507	0.0588	0.0286	0.0198
	Accuracy%	9.14 %	89.94 %	90.62 %	90.63 %	82 %
	Error %	8.53 %	10.06 %	9.37 %	9.36 %	18 %

c. It is tolerant to misspelling.

d. It allows taking into account the compound words. This decision can change for other dataset.

e. It provides more discriminating terms.

2. **Influence of Distance Measure:** The distance measure is the basic ingredient for each algorithm. The variation in the choice of this element was a very interesting experience in order to select the most coherent metrics with the dataset Spam V.0.1. By observing the previous tables and figures, we see clearly that:

Figure 15. The best results of AHS algorithm with different combination of distances

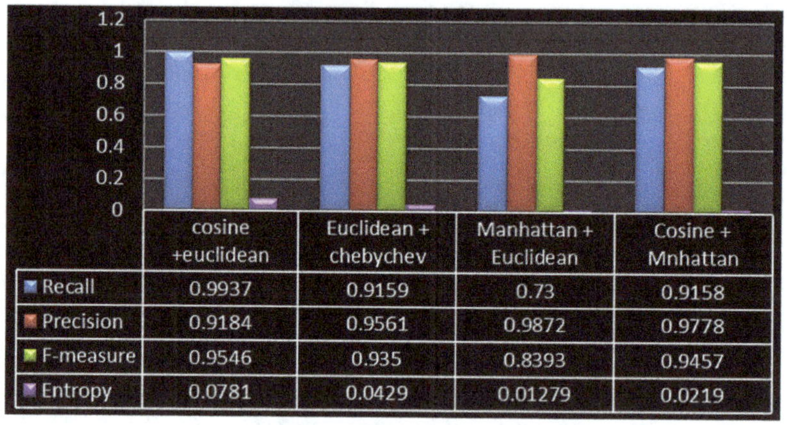

	cosine +euclidean	Euclidean + chebychev	Manhattan + Euclidean	Cosine + Mnhattan
Recall	0.9937	0.9159	0.73	0.9158
Precision	0.9184	0.9561	0.9872	0.9778
F-measure	0.9546	0.935	0.8393	0.9457
Entropy	0.0781	0.0429	0.01279	0.0219

Figure 16. The best results of AHS algorithm with different text representation technique

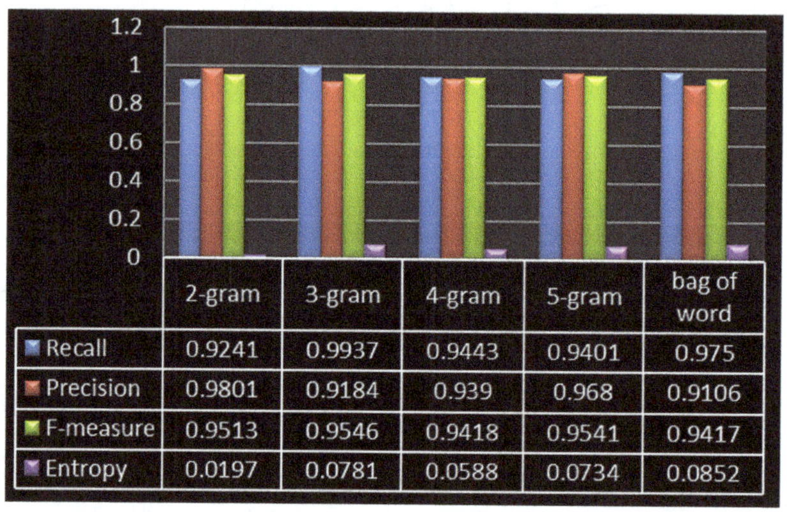

	2-gram	3-gram	4-gram	5-gram	bag of word
Recall	0.9241	0.9937	0.9443	0.9401	0.975
Precision	0.9801	0.9184	0.939	0.968	0.9106
F-measure	0.9513	0.9546	0.9418	0.9541	0.9417
Entropy	0.0197	0.0781	0.0588	0.0734	0.0852

a. The most ideal distance measure for ASC algorithm is cosine because it allows to the artificial cockroaches to better identify their congeners and make the best choice of shelter.

b. The ideal combination distances for the AHS and AHLs algorithms is (cosine + Euclidian) because they are complementary and allows to have a best cooperation between the operators of each algorithms.

3. **Influence of the Parameter N:** Figure 17 presents the relation existed between the variation of parameter N (gram number) and the quality of spam detection given by each algorithm (ASC, AHS and AHLS).

The previous curves show that the performance of algorithms (ASC, AHS and AHLS) depends on the parameter N. The question that arises; what is the value of N that gives the best result for each algorithm?

Figure 17. Results of spam detection results using with variation of text representation technique (parameter N) and best distance measure for each algorithm: a) AHLS algorithm with (Cosine + Euclidean), b) ASC algorithm with cosine distance, c) AHS algorithm with (cosine + Euclidean)

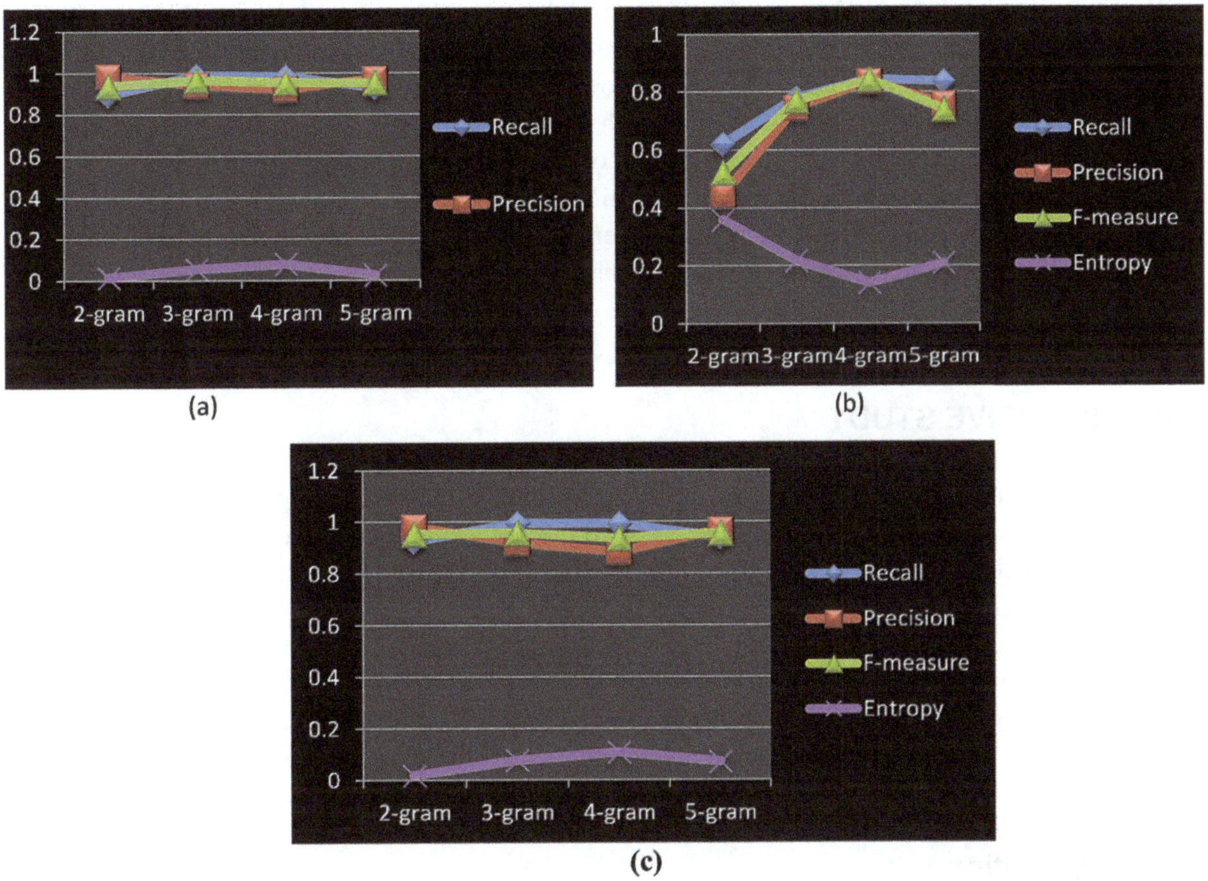

The two concepts (recall and precision) are very important to the user. If the precision is low the user will be unsatisfactory because it will lose significant messages (message ham classified as spam).

A perfect algorithm must have a precision and recall of one (1), but these two requirements are contradictory and a very high precision can be obtained with a low recall and vice versa.

By observing the two metrics (f-measure and entropy) we find that the performance of each algorithms is improving by increasing the value of N for each algorithm until achieve the optimal value (N = 3 for the algorithms (AHS and AHLS) and N = 4 for the ASC algorithm).

4. **In Term of Weighting Method:** The weighting technique plays a very important role in the quality performance. The TFC gives best results for all the algorithms (ASC, AHLs and AHS) because the TF * IDF weighting does not take into account the fact that the lexical units (words) can be of different length but TFC applies the normalization of the length to solve this problem.
5. **In Term of Threshold:** This parameter is very difficult to fix. We remark that when we increase the threshold the recall will be increased and precision will be decreased until reach the best f-

measure for each algorithm which represents a combination between the recall and precision. The choice of this parameter is also related to the distance measure used.

6. **In Term of Loss of Information:** In our work, The Information loss is negligible validated by the entropy obtained by each algorithm because we don't use the dimension reduction. The table proved that the best performance in term of entropy is given by AHLs algorithm because the entropy is based on the precision and this algorithm allows the detection of more than 94% of really spam messages. The badly entropy is obtained with the algorithm ASC because the first cockroaches which reach the shelter influences the quality of classification of others cockroaches.
7. **Ideal Configuration:** Arguably a good detection of spam is a combination of several parameters. The different tests realised had allowing the identification of ideal configuration for each algorithm validated by the validation measures used as presented in Table 6.

5. COMPARATIVE STUDY

In order to give our algorithms a reference and to view the interest of modelling new bio-inspired algorithms and their advantages compared to the classical machine learning techniques. A comparison was realized between the results provided by the algorithms (ASC, AHS and AHLs) with the results obtained by the following algorithms:

Algorithm based on the behaviour of social worker bees (foraging guardian and cleaning) was proposed by Hamou in (Hamou, 2013).

We had implemented classical supervised learning algorithms (decision tree C4.5 and K-means) on the SMS spam V.0.1 using the Weka tool (Hall, 2009).

The results of this comparison are grouped in the accompanying Table 7.

5.1. Interpretation

The Table 7 and Figure 18 show clearly that the best results are obtained by the algorithm AHLs (highlighted in blue) because it is based on a selection step using naive Bayes algorithm and lungs filtering

Table 6. Ideal configuration for each algorithm: ASC, AHS and AHLs

	Ideal configuration				
Parameters / Algorithms	Distance measure	Text representation	weighting	Adjustments parameters	Threshold
ASC	Cosine	4-gram characters	TFC	$\alpha = 0.4$ $\beta = 0.6$ $\gamma = 0.4$	0.6
AHS	Cosine + Euclidean	3-gram characters	TFC	-	0.4
AHLs	Cosine + Euclidean	3-gram characters	TFC	$\alpha = 2\ \beta = 3$	0.6

Table 7. The results of comparison

Validation measure / Algorithms	Recall	Precision	F-measure	Entropy	Accuracy	Error
K-means	0.8956	0.967	0.9326	0.0324	0.9083	0.0917
AHS	0.9937	0.9184	0.9546	0.0781	0.9231	0.0769
AHLS	0.9885	0.9407	0.964	0.057	0.9408	0.0592
Social worker bees	074	0.93	0.82	0.067	0.87	0.13
Decision treeC4.5	0.65	0.865	0.803	0.2164	0.851	0.149
ASC	0.8454	0.8378	0.8432	0.1419	0.9316	0.0684

using combination of two distances. It allows to obtains an elevate precision in the detection of spam messages.

The bad results are obtained by C4.5 algorithm (red boxes) that is based on the construction of a rule-based model. This algorithm has a problem in the classification of textual data.

The other algorithms (AHS, ASC and K-means) give satisfactory results (yellow boxes).

6. RESULTS MINING TOOL (VISUALISATION)

One of the strengths of our work for spam detection is the visual of messages (Spam or Ham). This part of our system; is characterised by the functionality of zoom and rotation. It provides a visualization of spam detection result in:

6.1. 3D Cub Format

The Figure 19b represents a global view of the spam detection results. We observe two shelters where the messages (cockroaches) that joined the spam class (shelter) are presented by blue cases and the cockroaches that had joined the ham class are presented by green cases. We can know class of each

Figure 18. Podium of algorithms results

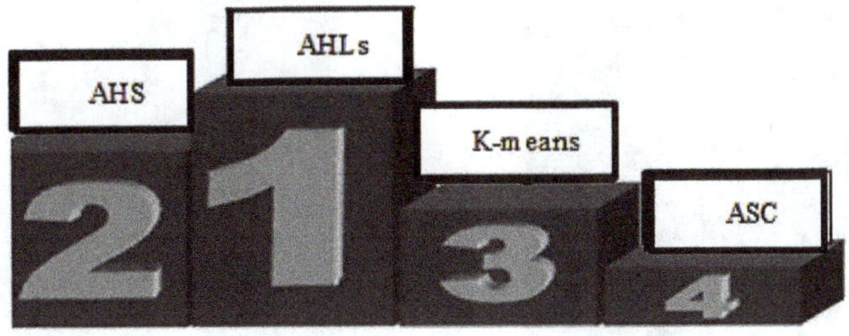

Figure 19. 3D visualisation of spam detection results as a 3D cub

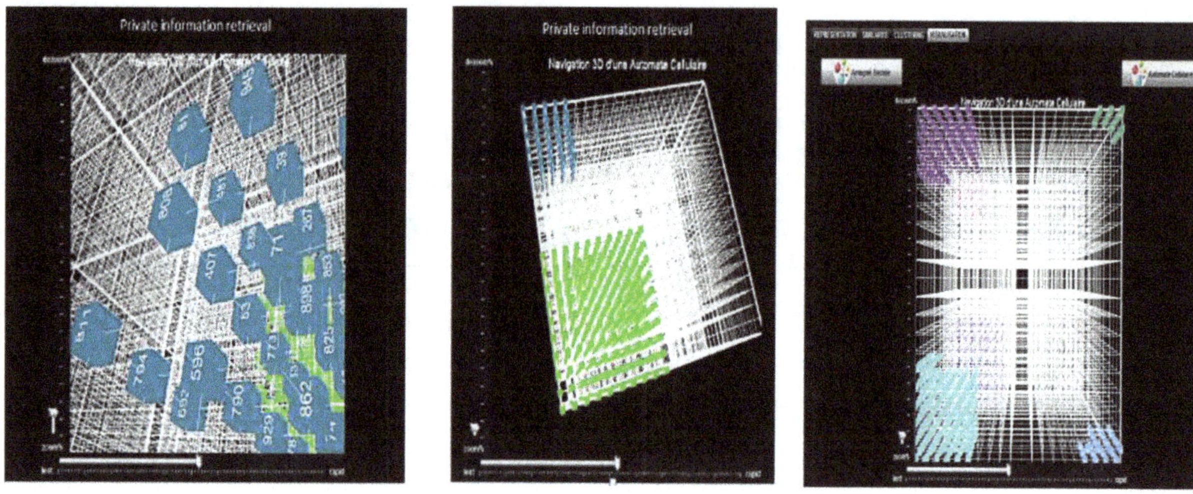

a) Detailed view b) Global view c) straight view

message by using the zoom functionality as shown in the Figure 19a, which designed the visualization of the identifier (message number) of each message by approaching the cursor to the box of the message

6.2. Cube Web Format

In a silky structure format (cobweb) where the white point represents message. For each new message detected as spam a link going to be woven between this message and the messages already spam (respectively Ham messages). The red path connects the spam messages and the green path connects the ham messages as depicts the next Figure 20.

Figure 20. Visualization of spam detection results as a cube web

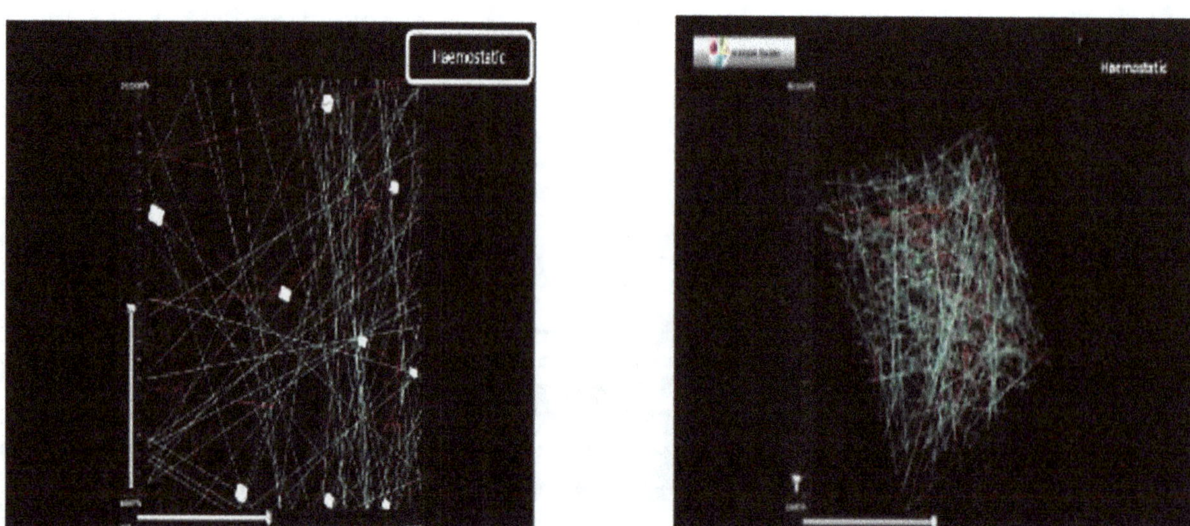

CONCLUSION

With the arrival of the World Wide Web the spam detection field has proven over the last few years as a major area looking for companies and for individuals. This dynamism is especially view to the high demand of users and entreprises for this technology. It is becoming increasingly essential in many situations where the Internet and the incredibly rapid development of telecommunication technology makes the spam messages as the major stake in our modern electronic world. Our work focuses on two aspects directly related to the modern user needs (the precision and the relevance of the detection and the representation of results).

The bio-inspired techniques and especially the family of swarm intelligence (ant colony optimisation, particle swarm optimization, and bees' colony algorithm) was essentially progressed these last few years. They have known a great success face to the optimization problem. This article deals on the introduction of a fresh bio-inspired algorithms and their adaptation for spam detection problem: AHS algorithm based on the phenomenon of stopping the loss of blood and physiological mechanism of homeostasis. ASC algorithm based on the societal phenomenon of cockroaches and their capacity to choose the attractive and secure place for hiding. AHLs algorithm based on the blood oxygenation phenomenon and the work of heart lungs system to ensure the systematic circulation and pulmonary circulation.

The result obtained are satisfactory and proved that the algorithms (ASC, AHS and AHLS) are able to detect and filter spam. They give better results compared to other algorithms existed in literature validated by the evaluation measures (recall, precision and F-measure, entropy, accuracy and error). We have studied the impact of each parameter in the performance quality of each algorithm in order to identify the ideal configuration.

For the purpose to improve the information user interface (IUI) and meet the modern needs of users. The second part of our work is dedicated to the development of new visualization tool using the java 3D in order to see the result in graphical form with more representation size space.

Our work proposes an original coupling of several methods: natural language processing (NLP), bio-inspired methods, text mining, and visualization methods. Finally, we can conclude that our satisfaction is full because the objectives fixed in the beginning have been reached.

FUTURE WORKS

As perspective, we plan to apply the algorithms (ASC, AHS and AHLS) for different problems such as: image processing, preservation of private life in big data. We can also combine these algorithms with other meta-heuristic methods (simulating annealing, taboo search and genetic algorithm). We can increase the quality of results by using the dimension reduction techniques (chi two and mutual information). Applying the conceptual representation technique:

Traveling Salesman Problem (TSP)

The functioning of The ASC algorithm gives us the opportunity to solve the TSP. The city represents an artificial shelter, and the choice of displacement of each cockroach from a shelter i to a shelter j depends on the decision of the group. First, the cockroaches will move randomly and after a moment the cockroaches will be grouped under the same shelter j that represents the nearest city to the city i and so on until city arrived. As a result, we can say that if we have a multiple paths to reach the arrived city. The shortest path is the path which connects the cities (shelters) where cockroaches have been grouped.

Information Retrieval Problem (IR)

The information retrieval problem is a big problem. We can solve this problem using:

The ASC algorithm: The query represents the shelter. The documents that joined this shelter are pertinent and the rest of documents are no-pertinent.

The AHS algorithm: the query represents the external haemorrhagic and each document must pass through two filters (first and second filter) using primary retrieval and secondary retrieval. The final decision if a document is retrieval or not is realised using the result of the previous filters.

Bi-Class Problems

The algorithms (ASC, AHS and AHLS) can solve the bi-classification problems such as plagiarism detection, intrusion detection, and suspicious person detection because they have the same structure as the spam detection problem.

REFERENCES

Applications (AINA). (2012). In *IEEE 26th International Conference on* (pp. 868-873). IEEE.

Aviles, L. (1997). Causes and consequences of cooperation and permanent-sociality in spiders. In J. Choe & B. Crespi (Eds.), *Evolution of social behaviour in insects and arachnids* (pp. 467–498). Cambridge, UK: Cambridge Univ. Press.

Back, T., & Schwefel, H. P. (1996, May). Evolutionary computation: An overview. In *Evolutionary Computation, 1996., Proceedings of IEEE International Conference on* (pp. 20-29). IEEE.

Bell, W. J., Roth, L. M., & Nalepa, C. A. (2007). *Cockroaches: Ecology, behaviour, and natural history.* JHU Press.

Benczur, A. A., Csalogany, K., Sarlos, T., & Uher, M. (2005, May). SpamRank–Fully Automatic Link Spam Detection Work in progress. *Proceedings of the First International Workshop on Adversarial Information Retrieval on the Web.*

Bourjot, C., Chevrier, V., & Thomas, V. (2003). A new swarm mechanism based on social spiders colonies: From web weaving to region detection. *Web Intelligence and Agent Systems, 1*(1), 47–64.

Cheng, L., Wang, Z. B., Song, Y. H., & Guo, A. H. (2011, September). Cockroach swarm optimization algorithm for TSP. In *Advanced Engineering Forum* (Vol. 1, pp. 226-229). doi:10.4028/www.scientific.net/AEF.1.226

Chirita, P. A., Diederich, J., & Nejdl, W. (2005, October). MailRank: using ranking for spam detection. In *Proceedings of the 14th ACM international conference on Information and knowledge management* (pp. 373-380). ACM.

Cormack, G. V., Gómez Hidalgo, J. M., & Puertas Sánz, E. (2007a). Feature engineering for mobile (SMS) spam filtering. In *Proceedings of the 30th Annual international ACM SIGIR Conference on Research and Development in information Retrieval*. ACM. doi:10.1145/1277741.1277951

Cormack, G. V., Gómez Hidalgo, J. M., & Puertas Sánz, E. (2007b). Spam filtering for short messages. In *Proceedings of the Sixteenth ACM Conference on Conference on information and Knowledge Management*. ACM. doi:10.1145/1321440.1321486

Dasgupta, D., & Forrest, S. (1999, July). Artificial immune systems in industrial applications. In *Intelligent Processing and Manufacturing of Materials, 1999. IPMM'99. Proceedings of the Second International Conference on* (*Vol. 1*, pp. 257-267). IEEE.

Dorigo, M., Maniezzo, V., & Colorni, A. (1996). Ant system: optimization by a colony of cooperating agents. *Systems, Man, and Cybernetics, Part B: Cybernetics. IEEE Transactions on*, *26*(1), 29–41.

Drucker, H., Wu, D., & Vapnik, V. N. (1999). Support vector machines for spam categorization. *IEEE Transactions on Neural Networks*, *10*(5), 1048–1054. doi:10.1109/72.788645 PMID:18252607

Elssied, N. O. F., Ibrahim, O., & Osman, A. H. (n.d.). Enhancement of spam detection mechanism based on hybrid\ varvec {k}-mean clustering and support vector machine. *Soft Computing*, 1-12.

Firte, L., Lemnaru, C., & Potolea, R. (2010, August). Spam detection filter using KNN algorithm and resampling. In *Intelligent Computer Communication and Processing (ICCP), 2010 IEEE International Conference on* (pp. 27-33). IEEE. doi:10.1109/ICCP.2010.5606466

Fogel, L. J. (1962). Autonomous automata. *Industrial Research*, *4*, 14–19.

Gómez Hidalgo, J. M., Cajigas Bringas, G., Puertas Sanz, E., & Carrero García, F. (2006). Content Based SMS Spam Filtering. In *Proceedings of the 2006 ACM Symposium on Document Engineering*. Amsterdam, The Netherlands: ACM Press.

González-Talaván, G. (2006). A simple, configurable SMTP anti-spam filter: Greylists. *Computers & Security*, *25*(3), 229–236. doi:10.1016/j.cose.2006.02.005

Gould, H. W. (1961). The $ q $-Stirling numbers of first and second kinds. *Duke Mathematical Journal*, *28*(2), 281–289. doi:10.1215/S0012-7094-61-02826-5

Grier, C., Thomas, K., Paxson, V., & Zhang, M. (2010, October). @ spam: the underground on 140 characters or less. In *Proceedings of the 17th ACM conference on Computer and communications security* (pp. 27-37). ACM. doi:10.1145/1866307.1866311

Hall, M., Frank, E., Holmes, G., Pfahringer, B., Reutemann, P., & Witten, I. H. (2009). The WEKA data mining software: an update. *ACM SIGKDD Explorations Newsletter*, *11*(1), 10-18.

Halloy, J., Sempo, G., Caprari, G., Rivault, C., Asadpour, M., Tâche, F., & Deneubourg, J. L. et al. (2007). Social integration of robots into groups of cockroaches to control self-organized choices. *Science*, *318*(5853), 1155–1158. doi:10.1126/science.1144259 PMID:18006751

Hamou, R. M., Amine, A., & Boudia, A. (2013). A New Meta-Heuristic Based on Social Bees for Detection and Filtering of Spam. *International Journal of Applied Metaheuristic Computing*, *4*(3), 15–33. doi:10.4018/ijamc.2013070102

Hamou, R. M., Amine, A., & Boudia, A. (2013). A New Meta-Heuristic Based on Social Bees for Detection and Filtering of Spam. *International Journal of Applied Metaheuristic Computing*, *4*(3), 15–33. doi:10.4018/ijamc.2013070102

Havens, T. C., Spain, C. J., Salmon, N. G., & Keller, J. M. (2008, September). Roach infestation optimization. In *Swarm Intelligence Symposium, 2008. SIS 2008. IEEE* (pp. 1-7). IEEE. doi:10.1109/SIS.2008.4668317

Holland, J. (1962). Outline for a logical theory of adaptive systems. *Journal of the Association for Computing Machinery*, 3.

Holland, J. H. (1973). Genetic algorithms and the optimal allocation of trials. *SIAM Journal on Computing*, *2*(2), 88–105. doi:10.1137/0202009

Jain, K., & Agrawal, S. (2014, August). A hybrid approach for spam filtering using support vector machine and artificial immune system. In *Networks & Soft Computing (ICNSC), 2014 First International Conference on* (pp. 5-9). IEEE. doi:10.1109/CNSC.2014.6906699

Jeanson, R., Rivault, C., Deneubourg, J. L., Blanco, S., Fournier, R., Jost, C., & Theraulaz, G. (2005). Self-organized aggregation in cockroaches. *Animal Behaviour*, *69*(1), 169–180. doi:10.1016/j.anbehav.2004.02.009

Jindal, N., & Liu, B. (2007, May). Review spam detection. In *Proceedings of the 16th international conference on World Wide Web* (pp. 1189-1190). ACM. doi:10.1145/1242572.1242759

Joachims, T. (1998). *Text categorization with support vector machines: Learning with many relevant features*. Springer Berlin Heidelberg.

Karaboga, D., & Basturk, B. (2007). A powerful and efficient algorithm for numerical function optimization: Artificial bee colony (ABC) algorithm. *Journal of Global Optimization*, *39*(3), 459–471. doi:10.1007/s10898-007-9149-x

Karaboga, D., & Basturk, B. (2007). A powerful and efficient algorithm for numerical function optimization: Artificial bee colony (ABC) algorithm. *Journal of Global Optimization*, *39*(3), 459–471. doi:10.1007/s10898-007-9149-x

Katriel, J. (2000). Bell numbers and coherent states. *Physics Letters. [Part A]*, *273*(3), 159–161. doi:10.1016/S0375-9601(00)00488-6

Kennedy, J., & Eberhart, R. (1995). Particle Swarm Optimization. *Proceedings of IEEE International Conference on Neural Networks, 4*, 1942–1948. doi:10.1109/ICNN.1995.488968

Koza, J. R. (1992). *Genetic Programming: On the Programming of Computers by Means of Natural Selection*. Cambridge, MA: The MIT Press.

Krasser, S., Tang, Y., Gould, J., Alperovitch, D., & Judge, P. (2007, June). Identifying image spam based on header and file properties using C4. 5 decision trees and support vector machine learning. In *Information Assurance and Security Workshop, 2007. IAW'07. IEEE SMC* (pp. 255-261). IEEE.

Lai, C. C., & Wu, C. H. (2007). *Particle swarm optimization-aided feature selection for spam email classification*. Kumamoto: IEEE. doi:10.1109/ICICIC.2007.442

Li, X., Shao, Z., & Qian, J. (2002). Anoptimizing method base on autonomous animates: Fish- swarm algorithm. *Systems Engineering Theory and Practice, 22*, 32–38.

Lihoreau, M., Costa, J. T., & Rivault, C. (2012). The social biology of domiciliary cockroaches: Colony structure, kin recognition and collective decisions. *Insectes Sociaux, 59*(4), 445–452. doi:10.1007/s00040-012-0234-x

Lihoreau, M., & Rivault, C. (2009). Kin recognition via cuticular hydrocarbons shapes cockroach social life. *Behaviour al Ecology, 20*(1), 46–53. doi:10.1093/beheco/arn113

Mohammad, A. H., & Zitar, R. A. (2011). Application of genetic optimized artificial immune system and neural networks in spam detection. *Applied Soft Computing, 11*(4), 3827-3845.

Moore, A. J., Ciccone, W. J., & Breed, M. D. (1988). The influence of social experience on the behaviour of male cockroaches, Nauphoeta cinerea. *Journal of Insect Behavior, 1*(2), 157–168. doi:10.1007/BF01052235

Oda, T., & White, T. (2005). Immunity from spam: An analysis of an artificial immune system for junk email detection. In *Artificial Immune Systems* (pp. 276–289). Springer Berlin Heidelberg. doi:10.1007/11536444_21

Sahami, M., Dumais, S., Heckerman, D., & Horvitz, E. (1998, July). A Bayesian approach to filtering junk e-mail. In *Learning for Text Categorization: Papers from the 1998 workshop* (*Vol. 62*, pp. 98-105). Academic Press.

Sasaki, M., & Shinnou, H. (2005, November). Spam detection using text clustering. In *Cyberworlds, 2005. International Conference on* (pp. 4-pp). IEEE. doi:10.1109/CW.2005.83

Silva, T. P., Santos, I., Almeida, T. A., & Gómez Hidalgo, J. M. (2014). Normalização Textual e Indexacção Semântica Aplicadas na Filtragem de SMS Spam. *Proceedings of the Encontro Nacional de Inteligência Artificial e Computacional (ENIAC).*

Suykens, J. A., & Vandewalle, J. (1999). Least squares support vector machine classifiers. *Neural Processing Letters, 9*(3), 293–300. doi:10.1023/A:1018628609742

Taweesiriwate, A., Manaskasemsak, B., & Rungsawang, A. (2012, March). Web spam detection using link-based ant colony optimization. Advanced Information Networking. doi:10.1109/AINA.2012.118

Wang, H. B., Yu, Y., & Liu, Z. (2005). SVM classifier incorporating feature selection using GA for spam detection. In *Embedded and Ubiquitous Computing–EUC 2005* (pp. 1147–1154). Springer Berlin Heidelberg. doi:10.1007/11596356_113

Chapter 12
Information Attacks and Defenses on the Social Web

Agostino Poggi
University of Parma, Italy

Michele Tomaiuolo
University of Parma, Italy

ABSTRACT

Social web sites are used daily by many millions of users. They have attracted users with very weak interest in technology, including absolute neophytes of computers in general. Common users of social web sites often have a carefree attitude in sharing information. Moreover, some system operators offer sub-par security measures, which are not adequate for the high value of the published information. For all these reasons, online social networks suffer more and more attacks by sophisticated crackers and scammers. To make things worse, the information gathered from social web sites can trigger attacks to even more sensible targets. This work reviews some typical social attacks that are conducted on social networking systems, describing real-world examples of such violations and analyzing in particular the weakness of password mechanisms. It then presents some solutions that could improve the overall security of the systems.

INTRODUCTION

Nowadays, millions of people of any age and gender regularly access Online Social Networks (OSNs) and spend most of their online time social networking. In fact, if we have to chose among the innovations of the past decade just a single phenomenon because of its outstanding social impact, that would be the diffusion of online social networks. While some social networking services were already active in the nineties, the capillary diffusion and the sheer number of people involved transformed online social networking in an unprecedented revolution only recently.

Social web sites have attracted also users with very weak interest in technology, including people that before the social networking revolution were not even regular users of other popular Internet services and computers in general (Stroud, 2008). The phenomenon is so widespread that many people started

DOI: 10.4018/978-1-5225-4944-4.ch012

using social networking systems to ask questions to people instead of querying search engines and in place of regular email. Moreover, some of the largest social networking sites constitute a separate and closed network (Sabbag, 2011). After the huge success of the early social networking systems, many other players came in the social networking market and nowadays hundreds of different social networking systems exist.

According to Boyd and Ellison (2008), teenagers have a clear understanding of privacy related issues; however, the same does not apply to some adults that: (i) did not even use email and other basic Internet services before the social networking revolution (Stroud, 2008) and (ii) not only have limited computer-related technical skills, but they also lack risk consciousness about privacy issues. Moreover, many people are becoming uncomfortable with the presence of their employers in the same social networking systems, because some personal data may leak in their corporate environment due to privacy configuration errors (Skeels & Grudin, 2009).

Unfortunately, online social networks are becoming an interesting target for crackers and scammers alike. In fact, many factors concur to attract malicious actions: (i) the users' carefree attitude in sharing information, (ii) the often sub-par security measures from the part of the system operators, and (iii) the high value of the published information. In particular, the information available through social media can be used to trigger attacks to even more sensible targets. Various forms of social engineering inside the system, including sophisticated and long-term attacks, may be facilitated by the general sense of sociability shared by users, which per se is an intrinsic objective of social media. However, while the only lasting solution to privacy and security issues would be increasing the users' awareness, much can and shall be done at the system level in order to protect the data with cryptography and to decrease the impact of wrong choices and mistakes on the user's part.

SECURITY THREATS ASSOCIATED WITH SOCIAL MEDIA

Nowadays, online social networks involve people from the entire world, of any age and with any kind of education. They also helped to increase computer usage among categories that previously showed little interest for it (Angiani et al., 2016; Franchi, Poggi & Tomaiuolo, 2016; Stroud, 2008). The users of information systems have various types of security requirements, including confidentiality, integrity, accountability, availability and anonymity. The same security requirements can be applied to social networking platforms, as well.

Unfortunately, while most users are aware that their profile and the information they publish is essentially public, they usually strengthen their privacy settings only after problems arise and tend to overlook the actual impact of the information they disclose (Stroud, 2008). Apparently, harmless information can be exploited, and the more information the attacker has, the more severe and sophisticated the attack can be. For example, name, location and age can be used to connect a profile to a real-world identity for more than half of the residents in the USA (Irani et al., 2011).

In fact, social networking platforms are susceptible to different types of attacks, targeting different components, conducted from different domains, using different techniques. For better analyzing these attacks, it is useful to identify the main components of a generic social networking platform, corresponding to different functional aspects of those systems. Attackers can target each of the different components, or they can target different levels, possibly with roughly the same logic. We identify four main components:

1. The social networking component. It manages and protects access to the users' personal profiles and the social relationships among users.
2. The content management component. It manages and protects access to all user-generated content, including personal status updates, comments, content links, photos and multimedia galleries.
3. The infrastructure services component. It provides the basic infrastructure services needed to run the social networking platform, including storage and replication services for content and profiles, information indexing and routing, management of users' online presence.
4. The communication and transport component. It encapsulates basic inter-networking and ad-hoc networking functionalities.

Moreover, we can distinguish two different kinds of attackers:

1. Intruders. An attack can be conducted by users accessing the system without proper authorization, or with accounts created for conducting the attack, purposely.
2. Insiders. Legitimate users or entities participating in the systems operations can assume malicious behaviors. From the users' point of view, malicious behavior can also be attributed to the service provider.

All the various classical kinds of security attacks may be adapted and applied to online social networks, too. The most typical threats against OSNs include:

- Unauthorized Access. Users, who have not been granted adequate permissions for accessing some services and resources, may attempt to circumvent the security mechanisms and policies of the system and gain unauthorized access. In a social networking platform, any user who has access to some profiles and messages can harm their legitimate owners. The collection of existing data is the basis of profiling attacks. These data may also supply some knowledge for secondary data collection from a wide range of different sources, including other OSNs. Remote access can also occur at system level. In this case, the attacker may directly gain control of all resources.
- Social engineering. In a social networking application, a common attack is to psychologically manipulate a user into performing misguided actions. It is similar to a confidence trick or a traditional fraud, but by means of computer-based communications and online social networking, typically to gain access to confidential information. In a phishing scheme, the attacker masquerades as a trustworthy entity to obtain the desired information. In most cases, the victim and the attacker never acknowledged each other directly in real life.
- Masquerading. When a rogue user disguises his identity and claims the identity of another user, the former is said to be masquerading. Masquerading may be attempted by an attacker either during a conversation or while registering his own profile, for deceiving other users or the whole social networking platform. Sometimes, masquerading is the first step to gain access to infrastructure services and resources to which the attacker is not entitled. Simple impersonation, by cloning the victim's profile from the same platform or by porting profile data from a different platform, may easily lead the attacker to gain trust from the victim's contacts. This way, it can damage other users eventually deceived. Especially in communities where reputation is valued, masquerading can also damage the user whose identity has been stolen. In fact, the attacker may pretend to be another user in order to shift the blame for any liable action.

We will provide some examples and a discussion of these threats in the following subsections. However, also other traditional security attacks can be conducted against OSNs, including:

- Denial of Service (DoS). The services and communications at the infrastructure level can be disrupted by common denial of service attacks. Social networking platforms are also susceptible to all the conventional denial of service attacks aimed at the underlying operating system or communication protocols. In addition to attacking the whole infrastructure of a social networking platform, users can also launch denial of service attacks against specific users, especially in a distributed platform. For example, repeatedly sending messages or other spam may place undue burden on the recipient users and their systems. Malicious users can also intentionally distribute false or useless information to prevent other users from completing their social activities.
- Repudiation. In general, repudiation occurs when a user, after having performed some action, later denies that action having happened. Repudiation can be intentional or even accidental. It can also be the result of a misunderstanding, when users have a different view of events. In any case, it can generate important disputes. In a sense, nothing can prevent a user from repudiating one of his actions. However, a social networking platform can eventually help resolving disputes by providing needed evidence, if it maintains a sufficiently detailed log of events. For users who value their reputation, the availability of such evidence may constitute a valid deterrent.
- Eavesdropping. The attempt to observe the flow and possibly the content of confidential messages is one of the most classical security threats. Apart from reading the content of messages, which may require cryptanalysis, an eavesdropper may gather useful information by simply observing the pattern of messages and their recipients, for example inferring the type of services being requested. To eavesdrop on other users, an attacker may also exploit the infrastructure and communication services of the platform, e.g. through unauthorized access.
- Alteration. When a user signs up a social networking service, he starts exposing his profile and content to the platform. An attacker may tamper with the profile and content data published by the victim, with all the messages he communicates to other users and all data used on the infrastructure services. Alteration can also be conducted by the service operator, which provides the facilities for online social networking and may take control of published data. Alteration may take the particular form of filtering, or censorship, when applied systematically for removing undesired content from the OSN.
- Copy and Replay. Each action in a social network may be subject to copy and reply. In this type of security threat, an attacker attempts to intercept some data and clone it, for retransmitting it later. The interceptor may successfully copy and replay a message, a complete profile or any other data. If those data are not associated with a signature and a timestamp, the repeated reception of such copies may pass unnoticed and accepted as a legitimate action.

Types of Information Leaks From OSNs

According to Li, Li and Venkatasubramanian (2007), there are two types of privacy attacks in online social networks:

1. Identity disclosure. It occurs when the adversary is able to determine the mapping from an anonymous profile to a specific real-world entity.

2. Attribute disclosure. It occurs when an adversary is able to determine the value of a user attribute that the user intended to stay private.

Moreover, attackers may gather data from:

1. A single social networking platform.
2. Multiple platforms, services or extension apps.

Sensible data may be obtained both through:

1. Direct access, either through an authorized service or through some breach.
2. Attribute inference, based on public data available about a user's contacts, and the correlation usually existing among attributes of linked profiles.

Either about identity or attributes, privacy in online social networks is mostly intended as user-to-user privacy: even when the relative settings are set correctly, so that no other user in the system can access information not intended for his eyes, the system itself has full access to information. In fact, in most online social networks, the system owners actually rely on such information to make a profit, for example to improve the accuracy of target advertisement. Unfortunately, as long as they have full access to the information – i.e., the information is available in the system in the clear – any security issue or naivety results in privacy violations.

Even innocuous features can be easily become serious issues. For example, in order to promote the service, Facebook had a public directory containing the names of the users that used to be presented along with 10 random friends of theirs. Such feature allowed web-spiders to repeatedly request pages from such directories and essentially discover the structure of the network. Subsequently, the number of friends was reduced to eight and the selection of friends became a deterministic function of the IP address of the requester. However, even the greatly reduced information can still be used to violate privacy (Bonneau, 2009).

Another very relevant problem is the amount of integration we expect from digital services. Online social networks are relatively open system that can be enhanced by third party widgets and games, usually called apps, which are embedded in the user's pages and typically have access to some amount of user information. Moreover, the users often want different services to interoperate for various reasons, so that, for example, a new tweet is also notified to the Facebook friends or pictures stored on Flickr are accessed by an online printing service. Eventually, the credentials of the online social network can be used to log in a different system without needing a separate account. The problem is that when any of these external services, systems or apps is violated, private information can be stolen. For example, in June 2012, although the main Twitter servers were not compromised, sensible data was nonetheless stolen from a third party widget, TweetGif (Robertson, 2012).

Similarly, we expect our mobile devices to interact cleverly with the online social networks. Service providers create mobile applications, which, in turn, have some degree of access to the data stored on the mobile device. Although such applications usually do not maliciously access user's data for purposes different from those stated, sometimes privacy is neglected. For example, the iPhone LinkedIn App used to send all the user's calendars to central LinkedIn servers, including phone numbers, call details and passcodes, while only the relevant information should have been sent.

Even when the social networking platform does not provide data directly, yet many supposedly private properties may be inferred about a user. In fact, in social networks the attributes of users who share some kind of link are often correlated. Zheleva and Getoor (2009) introduce different models of attacks, to infer the hidden sensitive values using friendship links or public group membership data. Specifically:

- Friend-aggregate. It looks at the sensitive attribute distribution amongst the friends of the person under question.
- Collective classification. It aims at inferring class labels of linked objects together, instead of classifying each instance independently of the rest. It iteratively uses inferred values for connected private profiles, in addition to public ones.
- Flat-link. It deals with links by flattening the data and considering the adjacency matrix of the graph. Each user has a list of features, corresponding to the size of the network. The public profiles are used as a training set for the classifier.
- Block modelling. It is a stochastic model, supposing that users form natural clusters or blocks. Then, their interactions can be explained by the blocks they belong to.
- Groupmate-link. In this model, groupmates are considered as friends to whom users are implicitly linked. It assumes that each group is a clique of friends. Thus, it creates a friendship link between users who belong to at least one group together, without representing the strength of the link. The resulting network can then be analyzed using one of the previous models.
- Group-based classification. It considers each group as a feature in a classifier, and sensitive attributes are inferred from the groups a user belongs to.
- Basic. It calculates the overall marginal distribution of public data and uses it to infer sensitive attributes of private profiles. This is the simplest model, which can be applied also in the absence of relationship and group information.

Phishing and Impersonation Attacks

Social engineering has always been a major security threat to information systems. Two of the more frequent kinds of social engineering attacks, i.e., pretexting and phishing, require some background information, but such information was not always easy to obtain. In fact, some of the early social engineering attacks actually involved going through company or people's trash bins in order to find scraps of paper containing relevant information. Nowadays, many interesting pieces of information are publicly available in social networking systems or accessible only to victim's friends at best (or at worse, from the attacker's perspective).

Traditional phishing consisted in sending electronic communications to a huge number of individuals with the intent to have them disclose some relevant information, such as passwords or credit card numbers. Since the large scale of such attacks, even if only few users actually trusted the communication and disclosed information, it was nonetheless enough to make the attack profitable. Such communications typically used very generic ways to address the victim and typically avoided any specific detail so that the same message could be sent to all the recipients. Nowadays, people have been instructed not to trust such generic messages, and other forms of phishing were created.

Spear phishing is a form of phishing where the attack is directed against a single individual and the offending communication contains large amounts of information on the victim himself, in order to convince him of the communication authenticity. This strategy is very effective both if the information employed is public but the victim somewhat thinks it is not or if the attackers gained access to private information by other means. Another popular form of phishing is cloning: a cloning attack requires an original legitimate message that is subsequently tampered and sent to the victims. These communications look extremely authentic, but nonetheless they contain ways to direct the victims towards the attackers website (where he is usually requested his credentials). Data-leaks offer an even more dangerous variant: when the data leak becomes public, users expect their service providers to send them emails notifying them to change their password. If such mails are instead well-crafted phishing emails, that perhaps also use some private information about the user, many more people fall for the fraud. The kind of attack just described actually happened in the days immediately after the LinkedIn breach.

A serious weakness in many online services regards the "security questions". In fact, while phishing can be avoided by paying attention to the actual identity of the communications initiators, not much can be done against "security questions" attacks. Some services allow the users to choose the security questions so that the information required to answer it correctly is not publicly available, however, for many other sites only a closed list of security questions is available. Researchers estimated that more than 33% of the security questions involve names of relatives or friends and about 16% involve personal preferences ("favorite something"). Popular security questions are also the name of a pet (typically the current or first one), ZIP code or social security number. The victim in online social networks often directly or indirectly divulges such pieces of information. Even apparently, safer information, such as social security numbers, may be guessed with high confidence from publicly available data (Acquisti & Gross, 2009).

This kind of attacks is usually not directed towards online social networks, at least initially, but usually targets email service providers, because, when the security question is answered correctly, most systems send a password-reset email, and, consequently, controlling the main email account is necessarily the first step. Similarly, to spear phishing, security-question related attacks target only specific individuals. Popular examples are the violation of Sarah Palin's email account in 2008, using information publicly available on Wikipedia and on official websites and, more recently, the attack against Mat Honan. In the latter case, the tech journalist had his mail, iCloud and Twitter accounts compromised after an elaborate chain of social attacks, during which also his mobile phone, tablet and laptop have been remotely wiped (Honan, 2012).

Another typical attack on social networking systems is impersonation (identity theft), which can occur either on a site where the victim already has a profile (profile cloning) or on different site (cross-site profile cloning). In both cases, it is easy to convince the victim's acquaintances to accept friendship from the fake profile and subsequently to disclose confidential data. Experiments on the effectiveness of impersonation have been conducted and the results show that such attacks mostly succeed (over 50% success rate) even in the case of simple non-cross-profile cloning (Bilge et al., 2009).

When the attackers obtain credential for an online social network profile, additional attacks can be performed. If the attackers are interested in keeping a private database containing the violated accounts data, they can create a widget that copies the relevant resources or simply scrapes the web pages. Moreover, they can try to use the same credential to attack the user's accounts in other online services. Since users do not often use different passwords for different accounts, this strategy is rather effective. For example, during a period of roughly a week in June 2012, several services were compromised. About a month after this so-called "breach week", a number of Dropbox accounts have been hacked, using the passwords stolen from the already compromised accounts (Agarwal, 2012).

Another possibility is using the violated account to accept friendships from profiles controlled by the attackers. If the victim has already enough friends (on average, a Facebook user has 1000 friends), it is not likely that he notices the new friendship, especially because the attackers can remove any notification before the account owner actually sees that. Such friendship can be subsequently used as a Trojan horse to access the user's data even after the user changes the password and, moreover, is also likely to attract friendship from the victim contacts, considering that mutual friendships usually increases the confidence that an accounts should be trusted.

Controlling many profiles in a social network, either real ones or fake, can be used for spam purposes, to influence opinions in general or simply to gather huge amounts of information. In fact, in order to either access the data of most users in the network or deliver them a message appearing to come from a friend, it is not necessary to control a large number of accounts, but it is sufficient to control accounts that approximate a dominating set, i.e. a set D such that $V = D \cup friends(D)$, where V is the set of all users in the network. Although computing a minimal dominating set is an NP-complete problem and requires full knowledge of the networks, decent approximations can be made using greedy algorithms and using only sampled knowledge.

THE WEAKNESS OF PASSWORDS

Most of the attacks we discussed so far are forms of social engineering. However, online social networks are not, from a technical point of view, different from any other password-protected service: if the passwords are not chosen judiciously and stored in a safe way, then the violation of accounts becomes very easy. In fact, the analysis of available records about the robustness of users' passwords, as well as the security mechanisms and policies deployed in online systems, does not shape a reassuring scenario. Table 1 summarizes some major password leaks suffered by social websites in recent years.

The "Breach Week"

The first week of June 2012 has earned the fame of "Breach Week". Some million passwords were stolen, in order, from LinkedIn, eHarmony and last.fm accounts. At least the first two incidents appear

Table 1. Major password leaks

Date	Target Site	Size of Leak	Password Storage
2016, May	MySpace	360 million passwords	SHA1, salted
2016, May	LinkedIn	117 million passwords	SHA1, not salted
2013, Aug.	Yahoo	1+ billion passwords	MD5, not salted
2012, June	LinkedIn	6,5 million passwords	SHA1, not salted
2012, June	eHarmony	1,5 million passwords	MD5, not salted
2012, June	last.fm	2,5 million passwords	MD5, not salted
2012, June	League of Legends	32 million passwords	MD5, not salted
2011	PlayStation Network	77 million accounts	Hashed, not salted
2011	Sony Online Entertainment	25 million accounts	Hashed, not salted
2009	Rockyou	32 million accounts	Clear text

correlated, as the passwords were published by the same user on a Russian forum about cracking. The first file, leaked from the professional social site LinkedIn, totaled 6,5 million unique passwords. It is not clear how many other passwords were really leaked, but the episode has certainly involved at least 117 million users, as it has been later revealed (Franceschi-Bicchierai, 2016). The second file, leaked from the popular online dating site eHarmony, contained around 1,5 million passwords. Finally, the internet radio platform last.fm notified all its users of a possible leak of passwords, requesting to update their login information. Administrators announced ongoing investigations. A published file accounts for at least 2,5 million passwords leaked form the radio platform. However, these episodes are not isolated. In fact, few days later, an alert appeared on the online gaming platform League of Legends, notifying its 32 million users of a breach in some of its databases (Merrill & Beck, 2012). The site administrators revealed that more than half of the passwords would not resist to crack: "We compared encrypted password hashes and discovered that 11 passwords were shared by over 10,000 players each... A double-digit percentage of individuals had the same password as at least one other person." Apart from revealing a widespread lack of awareness among users about basic password security, the announcement may imply that passwords, like on LinkedIn and eHarmony, were hashed but not salted (Yin, 2012).

But other major leaks have appeared both before and after the "Breach Week" (Franchi, Poggi & Tomaiuolo, 2015). In fact, Yahoo suffered a huge password leak in August, 2013. However, a measure of the extent of the leak was only made public in December 2016. On the basis of Yahoo messages to its own users and to some early analysis, more than a billion users have been affected. The stolen data associate emails with their corresponding hashed passwords. Passwords are simply hashed, without salt, with the obsolete and ineffective MD5 algorithm (Pauli, 2016).

In May 2016, a large database of MySpace users' credentials appeared online. The database contains 360 million passwords, which are hashed in two different forms: in the first field, the password is lowercased, truncated to 10 characters, and then hashed with the SHA-1 algorithm; in the second field, the full password is salted with the user-id and then hashed, with the same SHA-1 algorithm. Though the first field does not allow to recover the full password, directly, it largely facilitates an attack against the second field. Some analysts claim to have recovered 45 million passwords, quite easily (Cynosure Prime, 2016).

Sony, in 2011, suffered from breaches in its PlayStation Network and Qriocity media streaming service. The services were suspended for almost a month. Personal data from around 77 million accounts were stolen, possibly including credit card numbers. During the same period, Sony Online Entertainment announced that another breach led to leaking personal data from 25 million accounts. These breaches compare to previous massive data leaks from TJX Companies, a fashion retail network, in 2007, affecting over 45 million customers, including details about credit card and in some cases also social security numbers and driver's license numbers. In 2009, Heartland Payment Systems, currently the fifth largest credit card processor in the United States and the 9th in the world, was affected by a breach, which costed something in the tens of million dollars. In 2005, a similar breach at CardSystems Solutions, another payment card processor, jeopardized roughly 40 million credit and debit card accounts (Vijayan, 2007).

One of the best-known episodes, finally, is the breach suffered by Rockyou. After an SQL injection attack in 2009, the intruders gained access to the website database, which included the full list of unprotected clear text passwords of all 32 million users. In fact, those passwords are still readily available over the BitTorrent network and are being used for dictionary attacks against other websites (Cubrilovic, 2009).

All these attacks have allowed crackers to aggregate a large quantity of credentials, some of which have also been made public. Some of the largest published database of credentials amount to around half a billion of records, including username and passwords, together with other sensible data. In spring 2017, two of such databases have attracted large interest, also from traditional media. In particular, the first release of data, codenamed "Anti Public", comprises 458 millions of records. After few days, a second database with 593 millions of records, codenamed "exploit.in", has been released. In this second database, 45% of data are new with respect to "Anti Public". Thus, the two database together amount to more than 700 thousands of leaked credentials, in total (CERT-NZ, 2017).

Common Password Storing and Cracking Techniques

The safe management of passwords is an old problem. Various password-storing mechanisms can be considered.

1. Clear text. Though used in practice, the storage of passwords as clear text should out of the question, as it offers no protection against intruders.
2. Encryption. The use of traditional encryption schemes is also discouraged. In fact, by knowledge of the decryption key, all passwords may be subverted in a single shot. If an intruder acquires the control of a machine, then the possibility of losing a decryption key is quite concrete.
3. Hashing. The solution adopted since decades in UNIX systems is based on cryptographic one-way functions, which can only be inverted by guessing the original clear text password (Morris & Thompson, 1979). However, common hashing algorithms are often designed for efficiency, which allows attackers to try many combinations in short time. Moreover, the effort to guess users' passwords can be reduced by attackers, if they generate the hash of a tentative password and confront it with each one of the actual password hashes of the attacked system.
4. Salting. If some unique value (a salt) is added to each password before hashing it, the result is unique for each user. If two users use the same password, two different hashes are obtained, since that password is combined with two different salts. Then, in the database, both the hash and the salt, in the clear, need to be stored. Thus, it is not possible to pre-compute hashes for all popular and simple passwords, or for all combinations generated through brute force (Morris & Thompson, 1979).
5. Password hashing algorithms. While common hashing algorithms are designed to be as fast and efficient as possible, password-hashing algorithms are designed to require a significant amount of computational resources. Bcrypt, one of the best options among password hashing algorithms uses the Blowfish algorithm and allows developers to decide the number of iterations of its main function, possibly requiring various orders of magnitude more time than generic hashing algorithms. The exact choice depends on the desired balance of password security and needed computational resources for normal operation, in particular for handling the regular number of logins (Provos & Mazieres, 1999).

Though password-storing mechanisms are well known and documented, they are not always used in existing systems, including some popular services, with large user bases. In fact, some lessons can be learned about implemented mechanisms for password security in real cases.

A number of sites adopts techniques that are far from the best practices in this field. We will leave the Rockyou case apart. LinkedIn, for example, avoided storing passwords in clear text, but used a suboptimal algorithm for hashing. In fact, it used a good generic hashing algorithm (SHA-1, namely), instead of a password hashing algorithm, like bcrypt. On moderate hardware, SHA-1 can be computed over almost 200MBs of data per second, and MD5 over more than 300 MB of data per second (Dai, 2009). With these algorithms, a password of 6 lowercase alphanumeric characters can be easily obtained through a brute force attack in less than a minute. Moreover, this is without using the potential of parallel GPU computing, which can obtain results which are at least an order of magnitude better. Exploiting four HD 5970 cards and some pre-calculations for the latest steps of MD5, the Whitepixel tool may achieve 33.1 billions MD5 hash/s, on a system costing 2.700 $ at the end of 2010 (Bevand, 2010).

Another lesson that can be learned is that many websites simply skip password salting, even if it is a well established technique (Morris & Thompson, 1979). LinkedIn and eHarmony are not isolated examples, though emblematic given their huge user bases. For example, it took many years and versions for the popular blogging platform Wordpress to finally add salt to its user passwords, in 2008 at version 2.5.

In all those careless sites, simple attacks use dictionaries of popular passwords, together with mangling rules to obtain similar and derived passwords. Another possibility is to try all possible combinations of lowercase letters, uppercase letters, digits and punctuation symbols, in a brute force attack. Some tools, just like John the Ripper, can apply both attacks on a given list of hashed passwords. Starting from a dictionary or a combinatorial engine, the obtained password is hashed and then compared to all available hashes, possibly leading to the discovery of one or more users' passwords after a single hash operation. The effectiveness of the operation is greatly simplified by the fact that a single algorithm is applied against all passwords, without salt or additional parameters.

Moreover, if passwords are not salted, the attacks can be made even more effective by calculating in advance the hashes of all possible passwords, up to a certain length. Obviously, taking into account the needed disk space, this approach is feasible only for very short passwords. However, techniques are available to trade time for space, thus reducing the needed disk space but requiring more hash calculations at runtime. Among such techniques, some use the so-called rainbow tables. Oechslin (2003) shows how a previous technique, described by Hellman and refined by Rivest, could be further improved, halving the number of calculations during cryptanalysis.

Those methods use the iterative calculation of a hash function and a reduction function, in an alternating sequence, starting from a given password and repeating the cycle some thousands of times, depending on the desired balance between space and runtime processing time. For a given chain, only the starting password and the final hash are stored, while intermediate results are discarded. The number of chains to store depends on the desired success probability in decrypting a given hashed password. In the original paper, the method is applied to Windows LanManager passwords. With a space of 1.4GB for rainbow tables (and thanks to the weakness of the old LanManager scheme), a success rate of 99.9% can be achieved.

Given a certain hash, finding the corresponding password requires finding a rainbow chain in the table. If the original hash is not found, then one or more cycle of the reduction function and hash function are applied and then the search is repeated. Finally, when the relevant rainbow chain is found, starting from the first password in the chain, all calculations are repeated, until the password associated with the original hash is found.

ANALYSING SOME MILLION USERS' PASSWORDS

Unfortunately, users do not seem to choose passwords with care. A first glimpse of this fact comes from the security alert issued after the League of Legends intrusion (Merrill & Beck, 2012). In that alert, the system administrators pointed out that a high percentage of users had the same password of another user. The reported data of 10000 different users sharing the same 11 passwords is a glaring evidence. Therefore, in order to confirm the impression that users choose very weak passwords, we decided to analyze the publicly available list of leaked hashes from LinkedIn.

The first step is to crack the hashes to obtain the original passwords. Since the hashing procedure is essentially irreversible, the typical strategy is to "guess" which string generated the hash. One possibility is to brute-force attack the hash, i.e., simply hashing every possible string and comparing it with the desired hash. Such strategy is very slow and ineffective in practice whereas dictionary attacks are much more effective. In a dictionary attack, the strings are not chosen at random, but are extracted from a dictionary of common words. Additionally, the words from the dictionary are mangled simulating the usual strategies users employ to make passwords harder to guess (e.g., substitute numbers for alphabetic characters, appending values to the end or the beginning of the words).

Although more techniques exist, we decided to start cracking the passwords using the popular password cracker "John the Ripper", which performs both brute force and dictionary attacks with mangling rules and we reported our results in Figure 1. All the attempts were performed on a Core2 Duo laptop at 2.40 MHz. Our first attempt ("Basic" in Figure 1) used a small dictionary of around 3500 common password that is distributed with John itself and we also enabled simple mangling rules for generating common derivative passwords, in two hours the program was able to recover more than a million (1.05 M) LinkedIn users' passwords. The second attempt ("Full" in Figure 1) used a much larger dictionary, containing a few million words from more than 20 languages, also available from the program website. The larger dictionary provides some marginal gain.

Eventually, we fed John with the database of 32 million passwords that were stolen from the RockYou website in 2009 and subsequently published on the web. The results significantly improved. In two hours around 1.5 million passwords were revealed, 44% more than those obtained with the first attempts. In this case, the first million passwords were obtained in less than five minutes.

The most likely explanation is that the LinkedIn password set has a large overlap with the RockYou dataset, which, in turn, confirms the idea that many users tend to reuse passwords and are not extremely creative in deciding their own passwords. Moreover, in this case, of the 2 million passwords we cracked in 24 hours starting with the RockYou password dictionary, more than 25% are not longer than 6 characters and that less than 15% are longer than 8 characters. Moreover, almost the 30% consists only of alphabetic lowercase characters and another 30% consists of alphanumeric characters, with no special characters, and, eventually, almost 1 password over 10 consists of only digits. Roughly, 12% of the passwords end with the digit 1, 200.000 passwords end with 123 and about half as many with 1234. Years are also frequently included (e.g., numbers from 1975 to 2011, while 2012 is significantly less popular and "future" years are rare). The most popular base word that constitutes the password is a variation of the name of the website ("linked", "linkedin" and "link") and "love" and "password" still made it in the top 10. Analyses of other leaks of passwords show that variations on the service name are popular and that "password" is an evergreen (Dragusin, 2012).

In fact, the passwords that have not been cracked during our 24h timeframe are expected to be much stronger (otherwise, they would have been cracked as easily) and the fact that only 1/3 of the password

Figure 1. Dependence of the number cracked by using both dictionary (labeled as "Basic", "Full" and "RockYou") and brute force attacks, as functions of cracking time. In the result of the "RockYou" approach, it is particularly evident how the "brute-force" attack starts after exhausting the "dictionary" attack.

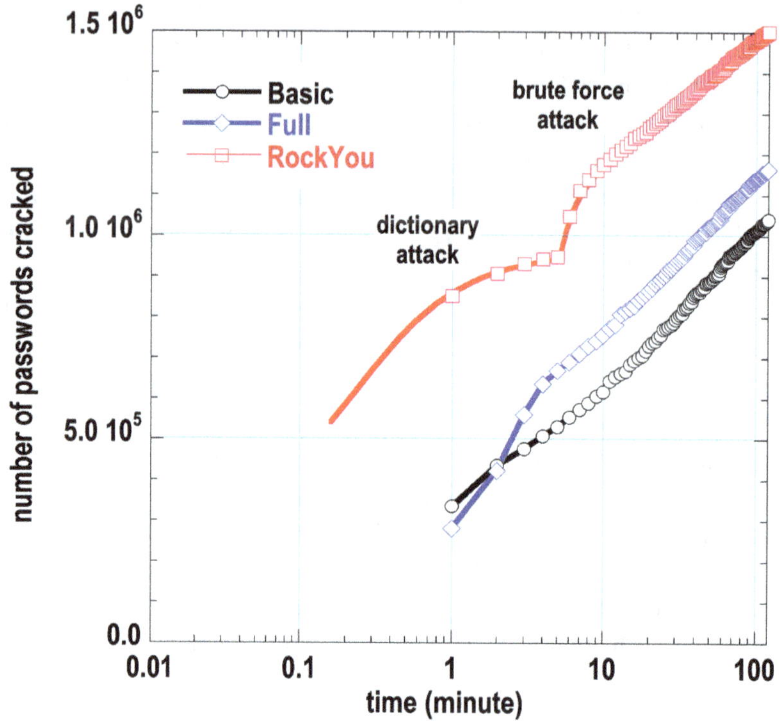

was cracked is somewhat encouraging. It is also interesting that, using the RockYou dataset as dictionary, the number of password cracked (N(p)) depends on the length of the passwords (L(p)) according to N(p) = A e-BL(p) where A = 3.15x106 and B = -0.786.

The last consideration regards the nature of the dataset: repeated passwords were not included, i.e., in the original system more than one user could have the same passwords, so that 2 million passwords could account for far more than 2 million of users, and, considering the remarks of the operators of League of Legends, they probably do.

HOW TO MAKE ONLINE SOCIAL NETWORKS MORE SECURE

Many solutions can be used to increase security. While the best long-term solution to the security problems is increasing the users' awareness, this may be unpractical because of the huge amount of people involved and their perception to be in a relatively safe environment, where friends are just few clicks away. However, there are many solutions that providers and software developers can use in order to improve the overall security.

For, example, most browsers and email clients offer some features to automatically detect phishing and one of the major weaknesses, i.e., security questions, can be fixed either improving the quality of the questions themselves or, better, sending reset messages via SMS, which at least would require the attacker to physically access the victim's mobile phone.

Zhou, Pei and Luk (2008) suggest some anonymization techniques to prevent privacy attacks in social networks. In particular, they suggest two different approaches, based respectively on generalization and perturbation:

1. Clustering. Vertices and edges are clustered into groups, so that a subgraph can be abstracted into a supervertex. This way, details about individuals are hidden.
2. Graph modification. Some vertices and edges of the graph are modified, by insertion or deletion. Such modifications may be applied using an optimization approach, randomly, or greedily, to match privacy requirements.

Felt & Evans (2008) suggest protecting data of social networks, especially from exposure to third-party developers. They present a privacy-by-proxy design, which mask data presented to external applications, using placeholders instead of actual values. However, it is not clear how resistant this simple approach is against de-anonymization techniques.

Li et al. (2007) discuss such de-anonymization techniques, under the conditions of k-anonymity. Under those conditions, each node is undistinguishable from other k-1 nodes, concerning attributes, which may potentially identify an individual. This group of node constitutes a class of equivalence. While k-anonymity can protect against identity disclosure, it fails to protect against attribute disclosure in the general case. Another discussed notion of privacy is the l-diversity, which requires each class of equivalence to contain at least l different values for a sensitive attribute. Authors propose instead t-closeness as a measure of privacy, which requires that the distribution of an attribute in a class of equivalence is close enough to that of the whole network, being less than a threshold t.

Weak passwords remain a major problem. Although, some service providers are either increasing strictness of the conditions to accept a password as valid, experience shows that "lazy" users always circumvent syntactical conditions (Yan et al., 2004). For example, in order to make passwords more secure the providers started to require that at least one character should be a number. Therefore, many users simply appended a single digit to their weak password, without neither substantially increasing the password security as a whole nor realizing they are failing to do so.

Part of the problem is that, when choosing passwords, users are facing the choice between easy to remember, easy to type passwords and secure passwords. As expected, users tend to choose the former. Even if they are instructed with good techniques to build secure and memorable passwords (e.g., pass-phrases), the large amount of services they subscribe creates the problem to remember which password is needed for which account and, consequently, users tend to reuse their passwords repeatedly, with terrible security implications. In addition, when longer pass-phrases are chosen, users prefer easier to remember sequences, which make sense to them. However, this motive also makes them vulnerable to more advanced dictionary attacks, which gather phrases from popular books and online sources.

Many solutions exist to improve password security, i.e., password managers and biometric identification. However, these solutions have serious drawbacks. Biometric identification is cumbersome because it may require specialized hardware and is often perceived as a privacy violation on its own right. Additionally, biometric data, when used for authentication, are roughly equivalent to strong and long passwords, which are hard to guess but which, on the other hand, are impossible to change. This way, if a service requiring biometric data at login is broken, or behaves in a rogue way, then the security based on those biometric data, on all sites, is made ineffective or completely broken. Password managers, on the other hand, create a single point of failure that, if violated, compromises the security of every single account the user had. In fact, this is not just a remote hypothesis (Slattery, 2011).

All considered, we think that one of the best strategies to improve the security of online social networks would be using strong asymmetric cryptography to protect all the data stored inside the system. Decrypting data protected with state-of-the-art cryptography is very hard. The system can be designed so that data travels only in encrypted format and is decrypted only on the client machine, so that, in essence no secret is revealed to a third party during regular usage. However, regular asymmetric cryptography has high computational costs when messages have to be sent to a restricted audience that varies its members during time, as in the case of a "group" or "circle" of friends (Canetti et al., 1999).

The solution we favor is using flexible attribute-based encryption, such as CP-ABE (Bethencourt et al., 2007). With attribute-based encryption, it is possible to create subordinate key-pairs associated with arbitrary attributes. Such attributes can indicate belonging to some group or time-dependent access rights. The system operators and other users have not necessarily access to the data so that, even if the system is compromised, the data remains secure.

The main problem is that, without remotely storing the private key, it becomes difficult to access the online social network from multiple devices, such as from home and from the office. In the specific case of mobile-centric online social networks, it becomes possible to keep the private key on the mobile device without impairing functionality. Backup and remote wipe procedures are widely implemented and they can be used to solve most issues regarding losing the physical device.

Another positive effect of using flexible cryptography, such as attribute-based encryption, is that it becomes possible to temporarily grant rights to another device from the main mobile device where the private key is stored. In fact, this essentially allows using the mobile device as a non-intrusive hardware key without actually having to copy the private key around.

The essential problem with encrypting data is that usually the system operators want to access such data in order to improve target advertisement that is their main source of revenue. Moreover, information quality, flow experience and trust in OSNs users' loyalty are correlated; thus, system operators have to pay close attention to them, to retain and increase their customers. Therefore, although technically feasible, an end-to-end encryption strategy would require either an entirely different, although similarly profitable, business model or a less expensive architecture, typically obtained moving from centralized to distributed (Franchi & Tomaiuolo, 2013).

There are two main categories of distributed social networks: federated and peer-to-peer (P2P). In a federated system, multiple entities cooperate to provide the service and each of them provides access to the whole system to a subset of the total users. Each user can choose the federated provider that he prefers, for example because he considers it worthier of trust. Service providers can also use high security and privacy standards as a way to attract more users and, perhaps even publish the code open source to increase confidence in their criteria. Direct access to the source code is possibly one of best way for user to discover whether their provider is using inadequate security standards, and, in general, more competition and no lock-in are positive factors for increasing quality.

On the other hand, in a P2P system, every participant is both a user and a system provider at the same time. However, purely P2P system can have issues with data availability, i.e., post from very badly connected users can be difficult to obtain. Consequently, P2P system may introduce "super-nodes" with a role akin to that of a federated provider. The distinction between the two categories is in practice blurred. The main advantage of P2P over federated approaches is that in the P2P scenario, the code runs on the user machine and, potentially, no information is transmitted outside with the explicit user content. The model of autonomous software agents has been often used in the development of such systems (Poggi & Tomaiuolo, 2010; Poggi & Tomaiuolo, 2013; Bergenti, Poggi & Tomaiuolo, 2014). Moreover, trust management and negotiation mechanisms among users can be used in order to improve the availability of the data even without introducing super-nodes (Bergenti, Rossi & Tomaiuolo, 2009; Tomaiuolo, Poggi & Franchi, 2013; Poggi, Tomaiuolo & Vitaglione, 2003).

Regardless of the actual form of distribution, the operative costs are essentially shared among multiple entities, and, in the case of P2P the required resources are already provided by the users, so that most costs are already covered.

Specifically in the field of social networking, various systems are being developed on the basis of peer-to-peer communications and DHT indexing, including Safebook (Cutillo, Molva & Strufe, 2009), PeerSoN (Bodriagov & Buchegger, 2013) and Blogracy (Franchi, Poggi & Tomaiuolo, 2016; Franchi, Poggi & Tomaiuolo, 2013). Safebook uses a network of socially close peers, defined Matryoshka. Peers in a user's Matryoshka are trusted and support the user by anonymizing communications and replicating content and profile information. PeerSoN proposes a Broadcast Encryption protocol, where each recipient has a different key, which can be used to decrypt data received by the broadcaster. Blogracy, on the other hand, adopts an Attribute-Based Encryption protocol for protecting access to users' content. It allows each user to assign credentials to various groups of followers, for accessing protected content.

In conclusion, it should be noted that cryptography does not outright prevent "mass" phishing, however it can make much more troublesome to discover enough data on users to conduct credible spear-phishing attacks. Cloning phishing attacks are also less likely to succeed in federated networks, since different federated providers would typically use different standards, and become meaningless in pure P2P scenarios. Similarly, impersonation attacks cannot be directly avoided, but cryptography still makes it harder to gather enough information to successfully conduct the attack. Another partial defense against impersonation is that the same cryptographic keys that are used to guarantee confidentiality can be used to guarantee authenticity, i.e., any message and any piece of information can be guaranteed to come from the owner of the key: a clone could be easily detected.

Security questions are used in order to allow users to change their password in case it is stolen. Depending on how the system is designed, there may not be a password to reset altogether. In these cases, a similar problem is losing the private key. Although, in theory, backups should avoid the problem, a failsafe mechanism may still be useful; however, because of the way the system works, it should probably be something completely different, hopefully not vulnerable to the classic problems of "security questions". Moreover, in P2P system there is not even the concept of a central authority that could "reset" any password or secret for the user and web of trust or similar strategies should be employed (Tomaiuolo, 2013).

Essentially, since all the pieces of information are encrypted, data published in online social networks cannot be used to gather information to violate other systems. Moreover, even if the online social network services were violated, the attackers would not access any data in clear.

CONCLUSION

Social web sites are used on a daily basis by many millions active users, including many complete neophytes of information and communication technologies. Give the values of the information they host and a combination of users and providers' lack of adequate protection, these systems are also attracting the attention of crackers and scammers, who may seek information for triggering attacks to even more sensible targets. In this work, we reviewed some typical social attacks that are conducted specifically on social Web sites, analyzing a number of widely known security violations and some corresponding public available set of leaked data. While any solution to privacy and security issues requires necessarily increasing the users' awareness, much can also be gained at the system level, using up-to-date cryptography schemes and effective security policies. This way, the impact of wrong choices and mistakes on the user's part would also be mitigated. In our study, we demonstrated that every system that is exploited makes every other system a bit less secure, because information leaks make information attacks more likely to succeed. Thus, security is going to be an increasingly important problem in the future and many systems will require an overall redesign, to provide good security standards. In this sense, the use of good encryption techniques will need to be accompanied by front-facing features, guides and well-documented procedures aimed at increasing the general users' awareness.

REFERENCES

Acquisti, A., & Gross, R. (2009). Predicting Social Security numbers from public data. *Proceedings of the National Academy of Sciences of the United States of America*, *106*(27), 10975–10980. doi:10.1073/pnas.0904891106 PMID:19581585

Agarwal, A. (2012). *Security update & new features.* Retrieved June 25, 2017, from https://blog.dropbox.com/bre/07/security-update-new-features/

Angiani, G., Fornacciari, P., Mordonini, M., Tomaiuolo, M., & Iotti, E. (2016). Models of participation in social networks. In Social Media Performance Evaluation and Success Measurements (pp. 196-224). Academic Press.

Bergenti, F., Poggi, A., & Tomaiuolo, M. (2014). An Actor Based Software Framework for Scalable Applications. *Lecture Notes in Computer Science*, *8729*, 26–35. doi:10.1007/978-3-319-11692-1_3

Bergenti, F., Rossi, L., & Tomaiuolo, M. (2009). Towards automated trust negotiation in MAS. *Proceeding 10th Workshop on Objects and Agents, WOA'09.*

Bethencourt, J., Sahai, A., & Waters, B. (2007). Ciphertext-policy attribute-based encryption. In *Symposium on Security and Privacy* (pp. 321-334). IEEE.

Bevand, M. (2010). *Whitepixel.* Retrieved June 25, 2017, from http://whitepixel.zorinaq.com/

Bilge, L., Strufe, T., Balzarotti, D., & Kirda, E. (2009, April). All your contacts are belong to us: automated identity theft attacks on social networks. In *Proceedings 18th International Conference on World Wide Web* (pp. 551--560). ACM. doi:10.1145/1526709.1526784

Bodriagov, O., & Buchegger, S. (2013). Encryption for peer-to-peer social networks. In *Security and Privacy in Social Networks* (pp. 47–65). Springer. doi:10.1007/978-1-4614-4139-7_4

Bonneau, J., Anderson, J., Anderson, R., & Stajano, F. (2009, March). Eight friends are enough: social graph approximation via public listings. In *Proceedings 2nd ACM EuroSys Workshop on Social Network Systems* (pp. 13--18). ACM. doi:10.1145/1578002.1578005

Canetti, R., Garay, J., Itkis, G., Micciancio, D., Naor, M., & Pinkas, B. (1999, March). Multicast security: A taxonomy and some efficient constructions. In *Proceedings 8th Annual Joint Conference of the IEEE Computer and Communications Societies* (Vol. 2, pp. 708--716). IEEE. doi:10.1109/INFCOM.1999.751457

CERT-NZ. (2017). *Anti Public and Exploit.in release of credentials*. Retrieved June 25, 2017, from https://www.cert.govt.nz/businesses-and-individuals/recent-threats/anti-public-and-exploit-in-release/

Cubrilovic, N. (2009). *RockYou Hack: From Bad To Worse*. Retrieved June 25, 2017, from http://techcrunch.com/2009/12/14/rockyou-hack-security-myspace-facebook-passwords/

Cutillo, L. A., Molva, R., & Strufe, T. (2009). Safebook: A privacy-preserving online social network leveraging on real-life trust. *Communications Magazine, IEEE*, *47*(12), 94–101. doi:10.1109/MCOM.2009.5350374

Cynosure Prime. (2016). *Myspace hashes, length 10 and beyond*. Retrieved June 25, 2017, from http://cynosureprime.blogspot.it/2016/07/myspace-hashes-length-10-and-beyond.html

Dai, W. (2009). *Crypto++ 5.6.0 Benchmarks*. Retrieved June 25, 2017, from http://www.cryptopp.com/benchmarks-amd64.html

Dragusin, R. (2012). *Data breach at IEEE.org: 100k plaintext passwords*. Retrieved June 25, 2017, from http://ieeelog.com/

Felt, A., & Evans, D. (2008). Privacy protection for social networking APIs. *2008 Web 2.0 Security and Privacy*.

Franceschi-Bicchierai, L. (2016). *Another Day, Another Hack: 117 Million LinkedIn Emails And Passwords*. Retrieved June 25, 2017, from https://motherboard.vice.com/en_us/article/78kk4z/another-day-another-hack-117-million-linkedin-emails-and-password

Franchi, E., Poggi, A., & Tomaiuolo, M. (2013). Open social networking for online collaboration. *International Journal of e-Collaboration*, *9*(3), 50–68. doi:10.4018/jec.2013070104

Franchi, E., Poggi, A., & Tomaiuolo, M. (2015). Information and password attacks on social networks: An argument for cryptography. *Journal of Information Technology Research*, *8*(1), 25–42. doi:10.4018/JITR.2015010103

Franchi, E., Poggi, A., & Tomaiuolo, M. (2016). Blogracy: A peer-to-peer social network. *International Journal of Distributed Systems and Technologies*, *7*(2), 37–56. doi:10.4018/IJDST.2016040103

Franchi, E., Poggi, A., & Tomaiuolo, M. (2016). Social media for online collaboration in firms and organizations. *International Journal of Information System Modeling and Design*, *7*(1), 18–31. doi:10.4018/IJISMD.2016010102

Franchi, E., & Tomaiuolo, M. (2013). Distributed social platforms for confidentiality and resilience. In Social Network Engineering for Secure Web Data and Services (pp. 114--136). Academic Press. doi:10.4018/978-1-4666-3926-3.ch006

Honan, M. (2012). *How Apple and Amazon Security Flaws Led to My Epic Hacking.* Retrieved June 25, 2017, from http://www.wired.com/gadgetlab/2012/08/apple-amazon-mat-honan-hacking

Irani, D., Webb, S., Pu, C., & Li, K. (2011). Modeling unintended personal-information leakage from multiple online social networks. *IEEE Internet Computing*, *15*(3), 13–19. doi:10.1109/MIC.2011.25

Li, N., Li, T., & Venkatasubramanian, S. (2007). t-closeness: Privacy beyond k-anonymity and l-diversity. In *IEEE 23rd International Conference on Data Engineering* (pp. 106-115). IEEE.

Merrill, M., & Beck, B. (2012). *League of Legends Account Security Alert.* Retrieved June 25, 2017, from http://euw.leagueoflegends.com/news/league-legends-account-security-alert

Morris, R., & Thompson, K. (1979). Password security: A case history. *Communications of the ACM*, *22*(11), 594–597. doi:10.1145/359168.359172

Oechslin, P. (2003). Making a faster cryptanalytic time-memory trade-off. In *Advances in Cryptology-CRYPTO 2003* (pp. 617–630). Springer. doi:10.1007/978-3-540-45146-4_36

Pauli, D. (2016). *Security! experts! slam! Yahoo! management! for! using! old! Crypto!* Retrieved June 25, 2017, from https://www.theregister.co.uk/2016/12/15/yahoos_password_hash/

Poggi, A., & Tomaiuolo, M. (2010). Integrating peer-to-peer and multi-agent technologies for the realization of content sharing applications. *Studies in Computational Intelligence*, *324*, 93–107.

Poggi, A., & Tomaiuolo, M. (2013). A DHT-based multi-agent system for semantic information sharing. *Studies in Computational Intelligence*, *439*, 197–213.

Poggi, A., Tomaiuolo, M., & Vitaglione, G. (2003). Security and trust in agent-oriented middleware. *Lecture Notes in Computer Science*, *2889*, 989–1003. doi:10.1007/978-3-540-39962-9_95

Provos, N., & Mazieres, D. (1999, June). A Future-Adaptable Password Scheme. In *USENIX Annual Technical Conference, FREENIX Track* (pp. 81-91). USENIX.

Robertson, A. (2012). *LulzSec hackers post data on 8,000 Twitter accounts, but your passwords are safe*. Retrieved June 25, 2017, from http://www.theverge.com/2012/6/12/3080534/lulzsec-reborn-twitter-tweetgif-hack

Slattery, B. (2011). *LastPass, Online Password Manager, May Have Been Hacked.* Retrieved June 25, 2017, from http://www.pcworld.com/article/227223/LastPass_Online_Password_Manager_May_Have_Been_Hacked.html

Stroud, D. (2008). Social networking: An age-neutral commodity—Social networking becomes a mature web application. *Journal of Direct Data and Digital Marketing Practice*, *9*(3), 278–292. doi:10.1057/palgrave.dddmp.4350099

Tomaiuolo, M. (2013). dDelega: Trust management for web services. *International Journal of Information Security and Privacy*, *7*(3), 53–67. doi:10.4018/jisp.2013070104

Tomaiuolo, M. (2013). *Trust management and delegation for the administration of web services.* Organizational, Legal, and Technological Dimensions of Information System Administration.

Tomaiuolo, M., Poggi, A., & Franchi, E. (2013). Supporting Social Networks With Agent-Based Services. *International Journal of Virtual Communities and Social Networking*, *5*(1), 62–74. doi:10.4018/jvcsn.2013010104

Vijayan, J. (2007). *TJX data breach: At 45.6M card numbers, it's the biggest ever.* Retrieved June 25, 2017, from http://www.computerworld.com/s/article/9014782

Yan, J., Blackwell, A., Anderson, R., & Grant, A. (2004). Password memorability and security: Empirical results. *Security & Privacy, IEEE*, *2*(5), 25–31. doi:10.1109/MSP.2004.81

Yin, S. (2012). *Last.FM Joins eHarmony, LinkedIn to Celebrate Breach Week.* Retrieved June 25, 2017, from http://www.pcmag.com/article.aspx/curl/2405492

Zheleva, E., & Getoor, L. (2009, April). To join or not to join: the illusion of privacy in social networks with mixed public and private user profiles. In *Proceedings 18th international conference on World Wide Web* (pp. 531--540). ACM. doi:10.1145/1526709.1526781

Zhou, B., Pei, J., & Luk, W. (2008). A brief survey on anonymization techniques for privacy preserving publishing of social network data. *ACM SIGKDD Explorations Newsletter*, *10*(2), 12–22. doi:10.1145/1540276.1540279

Chapter 13
Scaffolding the OEEU's Data-Driven Ecosystem to Analyze the Employability of Spanish Graduates

Andrea Vázquez-Ingelmo
University of Salamanca, Spain

Juan Cruz-Benito
University of Salamanca, Spain

Francisco J. García-Peñalvo
University of Salamanca, Spain

Martín Martín-González
Technical University of Madrid, Spain

ABSTRACT

This chapter outlines the technological evolution experimented by the Observatory for University Employability and Employment's information system to become a data-driven technological ecosystem. This observatory collects data from more than 50 Spanish universities and their graduate students (bachelor's degree, master's degree) with the goal of measuring the factors that lead to students' employability and employment. The goals pursued by the observatory need a strong technological support to gather, process, and disseminate the related data. The system that supports these tasks has evolved from a standard (traditional) information system to a data-driven ecosystem, which provides remarkable benefits covering the observatory's requirements. The benefits, the foundations, and the way the data-driven ecosystem is built will be described throughout the chapter, as well as how the information obtained is exploited in order to provide insights about the employment and employability variables.

DOI: 10.4018/978-1-5225-4944-4.ch013

1. INTRODUCTION

The Observatory of University Employability and Employment (also known as *OEEU* using the Spanish initials for *Observatorio de Empleabilidad and Empleo Universitarios*) http://www.oeeu.org/, is an organization composed by researchers and technicians who work together from different parts of Spain with a unified methodology. The purpose of the Observatory is to produce, analyze and spread information and insights regarding the graduates' employability and employment in Spain. The Observatory is under the direction of the UNESCO Chair in University Management and Policy (based in the Universidad Politécnica de Madrid, Spain) and it relies on the mentoring of an Expert Council composed of national and international academic and university experts. This project is also developed in collaboration with the "La Caixa" Foundation, the Conference of Rectors of Spanish Universities (CRUE) and the GRIAL Research Group of the University of Salamanca (Peñalvo et al., 2012).

The Observatory's vision is to become the information reference for understanding and exploiting knowledge about variables related to employability and university employment and its behavior. To reach this vision, the Observatory has the following goals (Michavila, Martín-González, Martínez, García-Peñalvo, & Cruz-Benito, 2015; Michavila, Martínez, Martín-González, García-Peñalvo, & Cruz-Benito, 2016):

- To understand the evolution of the employability and employment, and its characteristics related to university graduates.
- To develop a system and a uniform methodology for measuring indicators about employability and employment of graduates.
- To generate information on the employability and employment of university comparable between regions, branches of study (knowledge areas) and professional profiles, among others.
- To support the development of strategies and employment policies for universities, basing it on well-founded studies and information.
- To understand the mechanisms and actions that use the Spanish universities to promote employment and employability of their graduates.
- To provide information to individual universities to adjust their academic supply and training demands to the labor market based on reliable data.

These objectives seek to resolve the lack of public information (and its analysis) regarding employment and university employability. To achieve them, the Observatory is developing, implementing and exploiting a series of data-driven products (Patil, 2012).

This data-driven approach helps the Observatory gain knowledge and wisdom from the gathered data. The data gathering procedure of the Observatory and its storage are useless if there are no other procedures to generate knowledge from raw data. Large volumes of data do not provide knowledge and wisdom by itself, but data is the base of the taxonomy of knowledge (Zeleny, 1987), and that is why it is also the base of the Observatory's system. In order to act wisely, it is necessary to have knowledge, information and data about the tasks to be solved (Zeleny, 1987).

The Observatory's studies are supported by an information system implemented to accomplish the organization's main goals. As referenced before, the Observatory products, as well as the organization itself are data-driven (Patil & Mason, 2015). This means the information system built for the Observatory has a series of characteristics and components that facilitate the data collection, analysis, presentation and exploitation, as it will be explained below.

The first Observatory's study was released in 2015 and its goal was to collect and analyze information about degree students who graduated during the 2009-2010 course. Currently in 2017, the study is now in its second edition. The target of this second study are the university master's degree students (graduated in the 2013-2014 academic year).

However, the magnitude (a growing volume of data) and the continuity of these studies, as well as the broad vision of the Observatory, makes the traditional information systems not enough. As it will be detailed throughout this chapter, the challenges of the Observatory need the support of a more powerful system. That "powerful system" could be consider as a *data-driven* or *information-driven technological ecosystem*, because this kind of collaborative environment potentially fits the Observatory requirements.

A technological ecosystem can be seen as a set of different components connected through information flows in a physical environment that supports such flows (García-Holgado, García-Peñalvo, Hernández-García, & Llorens-Largo, 2015). Ecosystems make possible the provision of new and improved services that isolated tools or systems cannot be able to provide in order to solved knowledge and information management challenges inside any kind of institution or company (Garcia-Peñalvo & Garcia-Holgado, 2016). Moreover, technological ecosystems provide better support to the management of information and knowledge in heterogeneous environments (García-Holgado, Cruz-Benito, & García-Peñalvo, 2015; Garcia-Peñalvo & Garcia-Holgado, 2016).

In the case of the Observatory, the implementation of a technological ecosystem should be made considering a data-driven approach, resulting into a "data-driven ecosystem". This approach implies the incorporation of the characteristics of a technological ecosystem in addition to the data-driven philosophy. The purpose of a data-driven ecosystem is to create a collaborative environment to gather, process, analyze and disseminate data, in order to make evidence-based decisions by enabling the horizontal interaction of different users, stakeholders, systems and tools. These, among others, are the characteristics that make a data-driven ecosystem suitable for the Observatory.

In addition to the above, the data-driven ecosystem provides the adequate features to support the Observatory's data-as-a-service (DaaS) approach. This means that the components within the ecosystem themselves, as well as other users, can consume data from other components on-demand and also produce and communicating data to other components (by encouraging the independence between components, their decoupling and enabling interoperability (Terzo, Ruiu, Bucci, & Xhafa, 2013). According to this philosophy and working procedure, the data-driven ecosystem stablishes a data-as-a-service approach that enables the consumption of the Observatory's raw or processed data, independently of the platform or location of the consumers and only depending on the consumer's goals and the permissions available in the ecosystem.

In 2015, the first implementation of the Observatory's information system faced a series of technical issues with keeping in mind the meeting of requirements at that moment. Later, considering the vision of becoming a reference information source, was required to make important technological changes to accomplish it. As the OEEU evolves into a fully data-driven organization, it deals with a lot of difficulties regarding the collection and sharing of heterogeneous data from different information sources, as well as it experiences the challenges associated to knowledge management and scalability. These difficulties could be addressed through a collaborative environment like an information ecosystem that could allow the management of heterogeneous data and knowledge (Cruz-Benito, García-Peñalvo, & Therón, 2014;

García-Holgado, Cruz-Benito, et al., 2015) and could report higher levels of scalability because of the independence and low decoupling of the components involved in the ecosystem.

The purpose of this chapter is to outline the current Observatory's technological ecosystem, to describe the evolution process from a monolithic information system to a data-driven ecosystem and to present the benefits and results derived from this transition.

The content is organized following the next structure: section 2 provides a description of the different technical and organizational challenges associated to the OEEU mission and their system's goals; the third section presents the system's transition to a new version that improves previously identified issues and weakness, as well as the data structuration principles, the users involved and the technologies used; section 4 outlines the results of the Observatory and the outcomes of the designed data-driven ecosystem. Finally, the fifth section discusses the problems solved, the issues to be solved and the next challenges in the short to medium term, followed by the sixth and last section, where the conclusions of this work are presented.

2. THE PROBLEM

The Observatory of Employability and Employment faces a series of challenges and considerations to perform its work and carry out its mission.

The first challenge faced to become an information source reference are the design of data collection and data processing procedures. All the organization's purposes and products are backed by data, so data can be seen as the backbone of the Observatory. However, the possession and storage of large volumes of data are not enough to provide important outcomes. Data Analytics techniques are necessary to make these datasets valuable and to reach insights (LaValle, Lesser, Shockley, Hopkins, & Kruschwitz, 2011) regarding the employment landscape in Spain (in this case). In brief, the Observatory needs to be able to properly support the data collection and its analysis, i.e. having efficient and automatic gathering and computation methods that enable people to make evidence-based decisions regarding employability and employment.

Nevertheless, having large volumes of data brings great responsibilities, even more if the data is considered as sensitive. The main sources of information are Spanish universities and their students; and universities, as data providers and customers of the insights provided by the Observatory, rely on the organization to keep their (and their students') data safe and anonymized. If the Observatory does not meet these conditions, universities will not trust in its projects and they will not want to be involved in them (potentially leading to the project's death by *data-inanition*).

Considering these conditions, data should be collected and stored complying the Spanish Personal Data Protection Act – generally referred to as LOPD (Ley Orgánica 15/1999, 1999). Regarding the anonymization and the Spanish law, the Observatory cannot allow the identification of individuals and universities behind the results of the study, as they will be exposed to the general audience. Anonymization and encryption techniques are needed to stick to these requirements.

In addition to the foregoing, the amount of data managed by the Observatory follows a growth trend. Nowadays, it handles more than 700 variables related more that 180000 students (these amounts are an addition of the variables and students involved in the first and second study performed by the Observatory). This is a challenge to consider, because the success of the project and the organization itself mainly remains in the capacity of handling this information in a proper way. The Observatory should keep developing its work without became flooded by the amount of data held, and this require high levels of scalability. Scalability could be a difficult issue to manage in any organization or (eco)system, but the benefits in the medium-long term are worth it.

Finally, the Observatory needs to spread the results once completed the different studies carried out. Given the importance of the employment and employability fields for the society, it is necessary to show and exploit the knowledge produced by the organization's activities in a proper way to reach all kind of audiences and stakeholders. This knowledge must be disseminated to anyone that aims to understand the Observatory's mission as well as its studies' results. Visualization and interaction methods can help solving this challenge, as it will be shown below.

Moreover, not only the society, but another similar organizations could take advantage of the Observatory results and knowledge bank (the knowledge and extra information that support the Observatory's activities). In this case, the communication methods for this kind of outcomes are also essential. These issues will be addressed in a deeper way in the following sections.

3. TRANSITION TO A DATA-DRIVEN TECHNOLOGICAL ECOSYSTEM

3.1 The Transition Itself

The Observatory owns a significant amount of information needed to achieve the goals of analyzing and understanding the factors involved in to became more employable (employability) or in the fact of get a job (employment). Regarding that, it is necessary to build the Observatory's system focusing on the effective use of large volumes of data. Not only that, but also being capable of spread and disseminate the collected information and knowledge among the system's users as well as to other external systems, so they can interact with the information collected and processed by the Observatory to reach deeper insights about employability and employment. It is important to consider that these users will have different requirements and the Observatory needs to support all of them.

After the first implementation of the Observatory's information system in 2015, its main features and functionality were the following (Michavila et al., 2015):

- **Data Collector:** This feature allowed the data gathering from different sources. The two main information sources are described below:
  - **Universities:** The participant universities provided administrative data about their students. To accomplish the storage of the administrative data, the system allowed the universities to upload CSV files containing the information required. When uploaded, the files were validated -in a semi-automatic way- by the Observatory's staff before its definitive storage, fixing the possible errors incorporated in the initial files.
  - **Students:** The system included custom questionnaires based on the administrative data collected from the universities. The questionnaires were the method used to collect employability and employment information about the graduates involved in the study. The answers given by the students were stored in the system and completed the universities' administrative data previously described, as well as the Observatory's knowledge bank, obtaining a larger and more interesting dataset to analyze.

- **Data Analyzer:** The main goal of this functionality was to help the analysis processes. It provided automated methods to obtain results from the gathered data. The data analyzer retrieved records from the persistent storage and processed (in a batch way) it to obtain simple statistics, derived variables and other results.
- **Interoperability:** The first implementation of the system was not prepared to support high interoperability levels between its components. On the other hand, there were neither methods to exchange information between the Observatory's system and other information systems, as it was not a priority at that time.
- **Information Dashboards:** Other main feature of the Observatory's system, supported by the data collector and the data analyzer, was the implementation of information dashboards. These dashboards allow the different users of the system to have a broad view of the results of the study through information visualizations. To make this possible, the dashboards were implemented following some basic Visual Analytics and Data Visualization principles, concepts and techniques (Keim et al., 2008; Keim, Kohlhammer, Ellis, & Mansmann, 2010). On the other hand, universities also had access to a basic dashboard with information tables to keep track of their own student's participation and other basic results.

These features satisfied the requirements at that moment (2015), giving the system the capacities to support the first edition of the Observatory's study with a simple administrative data collector and customized questionnaires for the students, as well as dashboards to show the general results of the study.

One of the principal issues were the performance of the data analysis, which made the user experience less satisfactory. Although pre-calculated statistics were used to reduce the response times in the public portal and dashboards, this solution was not flexible and, even more, it was not scalable.

As explained before, the public portal held a significant amount of visualizations representing the general results of the study, which provided general employment and employability information to anyone who accessed the results' section. However, universities were not able to have these visualizations and individual results applied to their own administrative and graduates' data. They only had the possibility of downloading the data from the system and analyze it on their own. This was an important issue and a valuable feature to be added in the future, but the performance of the data analyzer and the low levels of interoperability between the system's components made it not an easily addressable challenge.

The absence of high levels of interoperability was also a problem due to the organization's mission of analyze the data gathered and share the produced knowledge to external stakeholders. Increasing these interoperability levels could allow other information systems to take advantage of the Observatory's data, creating a collaborative knowledge environment and reporting more benefits to the project and society.

After analyzing all the issues found behind the first implementation of the Observatory's system, it was clear that the system must be upgraded to prepare the environment for future studies and functionalities. This was when the transition to a data-driven ecosystem began, considering the similarities between the Observatory's requirements and the problems addressed by this approach.

It is important to consider the challenges and priorities behind the construction of a digital ecosystem focused in knowledge discovery through data analysis and exploitation in order to obtain valuable insights from the domain data (Chajri & Fakir, 2014; Touya & Fakir, 2014). These priorities especially involve the interoperability between the series of components within the ecosystem, as well as the flexibility to be used by different stakeholders, and the scalability regarding to data management and data analysis (Garcia-Peñalvo & Garcia-Holgado, 2016). As referenced before, if high levels of interoperability are achieved, the components could able to evolve individually and collectively, and they even could add or change their own functionalities without affecting the rest of the components. Moreover, additional components, both internal or external, could be connected to build up a broader technological ecosystem without major difficulties in a transparent way (García-Holgado & García-Peñalvo, 2014a, 2014b).

Those characteristics met the Observatory's philosophy; employment and university employability fields are constantly evolving, and the Observatory needs to adapt their studies to get the maximum amount of information and analyze it in the best possible way. The Observatory needs to collect, process, analyze and present data, and in order to support those activities, as presented above, a data collector, data analyzer and data dashboards functionalities have been implemented through a series of software components. These components consume information from the persistent storage of the Observatory and also consume data generated by each one of them. It is clear that every component should be independent and need to have well-defined tasks, but also they must collaborate to achieve the goals of the whole system. Information flows between the components are important in order to reach the collaboration of all them, and promoting high interoperability levels is a common solution to decouple the components while they continue collaborating in a transparent way.

For all these reasons, it has been concluded that the data-driven ecosystem approach would report very important benefits to the Observatory system purposes. By implementing a data-driven ecosystem, the Observatory's technological environment can evolve following new requirements and providing new functionalities or services by the addition or modification of components, thanks to the high cohesion of the components but also to their loose coupling.

Taking all these points into account, the system has been upgraded in order to begin a transition to a data-driven ecosystem and accomplish the Observatory's new requirements due next studies' editions, as well as to fix other issues identified after the 2015 study and with the authors' previous work in software architectures that gather and manage data in different contexts (Cruz-Benito, Borrás-Gené, García-Peñalvo, Fidalgo Blanco, & Therón, 2015; Cruz-Benito, García-Peñalvo, et al., 2014; J. Cruz-Benito et al., 2016; Cruz-Benito, Therón, et al., 2014; Cruz-Benito, Therón, & García-Peñalvo, 2016; Cruz-Benito, Therón, García-Peñalvo, & Pizarro Lucas, 2015; García-Peñalvo, Cruz-Benito, Maderuelo, Pérez-Blanco, & Martín-Suárez, 2014; García-Sánchez, Cruz-Benito, Therón, & Gómez-Isla, 2015).

This upgrade took place during the development and beginning of the second study conducted by the Observatory (2016-2017). Among the different feature additions and modifications, the following could be highlighted:

- **Data Collector:** The component regarding the data collection has been upgraded to solve the issues encountered after the first implementation of the system. Despite the technical changes, the information sources have not varied:
    - **Universities:** As in the 2015 edition, the data gathering starts with universities sending its administrative information. One of the issues of that first edition was the need of manual validations by the Observatory's staff before considering the data as correct. This manual validation procedure slowed down the gathering process, since it was not automatized and did not give guarantees of the proper correction of all the errors found. For that reason, the administrative data collector has been upgraded to include a fully-automated data validation stage before the information is persistently stored into the system. The validation stage avoids manual checking procedures by analyzing the CSV files just after the universities have uploaded it, and presenting them a detailed report containing the exceptions and its

explanations. Universities are responsible of fixing the errors found and upload their new and corrected files afterwards. Once the system receives a file without any errors, the data passes another stage of cleaning and structuring to keep the Observatory's bank of information organized, as it will be briefly explained in the next subsection.

  - **Students:** The student's data collector has also experienced an upgrade. Rather than fix issues from the first edition, some changes had been added in order to make the current component more valuable. The data is collected (again) through questionnaires, but one of the main upgrades includes the improvement of the questionnaires themselves by creating different versions that will be displayed depending on the student that fills the questionnaire (some kind of adaptivity), in pursuance of an increase of the completion rate. Another upgrade involves the collection of variables generated from the student's interaction with the questionnaires (technically referred to as *paradata* (Stieger & Reips, 2010). These adaptive questionnaires themselves could lead to future research about this topic (research using A/B methodologies, etc.) (Cruz-Benito et al., 2017). As outlined in the previous bullet, the students' answers are also structured and cleaned before saving it permanently.

- **Data Analyzer:** The data analyzer has been modified to increase its performance and support more complex operations. The data retrieval process is made through the same methods as in the old version, against the Observatory's database, excepting the data calculations are performed on-demand (and not in a batch way) due to the increased efficiency through an in-memory analysis approach. Once the data is retrieved, it is loaded into memory, where the processing and analysis take place. In-memory analysis allows the datasets to be processed more efficiently, reporting significant increases of performance (Zhang, Chen, Ooi, Tan, & Zhang, 2015). This enhancement opens the capacities of the whole system, as most components could consume information generated by the data analyzer.
- **Interoperability:** A data-driven ecosystem should be able to exchange and exploit data along the components that form part of it. Interoperability reports several benefits to the Observatory's system and improve scalability, communication and collaboration between its elements. To achieve these goals, an interoperability component has been introduced to the system. This component was implemented by the incorporation of a REST (Fielding & Taylor, 2002) API (Application Programming Interface). Different data flows can be found within (and against) the system, so the API is designed to contemplate and give support to all of them. The design of this API brings to the Observatory's database isolation properties, as transactions won't interfere with each other considering the secure behavior of the API requests. This component allows the data to be available for users and other components through network, making a data-as-a-service approach possible (Terzo et al., 2013). It also allows the creation of methods for the users to download their entire datasets in different open format files (like CSV, etc.) on-demand.
- **Information Dashboards:** In general, the system's information dashboards and visualizations are the main beneficiaries of the transition to a data-driven ecosystem. The poor performance level of the older version of the system led to difficulties regarding the provision of individual dashboards to each participant university. The enhancement of the communication between components and the improvement of the data analyzer made possible the sending of real-time data to the information dashboards (as the data is retrieved and processed on demand). In addition, the good performance of this procedure makes the user experience more satisfying.

The backbone of the transition from a monolithic system to a data-driven ecosystem is the interoperability component, not only because it can be seen as a nexus between several information (eco)systems, but also because it made possible the scalability of the information dashboards and other components of the system. This interoperability component enables the information dashboards to consume processed data from the data analyzer component, decoupling them from other system's components and supporting the data-as-a-service approach. Having high decoupling levels facilitates the evolution of every component individually, without interfering with each other, but also collectively, meeting the characteristics of a technological ecosystem (García-Holgado & García-Peñalvo, 2014a, 2014b; García-Holgado & García-Peñalvo, 2014a, 2014b).

All the changes and functionalities introduced in this transition are resumed in *Table 1*, where the 2015 information system components and characteristics are compared with the 2017 data-driven technological ecosystem.

Finally, Figure 1 shows the conceptual view of the Observatory's data-driven ecosystem.

To complete the transition overview, the next subsections briefly describe the data structuration referenced before, the system's users, and the different technologies involved in the new ecosystem.

3.2 Data Structuration

Considering that the data is the driver element of the system, it is required to organize and structure it to perform its processing and analysis properly, reporting benefits for all the Observatory consumers. This structuration procedure is transparent for the users and it takes place at the system's backend.

Table 1. Summary of changes in the transition from a standard information system to a data-driven technological ecosystem

	OEEU's 2015 Information System	OEEU's 2017 Data-Driven Ecosystem
Data collector	1. Procedures to collect administrative data from universities. 2. Subsystem to generate questionnaires and collect data from the students' answers.	1. New stages in the gathering procedures: on-demand data validation and data cleaning. 2. Subsystem to generate questionnaires and different versions of them that could be displayed depending on the student profile that participates (supporting A/B testing). 3. Collection of *paradata* from the students involved in the study.
Data analyzer	1. Low levels of performance 2. Simple analysis	1. Increase of performance due to the in-memory data computation 2. Support of more complex analysis
Interoperability	Not contemplated in requirements.	Addition of an interoperability component to communicate the different components within the ecosystem and external systems or users through data flows, following the data-as-a-service approach.
Information dashboards	1. Visualizations for general results in the public website and personalized data tables for each universities' results in the system intranet. 2. Pre-calculated results. 3. Coupling between presentation and analysis components.	1. Adaptation of the visualizations and general dashboards to show each universities' results in their private dashboards. 2. Results calculated on-demand based on the request. 3. Higher levels of decoupling due to the interoperability component developed.

Figure 1. OEEU's ecosystem's components and relations

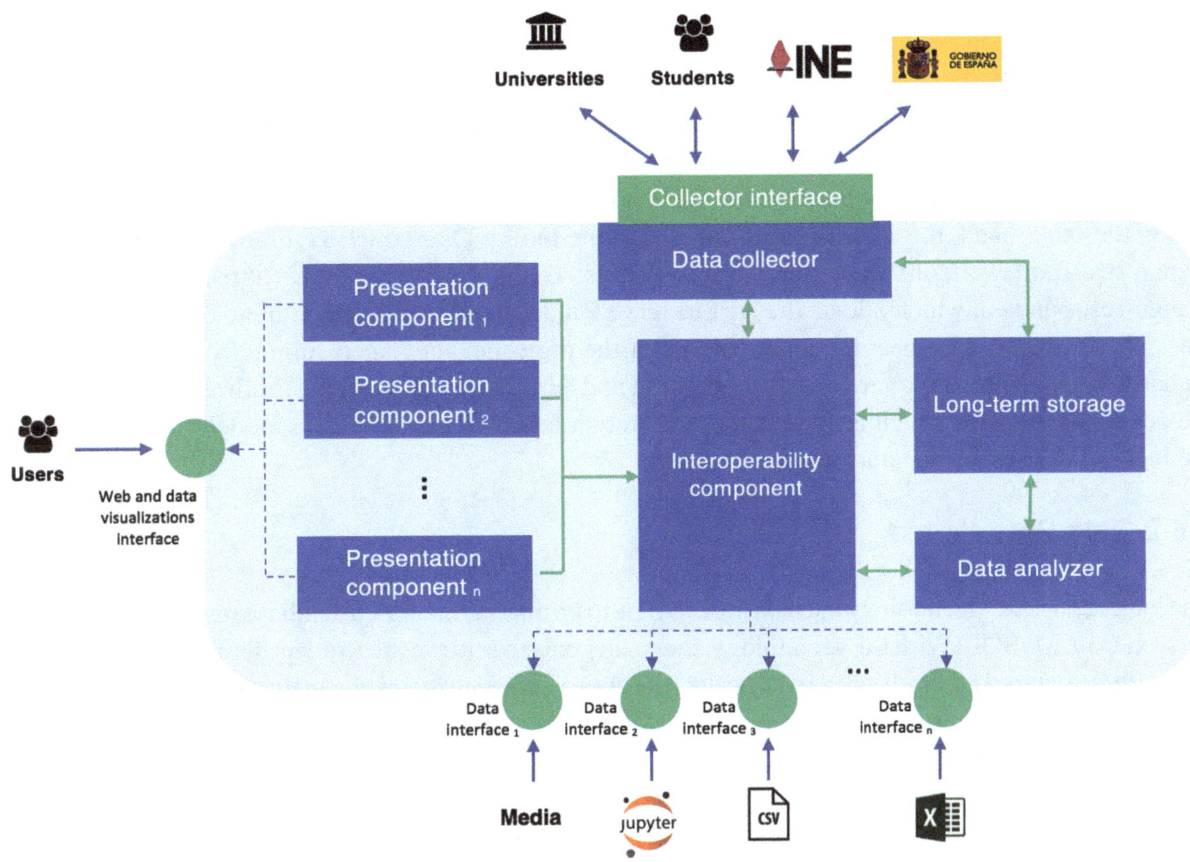

The principles followed to accomplish the data organization and structuration are technically named *tidy data* principles (Wickham, 2014). The Observatory's bank of information is organized in such a manner that, as Wickham points out, each variable forms a column, each observation forms a row and each type of observational unit forms a table. The raw data gathered from the different information sources is cleaned before its storage into the system's persistence layer sticking to these rules.

The principal benefit is the avoidance of messy datasets and the need of additional computations each time information is requested. For that last reason, tidy datasets also report performance gains as these unnecessary pre-processing operations are omitted.

3.3 System Users

There are a series of users of the Observatory's system, categorized by their needs. These users can be common users of the Observatory's public website (general audience), the graduates involved in the Observatory's studies, the Spanish universities that participate in the data gathering and consuming the knowledge and information generated, the data analysts (both external or internal to the organization) and other information systems that take advantage of the Observatory's components and information.

3.4 Technologies Used

3.4.1 Core System

The system has been implemented on Django, a Python web framework (Django Software Foundation, 2015; Holovaty & Kaplan-Moss, 2009). The election of Django has been backed up by the simplicity, flexibility and the community behind this framework.

On the other hand, the API has been built with the toolkit Django REST Framework (http://www.django-rest-framework.org/) and one of its variants: Django REST Pandas (https://github.com/wq/django-rest-pandas), which allows the API to serve Pandas dataframes (McKinney, 2011) to the client side. These frameworks meet the requirements of the project as they keep simplicity and good performance levels and provide a series of authorization and authentication methods for limiting and securing the access to the API, which is composed by different layers to keep the data model, the security and the logic of the system separate from each other.

3.4.2 Data Persistence

The long persistence technology used is MariaDB (https://mariadb.com/), a database management system derived from MySQL. With this technology, the main goals to achieve are to keep all of the Observatory's data safe, structured and well-organized, as well as obtain satisfactory response times on query processes.

3.4.3 Data Processing and Analysis

As already outlined above, the data analysis and processing take place at the backend of the system. These backend computations are made through Python Pandas (McKinney, 2011) because it allows in memory column analysis, reporting huge performance benefits.

All the computed data served by the Observatory's new engine is processed before sending it to the user who requested it, avoiding additional computational load on the user's side. Using this method, the system's information dashboards can show real-time data, unlike in the previous version of the system, where the statistics were pre-calculated, leading to flexibility and scalability problems.

3.4.4 Data Visualizations

The information dashboards have been built using D3, a Javascript library for visualizing data (Bostock, Ogievetsky, & Heer, 2011). D3 provide the appropriate methods to visualize the Observatory's bank of information, accomplishing the requirements and necessities of the users.

3.4.5 Data Exploitation

The interoperability component makes easy the external data retrieval and data exploitation through different tools independent of the system.

For example, data analysts can retrieve information from the Observatory's API and process the raw data using external tools like Jupyter notebooks (Kluyver et al., 2016), which could be very useful for publishing results, considering their capacity to show live code and outcomes even without being executed and the reproducibility feature it provides regarding to data analysis processes. These notebooks take advantage of the Observatory's API, facilitating the data retrieval process and hiding details of the system's persistent storage.

The ecosystem also has methods to allow universities to retrieve their students' data in CSV format, in order to use the downloaded files in other external tools like IBM SPSS, Microsoft Excel, etc.

The technologies involved in the Observatory's ecosystem are summarized in Figure 2.

4. RESULTS

4.1 Observatory's Results

The Observatory for University Employability and Employment has been running since 2013, and started to collect degree student's information between the end of 2014 and the beginning of 2015, merging both

Figure 2. Summary of components and technologies that compose the Observatory's data-driven ecosystem

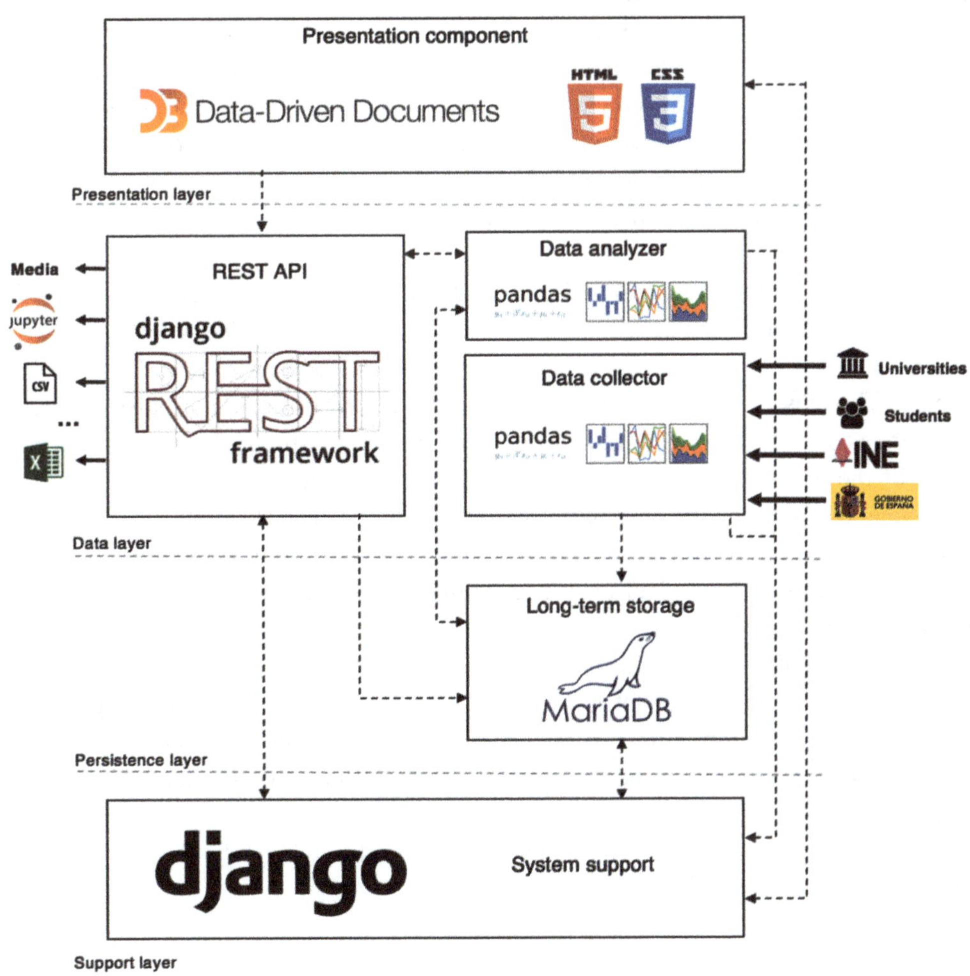

universities' administrative data and students' information collected through questionnaires. This data gathering corresponds to students that graduated in bachelor level in the 2009-2010 course.

At the end of 2016, a new period for data collection started, beginning with universities sending administrative information about students who finished their master's degree in the 2013-2014 academic year. As well as in the 2015 study, another data gathering process through questionnaires began in the early months of 2017, complementing the administrative information from universities.

The following results were reached during the Observatory's 2015 study (Michavila et al., 2016):

- The Observatory involved 49 Spanish universities, both public and private ones.
- The administrative data provided from all of these 49 universities reached at the end of the data collection stage 134,129 records about graduated students in the 2009-2010 season. Information about gender, sex, nationality, parents' studies, place where students live, average score in studies, mobility, internships, etc. are included among the total of data variables given (35 altogether).
- Moreover, of these 134,129 students, 13,006 started to answer their questionnaires, which is the 9,70% of the total registered students expressed as a percentage. Finally, 9,617 of them finished it (7,17% of the students in this case), giving a finalization rate of the questionnaires of 73,94%. The information provided by the completed questionnaires include around 400 variables regarding studies and employment. These variables keep information about the type of employment, relation with their jobs and their studies, wage, competencies, other studies done, opinions about the university and teaching methods, ratings about their studies, languages spoken, etc.
- Also, the Observatory keeps a knowledge bank with information related to Spanish degrees, information about branches of knowledge, as well as countries' data (economic, demographic, etc.), fields of employment and other relevant information about companies or employers. This information has been organized by the Observatory and can be useful to reach wider results data analysis.

On the other hand, the Observatory's 2017 study has reached other promising results during these last months of data gathering:

- In this case, The Observatory has involved 51 Spanish universities so far, also public and private. The difference on the number of participants between the two studies has its explanation in the fact that some of the universities that collaborated in the first study had decided not to participate in this last one and others that did not participate were enrolled in this new study. In the new study, universities added administrative data about its own master's graduated students.
- This time, the administrative data provided from these 51 universities involved 47,822 records about master students.
- Of these 47,822 students, approximately 6700 started questionnaires and 5200 finished them, making the questionnaire's finalization rate reach the more than 77%. The completion of the master's degree (MsC) questionnaires provides more than 200 variables related, again, ranging from information about their jobs, studies, competencies and other employment and university studies attributes.
- As in the 2015 study, variables given by the Observatory knowledge bank are used to allow deeper data analysis.

4.2 Results Regarding the System's Evolution

Regarding the results of the Observatory's information ecosystem itself, including the transition to a more powerful architecture supported by the concept of digital ecosystems and data-driven culture, the technological achievements reached are remarkable:

- The Observatory provides tools to help the universities to send its administrative information, as well as validation methods that give real-time feedback about possible errors within uploaded files.
- The Observatory has a system to generate personalized questionnaires for the involved students. The personalization is based on the student responses and profile. The personalization goes from the questionnaires' style to its interaction methods, among other changes. It also has procedures to gather information about the students' interaction with the questionnaires that could lead to future researches about usability and adaptivity.
- The Observatory has implemented components to analyze the collected data on-demand, achieving low response times even when computing large datasets.
- Finally, regarding the data presentation, the Observatory has a system for data visualization that handles the information for different stakeholders (universities, students or staff from the Observatory itself, among others). These presentations vary depending on the stakeholder.
    - The staff from the Observatory can access to all the data collected in order to obtain general results, as well as to manage individual universities' results or to monitor the questionnaires participation data.
    - Universities have access to personalized dashboards with information tables and visualizations adapted to its own data. This allow universities to consult effortless their results based on its administrative information and its students' answers.
    - The Observatory also provides public websites (http://datos.oeeu.org) that show the different studies overall results in an open way for any user. The websites make possible to disseminate all the knowledge discovered by the system.

The Figure 3 and Figure 4 show examples of the main Observatory's dashboards and its data visualizations, respectively. As explained before, these dashboards are also available for each participant University to check their own results.

5. DISCUSSION

Due to the upgrade of the Observatory's system, several issues associated to the complexity of the goals and technical challenges have been solved.

The Observatory's system has experienced an important performance increase in server's response, data analysis and data presentation, as this last one feature consumes data from the data analyzer. Since the nature of the project and the Observatory itself implies a continuous growth of the data amount, this was a priority issue to be resolved to avoid future problems regarding response and computation times.

Figure 3. Example of the administrative dashboard for the Observatory's staff users (contents in Spanish)

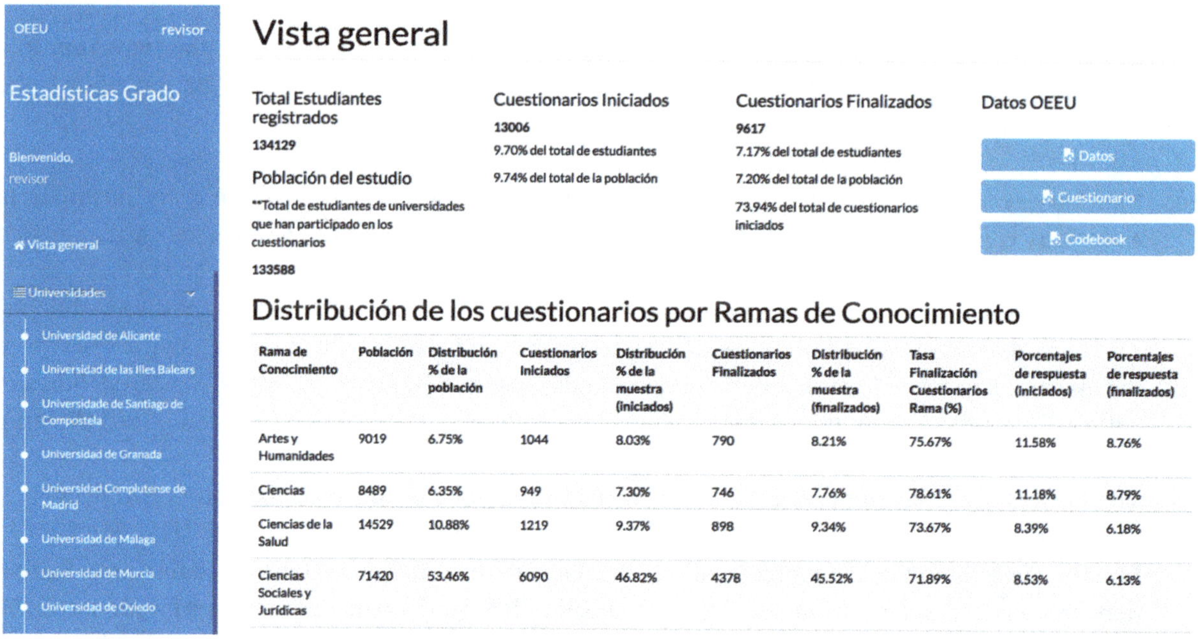

Figure 4. Example of information dashboards for the Observatory's data (contents in Spanish)

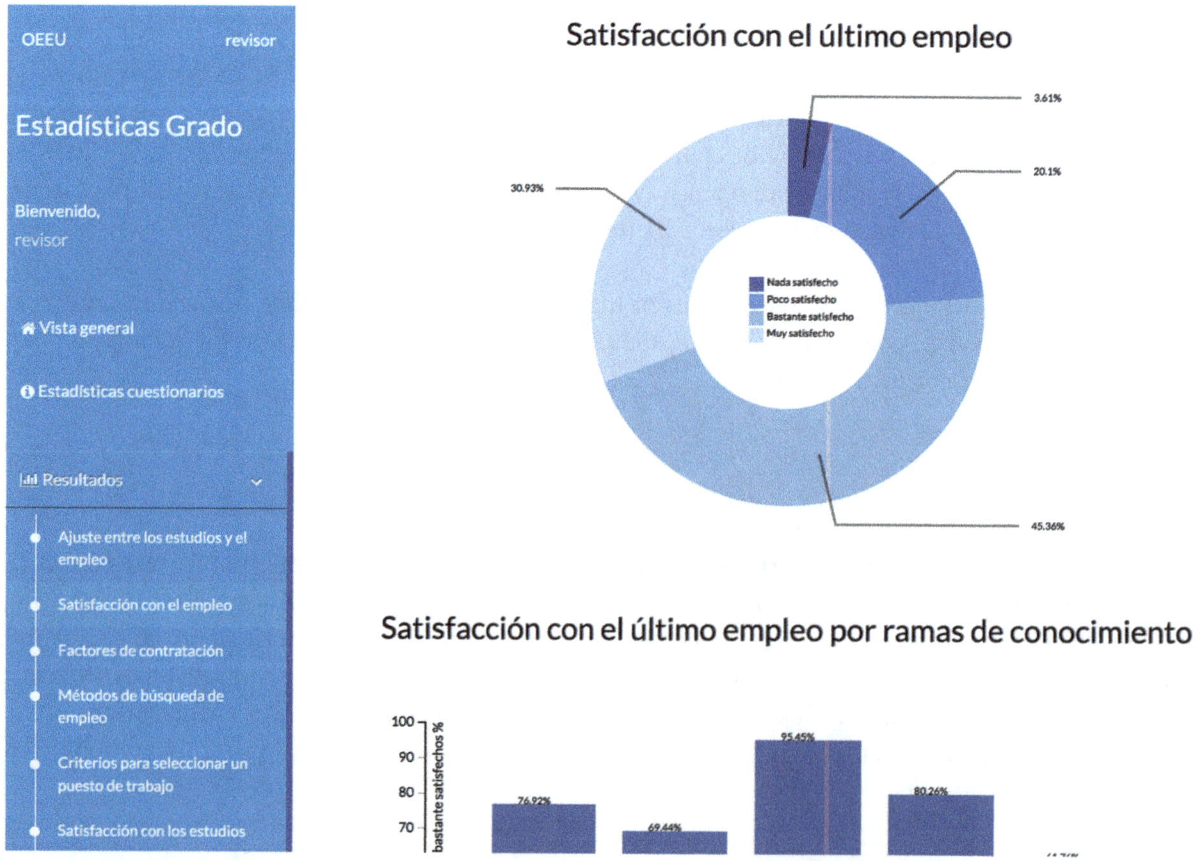

On the other hand, the data validation stage in the administrative data collector has provided great results, accelerating the first phase of the study and giving real-time feedback to the universities about the validity of their data. Before introducing an automatic data validator, manual fixes were essential and critical, because erroneous data could introduce noise in the studies. This manual validation consumed notable human and time resources from the Observatory, and it even could not be enough to fix all errors. Having an automatic validation stage, the errors are showed directly to the universities, making them responsible of the fixes and unburdening the Observatory from this task.

However, the main results come from the increase of interoperability of the system and the transition to a data-driven ecosystem. As it was introduced at the beginning of this chapter, the Observatory deals with high amounts of data coming from different sources, being the data the backbone of the system and the basis to make evidence-based decisions. The components within the system need to communicate with each other in a proper way to exchange the results of their tasks, but they also need to maintain high levels of independence, following technological ecosystems' guidelines and creating a data-as-a-service environment.

The data-driven ecosystem implemented allow software components to evolve individually and collectively, that made possible the addition of new functionalities and features without major difficulties. It reports remarkable benefits regarding communication between components and it reduces integration issues. Other of the results given by the interoperability and independence of the components is the enhancement of the information dashboards and the communication with third-party tools and systems. Data flows between visualizations and other components of the ecosystem enable dashboards to show more complex information, and even filter it depending on the user.

Although all the problems encountered after the first edition of the study have been covered, the needs of the Observatory are not static, and they are evolving continuously. The data-driven ecosystem implemented is a good basis to begin addressing more ambitious challenges. The decoupling of the components, especially the decoupling between the data analyzer and the dashboards opens many doors regarding information visualizations. Regarding that, some next challenges could include the evolution of the API responses to a graph approach for complex data (Vázquez-Ingelmo, Cruz-Benito, & García-Peñalvo, 2017), like it is being developed by main organizations as Facebook (i.e. with the GraphQL project https://facebook.github.io/graphql/).

One of the next challenges for the medium- or long-term is to develop linked data views for the visualization on the information dashboards. The philosophy behind linked visualizations lies in the creation of several simpler views (instead of creating one complex view) and their linkage, so that when the user interacts with one view, the other views will update and show the results of such an interaction (Wills, 2008). The Observatory owns a lot of information collected in each study, and it is difficult to reach valuable insights without the support reached through visualizations and interaction. Linked data views would make easier for the users to understand the Observatory's results. Also, could not be discarded the addition of complex data visualizations for the most complex problems regarding the information kept by the Observatory (multidimensional problems, etc.).

The data-driven ecosystem approach also allows the dissemination of information and knowledge gained by the Observatory among other external systems. Media, journals, data analysts etc., could link their own systems or tools to the Observatory's ecosystem to obtain results and insights of employment and university employability.

6. CONCLUSION

The Spanish Observatory for University Employability and Employment is generating, analyzing and disseminating information about the employability and employment of university graduates in Spain through a unified methodology. To support the data collection, as well as its analysis and dissemination, an information system was built in 2015 to accomplish the Observatory's goals and technological requirements. However, giving the challenges of the Observatory regarding its mission and information requirements, the traditional information system built has been transformed into a data-driven ecosystem. As it was presented throughout this chapter, this data-driven technological ecosystem presents a collaborative environment between software components and users that supports adequately the Observatory's vision and mission.

Technological ecosystems empower this kind of collaborative environments, enabling the connection of different components through data flows, resulting in a series of individual components with well-defined task working together to achieve the ecosystem's goals. They also allow the modification and update of components without affecting the rest of the ecosystem's elements, and makes easier the creation of new data flows, both internal and external to the ecosystem. On the other hand, data-driven philosophy focuses the attention on data in order to support decision making.

The combination of these two concepts make a data-driven ecosystem a powerful and suitable approach for the Observatory. This combination matches well the data-as-a-service approach implemented in the project, due to the interoperability, independence and decoupling gained through the transition to a data-driven ecosystem, allowing data to be served on-demand to different components and users.

With its data-driven ecosystem, the Observatory is now able to discover and understand better the variables that influence the employment and employability in graduate students. It also promotes the creation of data flows to disseminate the knowledge and wisdom gained during the analysis to internal and external users, different stakeholders and third-party information systems or tools, as well as it supports the evolution of its own components to meet new requirements.

ACKNOWLEDGMENT

The research leading to these results has received funding from "la Caixa" Foundation. Also, the author Juan Cruz-Benito would like to thank the European Social Fund and the *Consejería de Educación* of the *Junta de Castilla y León* (Spain) for funding his predoctoral fellow contract. This work has been partially funded by the Spanish Government Ministry of Economy and Competitiveness throughout the DEFINES project (Ref. TIN2016-80172-R).

REFERENCES

Bostock, M., Ogievetsky, V., & Heer, J. (2011). D³ data-driven documents. *IEEE Transactions on Visualization and Computer Graphics*, *17*(12), 2301–2309. doi:10.1109/TVCG.2011.185 PMID:22034350

Chajri, M., & Fakir, M. (2014). Application of Data Mining in e-Commerce. *Journal of Information Technology Research*, *7*(4), 79–91. doi:10.4018/jitr.2014100106

Cruz-Benito, J., Borrás-Gené, O., García-Peñalvo, F. J., Fidalgo Blanco, Á., & Therón, R. (2015). Extending MOOC ecosystems using web services and software architectures *Proceedings of the XVI International Conference on Human Computer Interaction* (pp. 1-7). Vilanova i la Geltrú, Spain: ACM. doi:10.1145/2829875.2829923

Cruz-Benito, J., García-Peñalvo, F. J., & Therón, R. (2014). *Defining generic data collectors for Learning Analytics: Facing up the heterogeneous data from heterogeneous environments.* Paper presented at the International Conference on Advanced Learning Technologies (ICALT) 2014, Athens, Greece. doi:10.1109/ICALT.2014.108

Cruz-Benito, J., Maderuelo, C., García-Peñalvo, F. J., Therón, R., Pérez-Blanco, J. S., Gómez, H. Z., & Martín-Suárez, A. (2016). Usalpharma: A Software Architecture to Support Learning in Virtual Worlds. *IEEE Revista Iberoamericana de Tecnologias del Aprendizaje, 11*(3), 194–204. doi:10.1109/RITA.2016.2589719

Cruz-Benito, J., Therón, R., & García-Peñalvo, F. J. (2016). Software Architectures Supporting Human-Computer Interaction Analysis: A Literature Review. In P. Zaphiris & A. Ioannou (Eds.), *Learning and Collaboration Technologies: Third International Conference, LCT 2016, Held as Part of HCI International 2016, Toronto, ON, Canada, July 17-22, 2016, Proceedings* (pp. 125-136). Cham: Springer International Publishing. doi:10.1007/978-3-319-39483-1_12

Cruz-Benito, J., Therón, R., García Peñalvo, F. J., Maderuelo, C., Pérez-Blanco, J. S., Zazo, H., & Martin-Suarez, A. (2014). Monitoring and feedback of Learning Processes in Virtual Worlds through analytics architectures: A real case. In Á. Rocha, D. Fonseca, E. Redondo, L. P. Reis, & M. P. Cota (Eds.), Sistemas y Tecnologías de Información. Actas de la 9ª Conferencia Ibérica de Sistemas y Tecnologías de Información (Vol. I Artículos, pp. 1126-1131). Barcelona, España: AISTI (Asociación Ibérica de Sistemas y Tecnologías de Información). doi:10.1109/CISTI.2014.6877097

Cruz-Benito, J., Therón, R., García-Peñalvo, F. J., & Pizarro Lucas, E. (2015). Discovering usage behaviors and engagement in an Educational Virtual World. *Computers in Human Behavior, 47*(0), 18–25. doi:10.1016/j.chb.2014.11.028

Cruz-Benito, J., Therón, R., García-Peñalvo, F. J., Sánchez-Prieto, J. C., Vázquez-Ingelmo, A., Martín-González, M., & Martínez, J. M. (2017). Improving Success/Completion Ratio in Large Surveys: A Proposal Based on Usability and Engagement. In P. Zaphiris & A. Ioannou (Eds.), *Learning and Collaboration Technologies. Technology in Education: 4th International Conference, LCT 2017, Held as Part of HCI International 2017, Vancouver, BC, Canada, July 9-14, 2017, Proceedings, Part II* (pp. 352-370). Cham: Springer International Publishing. doi:10.1007/978-3-319-58515-4_28

Django Software Foundation. (2015). *Django Web Framework.* Retrieved from https://www.django-project.com/

Fielding, R. T., & Taylor, R. N. (2002). Principled design of the modern Web architecture. *ACM Transactions on Internet Technology, 2*(2), 115–150. doi:10.1145/514183.514185

García-Holgado, A., Cruz-Benito, J., & García-Peñalvo, F. J. (2015). *Analysis of Knowledge Management Experiences in Spanish Public Administration.* Paper presented at the Third International Conference on Technological Ecosystems for Enhancing Multiculturality (TEEM'15), Porto, Portugal. doi:10.1145/2808580.2808609

García-Holgado, A., & García-Peñalvo, F. J. (2014a). *Architectural pattern for the definition of eLearning ecosystems based on Open Source developments.* Paper presented at the Computers in Education (SIIE), 2014 International Symposium on. doi:10.1109/SIIE.2014.7017711

García-Holgado, A., & García-Peñalvo, F. J. (2014b). *Patrón arquitectónico para la definición de ecosistemas de eLearning basados en desarrollos open source.* Academic Press.

García-Holgado, A., García-Peñalvo, F. J., Hernández-García, Á., & Llorens-Largo, F. (2015). *Analysis and improvement of knowledge management processes in organizations using the Business Process Model Notation.* Paper presented at the Annual Conference of the Global Innovation and Knowledge Academy. doi:10.1007/978-3-319-22204-2_9

García-Peñalvo, F. J., Cruz-Benito, J., Maderuelo, C., Pérez-Blanco, J. S., & Martín-Suárez, A. (2014). Usalpharma: A Cloud-Based Architecture to Support Quality Assurance Training Processes in Health Area Using Virtual Worlds. *The Scientific World Journal, 10.* doi:10.1155/2014/659364 PMID:24778593

Garcia-Peñalvo, F. J., & Garcia-Holgado, A. (2016). *Open Source Solutions for Knowledge Management and Technological Ecosystems.* IGI Global.

García-Sánchez, F., Cruz-Benito, J., Therón, R., & Gómez-Isla, J. (2015). *Designing and building systems and tools to analyze visual communications on social networks.* Paper presented at the Third International Conference on Technological Ecosystems for Enhancing Multiculturality (TEEM'15), Porto, Portugal. doi:10.1145/2808580.2808629

Holovaty, A., & Kaplan-Moss, J. (2009). *The definitive guide to Django: Web development done right.* Apress. doi:10.1007/978-1-4302-1937-8

Keim, D., Andrienko, G., Fekete, J.-D., Görg, C., Kohlhammer, J., & Melançon, G. (2008). Visual analytics: Definition, process, and challenges. *Information Visualization*, 154–175.

Keim, D., Kohlhammer, J., Ellis, G., & Mansmann, F. (2010). *Mastering the information age solving problems with visual analytics.* Eurographics Association.

Kluyver, T., Ragan-Kelley, B., Pérez, F., Granger, B., Bussonnier, M., Frederic, J., . . . Corlay, S. (2016). Jupyter Notebooks—a publishing format for reproducible computational workflows. *Positioning and Power in Academic Publishing: Players, Agents and Agendas*, 87.

LaValle, S., Lesser, E., Shockley, R., Hopkins, M. S., & Kruschwitz, N. (2011). Big data, analytics and the path from insights to value. *MIT Sloan Management Review, 52*(2), 21.

Ley Orgánica 15/1999. (1999). Ley Orgánica 15/1999, de 13 de diciembre, de Protección de Datos de Carácter Personal. *Boletín Oficial Del Estado, 298*(2), 43088-43099.

McKinney, W. (2011). pandas: a foundational Python library for data analysis and statistics. *Python for High Performance and Scientific Computing*, 1-9.

Michavila, F., Martín-González, M., Martínez, J. M., García-Peñalvo, F. J., & Cruz-Benito, J. (2015). Analyzing the employability and employment factors of graduate students in Spain: The OEEU Information System. *Proceedings of the 3rd International Conference on Technological Ecosystems for Enhancing Multiculturality*. doi:10.1145/2808580.2808622

Michavila, F., Martínez, J. M., Martín-González, M., García-Peñalvo, F. J., & Cruz-Benito, J. (2016). *Barómetro de Empleabilidad y Empleo de los Universitarios en España, 2015 (Primer informe de resultados)*. Academic Press.

Patil, D. (2012). *Data Jujitsu*. O'Reilly Media, Inc.

Patil, D., & Mason, H. (2015). *Data Driven*. O'Reilly Media, Inc.

Peñalvo, F. J. G., Conde, M. J. R., Pardo, A. M. S., González, M. Á. C., Zangrando, V., & Holgado, A. G. (2012). GRIAL (GRupo de investigación en InterAcción y eLearning), USAL. *IE Comunicaciones: Revista Iberoamericana de Informática Educativa,* (15), 85-94.

Stieger, S., & Reips, U.-D. (2010). What are participants doing while filling in an online questionnaire: A paradata collection tool and an empirical study. *Computers in Human Behavior*, *26*(6), 1488–1495. doi:10.1016/j.chb.2010.05.013

Terzo, O., Ruiu, P., Bucci, E., & Xhafa, F. (2013). *Data as a service (DaaS) for sharing and processing of large data collections in the cloud.* Paper presented at the Complex, Intelligent, and Software Intensive Systems (CISIS), 2013 Seventh International Conference on.

Touya, K., & Fakir, M. (2014). Mining Students' Learning Behavior in Moodle System. *Journal of Information Technology Research*, *7*(4), 12–26. doi:10.4018/jitr.2014100102

Vázquez-Ingelmo, A., Cruz-Benito, J., & García-Peñalvo, F. J. (2017). *Improving the OEEU's data-driven technological ecosystem's interoperability with GraphQL.* Paper presented at the 5th International Conference on Technological Ecosystems for Enhancing Multiculturality, Cádiz, Spain.

Wickham, H. (2014). Tidy data. *Journal of Statistical Software*, *59*(10), 1–23. doi:10.18637/jss.v059.i10 PMID:26917999

Wills, G. (2008). Linked data views. Handbook of data visualization, 217-241.

Zeleny, M. (1987). Management support systems: Towards integrated knowledge management. *Human Systems Management*, *7*(1), 59–70.

Zhang, H., Chen, G., Ooi, B. C., Tan, K.-L., & Zhang, M. (2015). In-memory big data management and processing: A survey. *IEEE Transactions on Knowledge and Data Engineering*, *27*(7), 1920–1948. doi:10.1109/TKDE.2015.2427795

Chapter 14
Exploring the Impact of Security Policy on Compliance

Winfred Yaokumah
Pentecost University College, Ghana

Peace Kumah
Ghana Education Service, Ghana

ABSTRACT

Extant studies on compliance with security policies have largely ignored the impact of monitoring, security operations, and roles and responsibilities on employees' compliance. This chapter proposes a theoretical model that integrates security policy, monitoring, security operations, and security roles to examine employees' security compliance. Data were collected from 233 IT security and management professionals. Using partial least square structural equation modelling and testing hypotheses, the study finds that information security policy has significant indirect influence on information security compliance. The effect of security policy is fully mediated by security roles, operations security activities, and security monitoring activities. Security policy strongly influences operations security activities and has the greatest effect on security roles and responsibilities. Among the three mediating variables, monitoring has the most significant influence on security compliance. Conversely, the direct impact of security policy on compliance is not significant.

INTRODUCTION

The failure of employee to comply with information systems security policies is a key concern for organizations (Puhakainen & Siponen, 2010). Information security breaches often result from employees' non-compliance with the security policy. A global survey profiles the nature of data breaches in 19 organizations from 27 countries. The study covers more than 47,000 reported security incidents and 621 confirmed data breaches. The findings reveal that over 50% of the insiders who committed sabotage were formal employees, 70% of Internet Protocol address (IP) theft cases were committed by internal people intended to resign their job, and 75% of attacks were opportunists with financial motives targeting no specific individual or organization (Data Breach Investigations Report [DBIR], 2013). This

DOI: 10.4018/978-1-5225-4944-4.ch014

report heightens the need for organizations to ensure that essential security controls are put in place and security policies are complied with. A recent study found that insiders (current and former employees, third parties) with trusted network access represent a major threat to information security, yet many organisations fail to implement processes and technologies to address internal incidents (PWC Report, 2015). Sometimes, even the organisational efforts to protect information assets from employee security threats may rather encourage the behaviors organizations are attempting to thwart (Lowry et al., 2015).

Compliance with information security policy remains a challenging task (D'Arcy & Greene, 2014). To ensure compliance with security objectives, legal, and regulatory requirements, organizations have established security policies to guide employees' behaviour. The information security policy contains intentions, principles, rules, and guidelines which the management wants the employees to adhere to (Sommestad et al., 2014). It provides management direction and support for information security (ISO/IEC, 2009). It generally describes the acceptable use of computer resources, information security roles and responsibilities, the type of training that employees should have, and the consequences of security policy violation (Sommestad et al., 2015). Providing adequate security to information security requires that technical information systems security and management personnel comply with security measures. For instance, critical data may be put at risk when the technical personnel fail to follow operational procedures, perform vulnerability assessment, check security in the third party products and services, perform regular backups, properly manage user accounts, secure mobile devices that are attached to the organization's productive networks, effectively control malware activities, protect data transfer and network services, monitor, log, and audit information systems regularly. Accordingly, Qing et al. (2011) suggest the development and deployment of more advanced protective technologies and enforcement of effective security policies and procedures.

Although organizations have instituted information security policy and compliance programs (D'Arcy & Herath, 2011; Sommestad et al., 2014), concise models to fulfil the organization's information security policy and compliance efforts are sparse. A major problem with security models is that there are just too many to follow. In just the ISO/IEC 27002:2013 framework, there are 14 major controls, which in itself a challenge for employees to comply with. Information security policy-compliance model for incorporating critical inter-related security measures is necessary for knowledge support in the management of information security in organizations where confidentiality, integrity, and availability are paramount. Therefore, the purpose of this study is to model the impact of information security policy on compliance, focusing on information systems security and management personnel, to understand how the activities of these personnel can influence security compliance. A recent study found that malicious database administrators can bypass the security mechanisms and make hidden modifications to databases, which may be untraceable (Kieseberg et al., 2013). Previous literature on security policy compliance focused on end-users (non-technical users). For instance, D'Arcy and Herath (2011) and Ifinedo (2016) applied deterrence theory; Hedström et al. (2013) applied social action theory (SAT) for management of information security; Sommestad et al. (2014) identified attitudes, intentions or actual behaviour having influence on compliance with information security policy. Also, Sommestad et al. (2015) found that anticipated regret and threat appraisal predict information security policy compliance. Also, organizational justice influences Internet use policy compliance (Li et al., 2014).

However, the actual technical and operations security activities are performed by the security and operations management personnel. This, therefore, suggests that there should be increasing focus on both technical and administrative controls to ensure information security compliance. The question still exists whether security and operations management personnel comply with technical guidelines and standards, and to what extent do security policies impact compliance. In order to address this issue, the study develops a model based on ISO/IEC 27002:2013 (International Organization for Standardization/International Electrotechnical Commission - security techniques - code of practice for information security management) to determine the effects of security policy on compliance, mediated by security roles and responsibilities, operations security activities, and security monitoring and reviews.

BACKGROUND

ISO/IEC 27002:2013 Framework

The focus of ISO/IEC 27002:2005 is to provide guideline for effective implementation of information security within an organization (ITGI, 2008). The framework is to improve information security practices through creating information security policies, procedures, assignment of roles and responsibilities, documentation of operational procedures, risk management (Myler & Broadbent, 2006), and to ensure compliance with legal and audit control (Saint-Germain, 2005). The current version of the standard, ISO/IEC 27002:2013 suggests that the effectiveness of organizational security compliance programs should involve establishment of security policies, procedures, work programs, and records. The standard contains 14 security control clauses collectively containing a total of 35 main security categories and 114 controls (ISO/IEC 27002:2013, 2013). The security control clauses are broadly categorized into administrative or organizational, technical or logical, and physical security (ISO/IEC 27002:2013, 2013) as listed below.

- Information security policies
- Organization of information security
- Human resources security
- Asset management
- Access control
- Cryptography
- Physical and environmental security
- Operations security
- Communications security
- Information systems acquisition, development and maintenance
- Supplier relationships
- Information security incident management
- Information security aspects of business continuity management
- Compliance.

This paper consists of five security controls drawn from technical security (access control, operations security) and organizational (organization of information security, information security policies, compliance) control clauses.

Information Security Threats and Controls

Various threats (potential dangers exploiting system vulnerabilities) are militating against information assets as a result of vulnerabilities in information systems. Vulnerabilities are weaknesses or lack of countermeasures in the information systems themselves (Shon, 2013). Threats may be intentional or unintentional and can come from both internal and external sources. Internal (insider) threats to information security are critical issues for organizations (Baracaldo & Joshi, 2013; Wang, Gupta, & Rao, 2015). Though several threats including browser exploits, data interception, malware, network exploits, spamming, spoofing, and theft/loss (US GAO, 2012) exist, technical (logical) controls are available to protect information systems. Technical controls such as firewalls, intrusion preventive and detection systems, antimalware, encryption, backup and restoration mechanisms, logging, monitoring, auditing, identification and authentication mechanisms were implemented in software configurations, hardware devices, and in procedures to protect information systems.

Apart from technical controls, administrative controls (such as security policies and procedures) and physical controls (such as cable locks, fencing, closed-circuit TV, and lighting) play a major role in security systems (Shon, 2013). These controls provide functionalities intended to secure information systems. These functionalities comprise of *deterrent* (intended to discourage a potential attacker), *preventive* (intended to avoid an incident from occurring), *corrective* (fixes components or systems after an incident has occurred), *recovery* (intended to bring the environment back to regular operations), and *detective* (identify an incident's activities and potential intruder) (Shon, 2013). Despite these security measures, the major threat is the members of the organization themselves who are entrusted to protect information systems (Willison & Warkentin, 2013) and are required to comply with the organization's security measures and policies.

Information Security Policy and Compliance

Information security policy can be defined as the formal guidelines normally established by top management that employees in the organization are required to follow in order to protect organizational information resources (Bulgurcu, Cavusoglu & Benbasat, 2010). The purpose of information security policy is to provide management direction and support for information security in accordance with business requirements and relevant laws and regulations (ISO/IEC 27002:2013, 2013). It addresses requirements imposed by business strategy, regulations, legislation and contracts within an increasing information security threat environment (ISO/IEC 27002:2013, 2013). Compliance with legal and contractual requirements is aimed at avoiding breaches of legal, statutory, regulatory or contractual obligations related to information security and of any security requirements. Compliance is a type of social influence that affects our everyday behavior (Aronson et al., 2010). It refers to an implicit or explicit response to a request (Cialdini & Goldstein, 2004). It is the effect that the words, actions, or mere presence of other people have on our thoughts, feelings, attitudes, or behavior (Aronson et al., 2010). Technical compliance review, which involves regular review for compliance with the organization's information security policies and standards (ISO /IEC 27002:2013, 2013), can be achieved through technical and administrative controls.

An enforceable information security policy ensures that vulnerabilities are identified and addressed (fostering business continuity), information is safely shared within the organization among the employees as well as with other stakeholders (mitigating enterprise risk), and heightening security awareness (increasing the likelihood of individual compliance) (Ungerman, 2005). For employees to follow these guidelines, previous works suggested that organizations should cultivate security culture. Improving security culture of the organization will positively impact compliance with security policies (Parsons et al., 2015). For effective cultural change towards information security policies and to better incorporate and enforce information security policy, strategic cultural change (cultural change at the top management) will be necessary Kolkowska & Dhillon, 2013). The work of Zhang, Reithel and Li (2009) attempted to incorporate perceived technical security protection into the theory of planned behavior by examining factors affecting end-user security behaviors, specifically, compliance with security policies. The findings indicated that perceived high technical protection leads to low intention to comply with security policy. The authors concluded that the power and capability of technical protection mechanisms should not be overstated; rather end-users should adopt more cautious security practices and adhere to the requirements of the organization's security policies. Regardless of this finding, without technical protection put in place, employees' actions could not be adequately restricted and could go unnoticed.

Operations Security, Roles and Responsibilities, and Monitoring and Reviews

Operations security, access control, and monitoring and reviews are generally technical and administrative means of security controls. Operations security is the day-to-day routine activities that keep communication networks, computer systems, operating systems and application software, and the systems environment up and running in a secure and protected manner (Shon, 2013). These activities entail interactions of people (performing various roles and responsibilities), software applications, servers, networks, and the resulting oversight responsibility through logging, monitoring, auditing, and reporting (Shon, 2013). Specifically, among others, operations security consists of activities such as managing operational procedures and assigning responsibilities, managing security in third-party product and services, protecting against malware, taking regular backup, logging, monitoring, auditing, and reporting. Failure to perform these activities on regular basis can make systems insecure. For instance, when malware is not updated regularly the system becomes vulnerable to virus activities that may cause destruction to data. Also, when backup is not taken regularly, critical data may be lost when, for example, the server clashes.

Roles comprise of a set of rules or norms that act as plans or blueprints to guide behavior (Biddle, 1986). Role theory proposed that human behavior is guided by expectations held both by the individual and by other people (Goffman, 1959). Roles come with formalized expectations as norms or rules which include what rewards will result when roles are successfully performed and what punishments will result when roles are not successfully performed (Goffman, 1959). Roles guide the behavior of the individuals. The individual, in turn, influences the norms, expectations, and behaviors associated with the roles (Goffman, 1959). Access control processes include policies pertaining to the roles and responsibilities of users (ISO/IEC 27002:2013, 2013). Two principles are often used to guide access control policy: *need-to-know* (granting access to information based on the need of the user to perform his tasks or roles) and *need-to-use* (granting access to the information processing facilities such as IT equipment, applications, procedures, and rooms based on the need to perform tasks) (ISO/IEC 27002:2013, 2013). *Need-to-use* principle may include access to facilities such as networks and network services by authorized users. Several user access management techniques are employed to ensure that only authorized users have access to resources (João et al., 2014), while unauthorized users are restricted or denied access to systems and services. User access management consists of user registration and de-registration of access rights, user access provisioning for assigning or revoking access rights to systems and services, management of privileged access rights for allocation and use of privileged access rights, review of user access rights at regular intervals, and user responsibilities to make users accountable for safeguarding their authentication information (ISO/IEC 27002:2013, 2013).

Another important aspect of information security is monitoring and reviews. Monitoring can be achieved through logging, observing, and recording a person's behaviour. New information equipments are available to help security personnel and system administrators enforce and monitor information security policies (Mei-Yu & Ming-Hsien, 2013). The devices are technical methods that can provide backup, logging, blocking and monitory functions. Communication monitoring technologies can filter, record, or block suspected activities in order to reduce the occurrence of security incidence (Mei-Yu & Ming-Hsien, 2013). Logging of data is important for monitoring and forensics purposes and should include network traffic monitoring. All packets that pass through the network could pass one determined traffic point where it is recorded and saved for later analysis (Rajdeep, Joshi, & Pilli, 2010). The logged data acts as forensic data sources on networks. The types of attacks on the network are stored in various files, depending upon the devices used and the network management policies (Grubor & Barac, 2014). Consequently, evidence for compliance with security policies can be ascertained from the logged files. For example, a logged data can be obtained from attacker's computer (log file, working files, ambient data), corrupted computer (log file, working files, ambient data), firewalls (log file), Internet Service Provider (client's traffic logged data), and victim's computer (log file, working files, ambient data change of configuration), and malicious files (Trojans, viruses, rootkit) (Grubor & Barac, 2014). In an organization, the logged data could provide evidence for accountability (Vance, Lowry & Eggett, 2013).

Research Model and Hypotheses

A key issue in information security policy is compliance. The proposed model is derived from five control objectives (security policy, access control, communications security, operations management, compliance) within ISO/IEC 27002:2013 framework. The model (see Figure 1) hypothesized that compliance with security measures is influenced by security controls as a result of the establishment of security policy. The model suggests that the establishment of information security policy can impact information security compliance directly or indirectly through assignment of security roles and responsibilities, operations security activities, and security monitoring and review activities.

Information security policy details the monitoring and review activities, operations security activities, and the roles and responsibilities of employees. Information security policy has effect on monitoring and review activities (H_1), operations security (H_2), roles & responsibilities (H_3), and security compliance (H_4) (Shon, 2013). Operations security tasks are performed by the individuals who are assigned security roles and responsibilities (H_5). Also, security operations generate activities that must be monitored and reviewed (H_6) (ISO/IEC27002:2013, 2013). Moreover, security compliance is impacted indirectly through assignment of security roles and responsibilities (H_9), performance of operations security tasks (H_8), and monitoring and review activities (H_7). On the other hand, security policy can directly impact compliance (H_4) through anticipated regret and threat (Sommestad et al., 2015) and deterrent (D'Arcy & Herath, 2011). Thus, the model aims at predicting security measures that have the greatest impact on security compliance. In order to determine the impact on compliance, the following hypotheses are proposed and tested.

Hypothesis One: Information security policy (POLICY) has positive and significant impact on information security monitoring and review activities (MONITOR).

Figure 1. Security policy-compliance conceptual framework

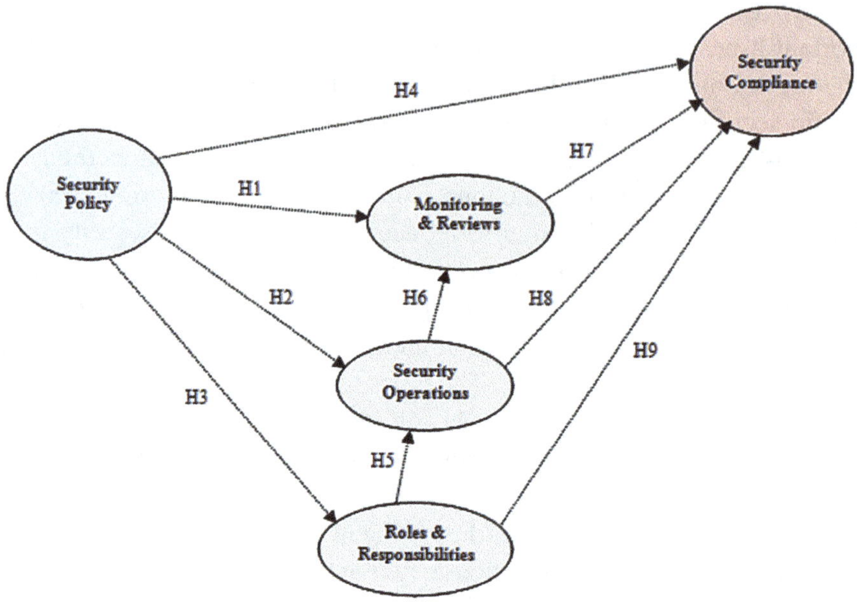

Hypothesis Two: Information security policy (POLICY) has positive and significant impact on operations security activities (OPERATE).

Hypothesis Three: Information security policy (POLICY) has positive and significant impact on information security roles and responsibilities (ROLES).

Hypothesis Four: Information security policy (POLICY) has positive and significant direct impact on information security compliance (COMPLY).

Hypothesis Five: Information security roles and responsibilities (ROLES) have positive and significant impact on information operations security activities (OPERATE).

Hypothesis Six: Information operations security activities (OPERATE) have positive and significant impact on information security monitoring and review activities (MONITOR).

Hypothesis Seven: Information security monitoring and review activities (MONITOR) have positive and significant impact on information security compliance (COMPLY).

Hypothesis Eight: Information operations security activities (OPERATE) have positive and significant impact on information security compliance (COMPLY).

Hypothesis Nine: Information security roles and responsibilities (ROLES) have positive and significant impact on information security compliance (COMPLY).

METHODOLOGY

Research Design

This study used quantitative research methodology with a cross-sectional explanatory survey research approach to assess the impact of information security policy on information security compliance. A structural model, made up of five constructs, was developed based on ISO/IEC 27002:2013 framework. Information security policy represents the exogenous variable while information security roles and responsibilities, monitoring and reviews, operations security, and compliance represent the endogenous variables. Hypotheses were formulated and tested to establish the existence of statistical significance and association between the constructs.

Research Samples

The participants in the study were security personnel, chief information officers and IT Managers (with responsibility of IT security) from five main industry sectors. The organizations were drawn from three large regional capitals in Ghana, including (a) government public service institutions, (b) public utility companies (water, electricity, and telecommunication), (c) financial institutions, (d) educational institutions, and (e) healthcare institutions. Five hundred questionnaires were self-administered to the participants.

Research Instrument and Measures

The survey instrument was developed based on security controls contained in the widely-accepted framework of ISO/IEC 27002:2013. The questionnaire was divided into four main sections consisting of 48 items. The sections were security policy (5 items), information security roles and responsibilities (2 items), monitoring and reviews (14 items), operations security (11 items), and compliance (13 items). All the items used a 6-point Likert scale (*Not performed = 0, Performed informally = 1, Planned = 2, Well defined = 3, Quantitatively controlled = 4, and Continuously improving = 5*) to measure the impact of information security policy on compliance. The scale generated continuous data on all the items. The demographic section consists of industry sector, job title, and the number of years of experience, which were categorical data.

Data Analysis

The structural model was analyzed using Partial Least Square Structural Equation Modelling (PLS-SEM) software. PLS-SEM works well with small sample sizes and complex models (Cassel, Hackl & Westlund, 1999). This method is suitable for this study because the objective of the research is to test the causal relationship between the predictor variables (information security policy, information security roles and responsibilities, operations security, and security monitoring and reviews) and the outcome variable (security compliance). Reliability by means of Cronbach's alpha and validity of the measurement model (Hair et al., 2014; Henseler et al., 2009) were conducted. Because Cronbach's alpha is sensitive to the number of items included and works on the assumption that all items load equally, composite reliability is more suitable (Hair et al., 2014) for this study.

Subsequent, the measurement model is assessed using convergent and discriminant validity. The convergent validity measures the internal consistency. It ensures that items measuring latent variables really measure the said variables and not any other latent variable (Hair et al., 2014; Urbach & Ahlemann, 2010). Convergent validity is the degree to which a measure correlates positively with alternative measures of the same construct (Hair et al., 2014). After verifying the reliability and validity of the constructs, the tests of structural relationships (by means of hypotheses) among the constructs were conducted using the coefficient of determination, R^2; the magnitude of the sign and significance of path coefficients; and the predictive relevance, Q^2.

RESULTS

Firstly, the tests of reliability and validity of the measurement instrument, assessment of the structural model, and the determination of the model fit were performed. Following, the tests of hypotheses were carried out to ascertain the impact of information security policy on security compliance, mediated through information security roles and responsibilities, operations security, and security monitoring and reviews.

Reliability and Validity

The rule of thumb for reliability of the constructs is that the values for Cronbach's alpha (Henseler, & Sarstedt, 2012; Nunnally & Bernstein, 1994) and the composite reliability (Hair et al., 2014) should be higher than 0.7. Table 1 shows that the Cronbach's alpha coefficient is reliable, with values ranging from 0.702 - 0.941. The composite reliability, which is the more preferred measure of reliability (Hair et al., 2014) also ranges from (0.866-0.948), therefore indicating that the constructs are reliable.

Next, the convergent validity was assessed with average variance extracted (AVE) and the factor loadings of the items. The AVE measures the amount of variance that the latent variable captures from the items it measures, relative to the amount of variance associated with the measurement errors (Hair et al., 2014). The rule of thumb is that AVE should be more than 0.5 (Hair et al., 2014), thus representing at least 50% of the measurement variance captured by the latent variables. In addition, factor loadings of all indicators should be above 0.7 (Hair et al., 2014). Table 1 shows that indicator loadings are all greater than the threshold of 0.7 or higher.

Also, Table 2 shows that all the latent variables have AVE greater than 0.5, therefore, convergent validity has been achieved. Discriminant validity is how a construct is truly different from other constructs both in terms of how much it correlates with other constructs and how distinctly measured variables represent only this single construct (Hair et al., 2014). Measuring discriminant validity requires that the Average Variance Extracted (AVE) of each latent construct should be greater than the highest squared correlations between any other construct (Fornell & Larcker, 1981) and the loadings of each indicator should be greater than all its cross loadings (Chin, 1998; Sarstedt et al., 2014). Table 2 shows the correlations between constructs with the square root of the AVEs on the diagonal. It can be seen that the square root of the AVE for all constructs are greater than correlation with other constructs, indicating sufficient discriminant validity. Based on these findings, it can be concluded that the measurement model exhibits acceptable reliability, convergent validity, and discriminant validity.

Structural Model Assessment

The structural model determines whether the structural relations in the model being tested are meaningful (Sarstedt et al., 2014). The quality of the structural model is assessed by the ability to predict endogenous constructs. This is achieved by using the coefficient of determination, R^2 and cross-validated redundancy, Q^2 (Sarstedt et al., 2014; Urbach & Ahlemann, 2010), and the strength of the path coefficients (Hair, Ringle, & Sarstedt, 2011).

Table 1. Loading and cross loadings with CA and CR

VARIABLES	ROLE	POLICY	MONITOR	OPERATE	COMPLY	CA	CR
ROLE	**0.874**	0.519	0.614	0.523	0.453	0.702	0.866
	0.874	0.552	0.516	0.534	0.437		
POLICY	0.374	**0.817**	0.537	0.566	0.634	0.885	0.916
	0.572	**0.838**	0.643	0.589	0.497		
	0.486	**0.826**	0.658	0.573	0.524		
	0.540	**0.820**	0.706	0.613	0.536		
	0.561	**0.838**	0.693	0.720	0.561		
MONITOR	0.525	0.664	**0.799**	0.566	0.612	0.941	0.948
	0.560	0.538	**0.814**	0.574	0.481		
	0.529	0.629	**0.793**	0.473	0.655		
	0.386	0.625	**0.784**	0.547	0.482		
	0.496	0.533	**0.756**	0.596	0.463		
	0.514	0.481	**0.746**	0.617	0.346		
	0.405	0.584	**0.754**	0.628	0.435		
	0.455	0.622	**0.736**	0.531	0.514		
	0.492	0.538	**0.725**	0.636	0.399		
	0.530	0.634	**0.758**	0.680	0.568		
	0.546	0.630	**0.730**	0.668	0.529		
	0.506	0.480	**0.647**	0.481	0.451		
	0.539	0.711	**0.806**	0.740	0.560		
	0.318	0.559	**0.668**	0.357	0.693		
OPEARATE	0.436	0.594	0.734	**0.753**	0.529	0.936	0.946
	0.409	0.476	0.522	**0.822**	0.482		
	0.545	0.574	0.607	**0.869**	0.382		
	0.457	0.497	0.576	**0.755**	0.421		
	0.542	0.632	0.520	**0.785**	0.412		
	0.481	0.620	0.570	**0.774**	0.309		
	0.355	0.465	0.624	**0.723**	0.499		
	0.423	0.698	0.562	**0.780**	0.471		
	0.464	0.610	0.674	**0.773**	0.471		
	0.605	0.649	0.602	**0.759**	0.390		
	0.484	0.559	0.659	**0.808**	0.396		
COMPLY	0.417	0.477	0.492	0.379	**0.775**	0.923	0.934
	0.324	0.425	0.389	0.437	**0.701**		
	0.479	0.455	0.539	0.356	**0.770**		
	0.497	0.483	0.478	0.438	**0.792**		
	0.333	0.447	0.440	0.351	**0.700**		
	0.265	0.506	0.515	0.378	**0.753**		

continued on following page

Table 1. Continued

VARIABLES	ROLE	POLICY	MONITOR	OPERATE	COMPLY	CA	CR
	0.085	0.395	0.471	0.359	**0.711**		
	0.462	0.604	0.613	0.429	**0.772**		
	0.362	0.476	0.490	0.550	**0.683**		
	0.315	0.489	0.487	0.471	**0.668**		
	0.407	0.492	0.406	0.421	**0.686**		
	0.508	0.485	0.466	0.211	**0.650**		
	0.312	0.513	0.607	0.399	**0.715**		

CA Cronbach's alpha; CR Composite Reliability, AVE Average Variance Extracted

Table 2. Descriptive statistics, correlation matrix, and AVE

VARIABLES	MEAN	SD	ROLE	POLICY	MONITOR	OPERATE	COMPLY	AVE
ROLE	2.940	0.986	0.874*					0.764
POLICY	3.337	0.906	0.612	0.828*				0.685
MONITOR	2.968	0.860	0.646	0.742	0.753*			0.566
OPERATE	3.264	0.933	0.605	0.740	0.750	0.783*		0.613
COMPLY	3.003	0.860	0.509	0.665	0.682	0.551	0.723*	0.522

* Square roots of average variances extracted (AVEs) shown on diagonal.

The R^2 is the measure of the variance of each of the endogenous constructs explained by exogenous constructs, thus measuring the predictive power of the exogenous constructs (Chin, 2010; Urbach & Ahlemann, 2010). The R^2 measure ranges from 0 to 1, with values closer to 1 indicating greater degree of predictive power. For variance explained by the endogenous variables to have practical and statistical significance, it is recommended that R^2 values be greater than 0.10 (Hair et al., 2014). As a guideline for assessing R^2, 0.67, 0.33 and 0.19, represent substantial, moderate, and weak respectively (Lee et al., 2011). A more stringent assessment with R^2 values of 0.75 representing substantial; values of 0.5 and 0.25 represent moderate and weak R^2 respectively (Hair et al., 2014).

From Table 3, R^2 for security monitoring and review activities was 0.703, indicating that about 70% of the variance in security monitoring and review activities is explained by security policy and operations security activities. The R^2 for operations security is 0.608, revealing that about 61% of the variance in

Table 3. R^2 and Q^2 coefficients

VARIABLES	R^2	Q^2
Security Roles and Responsibilities (ROLE)	0.376	0.377
Operations Security (OPERATE)	0.608	0.608
Security Logging, Monitoring, and Reviews (MONITOR)	0.703	0.704
Security Compliance (COMPLY)	0.151	0.635

operations security is accounted for by security policy and security roles and responsibilities. The R^2 for security roles and responsibilities is 0.376, showing that about 38% of the variance in security roles and responsibilities is accounted for by security policy. Moreover, the R^2 for security compliance is 0.151, indicating that only about 15% of the variance in security compliance is explained by the effects of security policy (not significant), security roles and responsibilities (significant), operations security activities (significant but negative), and security monitoring and review activities (significant).

Moreover, the Stone-Geiser Q^2 (Geisser, 1975; Lee et al., 2011) was used to assess the predictive accuracy of the proposed model. Q^2 measures how accurately the PLS-SEM model predicts the observed data points (Hair et al., 2014). The Q^2 is a nonparametric measure obtained using the blindfolding procedure with values larger than 0, indicating predictive relevance (Hair et al., 2011). Our proposed model (see Table 3) shows that Stone-Geisser Q^2 values are between 0.377 and 0.704, which is an indicative of predictive relevance.

Model Fit

Table 4 summarizes the model fit indices, namely Average R squared (ARS), Average Path Coefficient (APC), Average Variance Inflation Factor (AVIF), and Tanenhaus GoF. According to Tenenhaus et al. (2005), APC and ARS should be significant at 5% whiles AVIF should be less than 5. It can be seen from Table 4 that both ARS and APC are significant at 0.001 and AVIF is 2.156. These show evidence of good fit. Overall fit of the model is assessed with the Tenenhaus GoF, which is the Geometric mean of the average communality and average R^2 for endogenous variables (Tenenhaus et al., 2005). Tenenhaus GoF guidelines proposed that GoF of 0.1, 0.25 and 0.36 represent small, medium and large fit respectively (Kock, 2013). The model obtained a GoF of 0.538, which is impressively greater than the cut-off of 0.36 for large fit. Based on these results, it can be concluded that the proposed model has good fit with the data and measures well.

Hypotheses Testing

Table 5 shows the summary of the results of testing the hypotheses. The results show that security policy, POLICY (a) significantly predicts security monitoring and review activities, MONITOR (β=0.473, $p<0.001$), supporting H_1; (b) significantly impacts security operations security activities, OPERATE (β=0.583, $p<0.001$), lending support for H_2; and (c) is significant in predicting the establishment of security roles and responsibilities, ROLE (β=0.583, $p<0.001$), supporting hypothesis H_3. However,

Table 4. Model fit and quality indices

Index	Value/Thresholds
Average R Square (ARS)	0.459 ***
Average Adjusted R Square (AARS)	0.453 ***
Average Path Coefficient (APC)	0.369 ***
Average Block VIF	2.156 / acceptable if <= 5, ideally <= 3.3
Tanenhaus GoF	0.538 / small >= 0.1, medium >= 0.25, large >= 0.36

*** Significant at p=0.001

Table 5. Results of hypothesis testing

Hypotheses	Hypothesized Path	Path Coefficient	P-Value	Result
H_1	POLICY → MONITOR	0.473	<0.001	Supported
H_2	POLICY → OPERATE	0.583	<0.001	Supported
H_3	POLICY → ROLE	0.613	<0.001	Supported
H_4	POLICY → COMPLY	0.044	0.254	Not Supported
H_5	ROLE → OPERATE	0.277	<0.001	Supported
H_6	OPERATE → MONITOR	0.427	<0.001	Supported
H_7	MONITOR → COMPLY	0.328	<0.001	Supported
H_8	OPERATE → COMPLY	-0.348	<0.001	Not Supported
H_9	ROLE → COMPLY	0.223	<0.001	Supported

the direct effect of policies, POLICY on compliance to information security policy, COMPLY is not significant ($\beta=0.044$, $p=0.254$). Thus, hypothesis H_4 is not supported.

It can also be observed from Table 5 that security role and responsibilities, ROLE significantly impacts operations security activities, OPERATE ($\beta=0.277$, $p<0.001$), lending support for H_5. Moreover, the impact of security role and responsibilities, ROLE on compliance, COMPLY is significant ($\beta=0.223$, $p<0.001$), supporting hypothesis H_9. Observably, operations security activities, OPERATE have positive and significant impact on security monitoring and review control activities, MONITOR ($\beta=0.427$, $p<0.001$), supporting H_6.

On the other hand, the impact of operations security activities, OPERATE on compliance, COMPLY is significant, yet negative ($\beta = -0.348$, $p<0.001$). Thus, hypothesis H_8 is not supported. Security monitoring and review control activities, MONITOR has positive and significant impact on compliance, COMPLY ($\beta=0.328$, $p<0.001$). The path coefficients (in Figure 2) reveals that security policy, POLICY made the highest impact on security roles and responsibilities, ROLE, among the four security control activities (OPERATE, MONITOR, ROLE, COMPLY). Moreover, security monitoring and review control activities, MONITOR is the highest predictor of security compliance, COMPLY among the three security control measures (MONITOR, OPERATE, ROLE) explaining security compliance.

RECOMMENDATIONS

The purpose of this research is to model the impact of information security policy on security compliance and the pathway through which this effect is generated. To achieve this objective, this study investigates the relationships among security policy, mediated through information security roles and responsibilities, operations security, security monitoring and reviews, and compliance. The results reveal positive and significant effect of information security policy on monitoring and reviews, operations security activities, and security roles and responsibilities. The results also show that assignment of information security roles and responsibilities significantly influence both security operational activities and security compliance. Moreover, the findings provide support for positive and significant relationships between operations security and monitoring and reviews activities, and between monitoring and reviews and compliance.

Along the pathway from security policy to compliance (see Figure 2), it can be observed that about 70% of the variance in security monitoring and review activities is explained by security policy and operations security activities. The influence of security policy and operations security on monitoring and reviews were substantial (Lee, 2011). Moreover, 61% of the variance in operations security activities is accounted for by security policy and security roles and responsibilities. Accordingly, the impact of security policy and security roles and responsibilities on operations security activities is generally substantial (Lee, 2011). Also, 38% of the variance in security roles and responsibilities was accounted for by security policy. Thus, indicating that the effect of security policy on security roles and responsibilities is moderate (Lee, 2011). Overall, based on these findings and the path coefficients, the model reveals that the highest predictive power of security policy on security compliance should be through: POLICY → ROLE → OPERATE → MONITOR → COMPLY.

On the contrary, the direct impact of information security policy on security compliance was not significant. In addition, operations security activities have significant but negative impact on security compliance. Overall, about 15% of the variance in security compliance is accounted for by security policy (not significant), security roles and responsibilities (significant), operations security activities (significant but negative), and security monitoring and review activities (significant). This result shows that the combined effect of security policy, security roles and responsibilities, operations security activities, and security monitoring and review activities on security compliance is weak (Lee, 2011). This is as the result of the negative effect the operations security activities have on compliance and the insignificant effect of security policy on compliance. Thus, the model does not support a simple, direct effect of operations security on compliance. This means that the effect of operations security activities on compliance should be carried out through monitoring and review activities. Apparently, operations security activities provide data that must be logged, monitored, audited and reviewed for the purpose of ascertaining whether the organization complies with information security standards and guidelines.

Figure 2. Security policy-compliance model

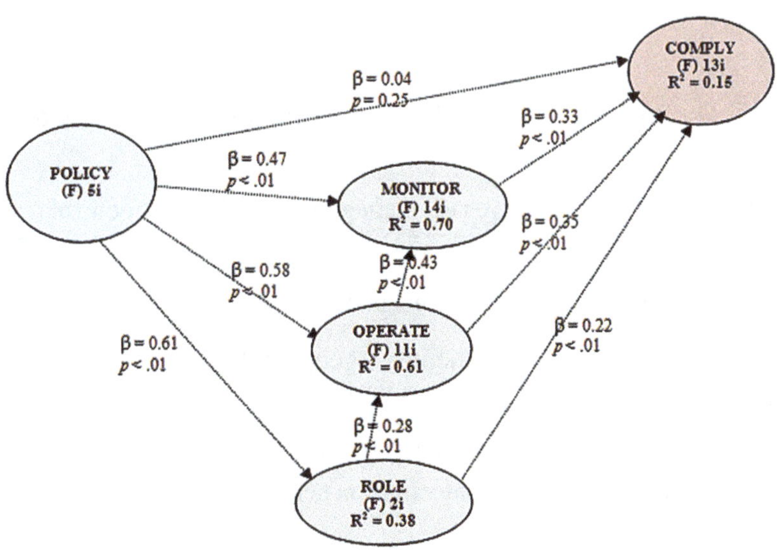

FUTURE RESEARCH DIRECTIONS

The study demonstrated that security policy, assignment of security roles and responsibilities, operations security, and monitoring and review activities have critical implications on compliance. The findings showed that security policy enhances the level of security compliance. However, direct impact of information security policy on security compliance was not significant. Therefore, security policy should be directed through security roles and responsibilities, operations security, and monitoring and review activities. This suggests that organizations investing in or implementing security compliance programs should pay attention to establishment of security policy, assignment of security roles and responsibilities, operations security, and monitoring and review activities. Thus, future work should involve extending the model to cover other control clauses of ISO/IEC 27002:2013 in order to ascertain their impacts on compliance.

CONCLUSION

This study developed a model that integrated information security policy, security roles and responsibilities, operations security, security monitoring and review activities, compliance, with the aim of providing better understanding of how security policy influences security compliance. The paper focused on the pathway through which security compliance generated its impact. Partial Least Square Structural Equation Modelling (PLS-SEM) was used to test hypotheses on the relationships among security policy and security compliance. While this study provided some very important understanding in security compliance, the specifics of examining five security controls of the ISO/IEC 27002:2013 framework could be deemed as a limitation. Though these five security controls might be satisfactory to understanding employee compliance, there might be other controls that could be significant to the phenomenon under investigation.

REFERENCES

Aronson, E., Wilson, T. D., & Akert, R. M. (2010). *Social Psychology*. Upper Saddle River, NJ: Prentice Hall.

Baracaldo, N., & Joshi, J. (2013). An adaptive risk management and access control framework to mitigate insider threats. *Computers & Security*, *39*, 237–254. doi:10.1016/j.cose.2013.08.001

Biddle, B. J. (1986). Recent Development in Role Theory. *Annual Review of Sociology*, 1267–1292.

Bovens, M. (2010). Two concepts of accountability: Accountability as a virtue and as a mechanism. *West European Politics*, *33*(5), 946–967. doi:10.1080/01402382.2010.486119

Bulgurcu, B., Cavusoglu, H., & Benbasat, I. (2010). Information security policy compliance: An empirical study of rationality-based beliefs and information security awareness. *Management Information Systems Quarterly*, *34*(3), 523–548.

Cassel, C. M., Hackl, P., & Westlund, A. H. (1999). Robustness of partial least-squares method for estimating latent variable quality structures. *Journal of Applied Statistics*, *26*(4), 435–446. doi:10.1080/02664769922322

Chin, W. W. (1998). The partial least squares approach to structural equation modelling. Modern Methods for Business Research, 295, 295–336. doi:10.1016/j.aap.2008.12.010

Chin, W. W. (2010). How to write up and report PLS analyses. Handbook of Partial Least Squares, 665–690. doi:10.1007/978-3-540-32827-8

Cialdini, R. B., & Goldstein, N. J. (2004). Social influence: Compliance and conformity. *Annual Review of Psychology*, *55*(1), 591–621. doi:10.1146/annurev.psych.55.090902.142015 PMID:14744228

D'Arcy, J., & Greene, G. (2014). Security culture and the employment relationship as drivers of employees' security compliance. *Information Management & Computer Security*, *22*(5), 474–489. doi:10.1108/IMCS-08-2013-0057

D'Arcy, J., & Herath, T. (2011). A review and analysis of deterrence theory in the IS security literature: Making sense of the disparate findings. *European Journal of Information Systems*, *20*(6), 643–658. doi:10.1057/ejis.2011.23

DBIR. (2013). *2013 Data Breach Investigations Report*. Retrieved from http://www.verizonenterprise.com/DBIR/

Fornell, C., & Larcker, D. F. (1981). Evaluating structural equation models with unobservable variables and measurements error. *JMR, Journal of Marketing Research*, *18*(1), 39–50. doi:10.2307/3151312

Geisser, S. (1975). The predictive sample reuse method with applications. *Journal of the American Statistical Association*, *70*(350), 320–328. doi:10.1080/01621459.1975.10479865

Goffman, E. (1959). *The Presentation of Self in Everyday Life*. Anchor Books.

Götz, O., Liehr-Gobbers, K., & Krafft, M. (2010). Evaluation of structural equation models using the partial least squares (PLS) approach. In Handbook of Partial Least Squares (pp. 691-711). Academic Press.doi:10.1007/978-3-540-32827-8_30

Grubor, G., & Barac, I. (2014). Integrated proactive forensics model in network information security. *Sinteza*, 693-699. doi: 10.15308/SInteZa-2014-693-699

Hair, J. F., Ringle, C. M., & Sarstedt, M. (2011). PLS-SEM: Indeed a silver bullet. *Journal of Marketing Theory and Practice*, *19*(2), 139–152. doi:10.2753/MTP1069-6679190202

Hair, J. F. J., Hult, G. T. M., Ringle, C., & Sarstedt, M. (2014). A primer on partial least squares structural equation modeling (PLS-SEM). *Long Range Planning*, *46*, 328. doi:10.1016/j.lrp.2013.01.002

Hedström, K., Karlsson, F., & Kolkowska, E. (2013). Social action theory for understanding information security non-compliance in hospitals. *Information Management & Computer Security*, *21*(4), 266–287. doi:10.1108/IMCS-08-2012-0043

Henseler, J., Ringle, C. M., & Sinkovics, R. (2009). The use of partial least squares path modeling in international marketing. *Advances in International Marketing*, *20*, 277–319. doi:10.1108/S1474-7979(2009)0000020014

Henseler, J., & Sarstedt, M. (2012). Goodness-of-fit indices for partial least squares path modeling. *Computational Statistics*, *28*(2), 565–580. doi:10.1007/s00180-012-0317-1

Ifinedo, P. (2016). Critical times for organizations: What should be done to curb workers' noncompliance with is security policy guidelines? *Information Systems Management*, *33*(1), 30–41. doi:10.1080/10580530.2015.1117868

ISO/IEC. (2009). Information technology – Security techniques – Information security management measurements. ISO/IEC 27004.

ISO/IEC 27002:2013. (2013). *Information technology security techniques - Code of practice for information security controls*. Retrieved from http://www.iso.org

ITGI. (2008). *Information security governance: Guidance for information security managers*. Retrieved from www.itgi.org

João, Z., Dulce, D., Mário, J. S., & Carlos, S. (2014). Group-Based Discretionary Access Control in Health Related Repositories. *Journal of Information Technology Research*, *7*(1), 1–17. doi:10.4018/jitr.2014010106

Kieseberg, P., Schrittwieser, S., Morgan, L., Mulazzani, M., Huber, M., & Weippl, E. (2013). Using the structure of B+-trees for enhancing logging mechanisms of databases. *International Journal of Web Information Systems*, *9*(1), 53–68. doi:10.1108/17440081311316389

Kock, N. (2013). *WarpPLS 4.0 User Manual*. Laredo, TX: ScriptWarp Systems.

Kolkowska, E., & Dhillon, G. (2013). Organizational power and information security rule compliance. *Computers & Security*, *33*, 3–11. doi:10.1016/j.cose.2012.07.001

Lee, L., Petter, S., Fayard, D., & Robinson, S. (2011). On the use of partial least squares path modeling in accounting research. *International Journal of Accounting Information Systems*, *12*(4), 305–328. doi:10.1016/j.accinf.2011.05.002

Li, H., Sarathy, R., Zhang, J., & Luo, X. (2014). Exploring the effects of organizational justice, personal ethics and sanction on internet use policy compliance. *Information Systems Journal*, *24*(6), 479–502. doi:10.1111/isj.12037

Lowry, P. B., Posey, C., Bennett, R. J., & Roberts, T. L. (2015). Leveraging fairness and reactance theories to deter reactive computer abuse following enhanced organisational information security policies: An empirical study of the influence of counterfactual reasoning and organisational trust. *Information Systems Journal*, *25*(3), 193–273. doi:10.1111/isj.12063

Mei-Yu, W., & Ming-Hsien, Y. (2013). Enterprise Information Security Management Based on Context-Aware RBAC and Communication Monitoring Technology. *Mathematical Problems in Engineering*, 1–11. doi:10.1155/2013/569562

Myler, E., & Broadbent, G. (2006). ISO 17799: Standard for security. *Information Management Journal, 40*(6), 43–52.

Nunnally, J., & Bernstein, I. (1994). *Psychometric Theory* (3rd ed.). New York: McGraw-Hill; Retrieved from http://scholar.google.com

Parsons, K. M., Young, E., Butavicius, M. A., McCormac, A., Pattinson, M. R., & Jerram, C. (2015). The Influence of Organizational Information Security Culture on Information Security Decision Making. *Journal of Cognitive Engineering and Decision Making, 9*(2), 117–129. doi:10.1177/1555343415575152

Puhakainen, P., & Siponen, M. (2010). Improving employees' compliance through information systems security training: An action research study. *Management Information Systems Quarterly, 34*(4), 767–A4.

Qing, H., Zhengchuan, X., Tamara, D., & Hong, L. (2011). Does deterrence work in reducing information security policy abuse by employees? *Communications of the ACM, 54*(6), 54–60. doi:10.1145/1953122.1953142

Rajdeep, A., Joshi, N.R.C., & Pilli, E.S. (2010). Generic framework for network forensics. *International Journal of Computer Applications, 1*(11), 0975 – 8887.

Report, P. W. C. (2015). *The Global State of Information Security Survey 2015*. Retrieved from http://www.pwc.com

Saint-Germain, R. (2005). Information security management best practice based on ISO/IEC 17799. *Information Management Journal, 39*(4), 60–65.

Sarstedt, M., Ringle, C. M., Smith, D., Reams, R., & Hair, J. F. Jr. (2014). Partial least squares structural equation modeling (PLS-SEM): A useful tool for family business researchers. *Journal of Family Business Strategy, 5*(1), 105–115. doi:10.1016/j.jfbs.2014.01.002

Shon, H. (2013). *All-In-One CISSP Exam Guide* (6th ed.). McGraw Hill.

Sommestad, T., Hallberg, J., Lundholm, K., & Bengtsson, J. (2014). Variables influencing information security policy compliance. *Information Management & Computer Security, 22*(1), 42–75. doi:10.1108/IMCS-08-2012-0045

Sommestad, T., Karlzén, H., & Hallberg, J. (2015). The sufficiency of the theory of planned behavior for explaining information security policy compliance. *Information & Computer Security, 23*(2), 200–217. doi:10.1108/ICS-04-2014-0025

Stone, M. (1974). Cross-validatory choice and assessment of statistical predictions. *Journal of the Royal Statistical Society. Series B. Methodological, 36*, 111–147. doi:10.2307/2984809

Tenenhaus, M., Vinzi, V. E., Chatelin, Y. M., & Lauro, C. (2005). PLS path modelling. *Computational Statistics & Data Analysis, 48*(1), 159–205. doi:10.1016/j.csda.2004.03.005

Ungerman, M. (2005). Creating and enforcing an effective information security policy. *Information Systems Control Journal, 6*. Retrieved from http://www.isaca.org

Urbach, N., & Ahlemann, F. (2010). Structural equation modelling in information systems research using partial least squares. *Journal of Information Technology Theory and Application, 11*, 5–40. Retrieved from http://iris.ebs.edu

US GAO. (2012). *Information security: Better implementation of controls for mobile devices should be encouraged.* Retrieved from http://www.gao.gov/products/GAO-12-757

Vance, A., Lowry, P. B., & Eggett, D. (2013). Using Accountability to Reduce Access Policy Violations in Information Systems. *Journal of Management Information Systems, 29*(4), 263–289. doi:10.2753/MIS0742-1222290410

Wang, J., Gupta, M., & Raghav Rao, H. R. (2015). Insider threats in a financial institution: Analysis of attack-proneness of information systems applications. *Management Information Systems Quarterly, 39*(1), 91–112. doi:10.25300/MISQ/2015/39.1.05

Willison, R., & Warkentin, M. (2013). Beyond deterrence: An expanded view of employee computer abuse. *Management Information Systems Quarterly, 37*(1), 1–20. doi:10.25300/MISQ/2013/37.1.01

Zhang, J., Reithel, B. J., & Li, H. (2009). Impact of perceived technical protection on security behaviors. *Information Management & Computer Security, 17*(4), 330–340. doi:10.1108/09685220910993980

KEY TERMS AND DEFINITIONS

Administrative Controls: Security measures or safeguards that are instituted by organizational leaders to protect information resources. This includes security policy, security awareness and training programs, and job rotation.

Compliance: An effort to ensure that organizations are abiding by their own security policies, standards, guidelines, industry security regulations, and government legislation.

Controls: Safeguards or countermeasures that are put in place to protect information resources from misuse, modification, and damage.

Information Resources: Data and information assets of an organization including databases, personnel, and equipment.

Monitoring: An ongoing process of regularly assessing information systems to ensure that they meet the organization's security requirements.

Physical Controls: Security measures or safeguards that are implemented to protect facilities (buildings, computer installations, etc.). This includes locks, CCTV cameras, smoke detectors, and fencing.

Policy: A document normally prepare by organizational leaders that states the rules and procedures for all employees accessing and using the organization's information resources.

Technical Controls: Security measures or safeguards that are included in hardware and software with the intention to protect information resources. This includes anti-malware, encryption systems, and access control systems.

Vulnerability: A weakness or flaw in hardware, networks and software systems, or in people that makes them open to attacks.

Compilation of References

Abelson, H. (2008). The creation of OpenCourseWare at MIT. *Journal of Science Education and Technology, 17*(2), 164–174. doi:10.1007/s10956-007-9060-8

Abras, C., Maloney-Krichmar, D., & Preece, J. (2004). User-centered design. In Encyclopedia of Human-Computer Interaction. Thousand Oaks, CA: Sage Publications.

Acquisti, A., & Gross, R. (2009). Predicting Social Security numbers from public data. *Proceedings of the National Academy of Sciences of the United States of America, 106*(27), 10975–10980. doi:10.1073/pnas.0904891106 PMID:19581585

Adell, J., & Castañeda, L. (2010). Los entornos personales de aprendizaje (PLEs): una nueva manera de entender el aprendizaje. In Claves para la investigación en Innovación y calidad educativas. La integración de las Tecnologías de la Información y la Comunicación y la Interculturalidad en las aulas (pp. 19-30) Alcoy: Marfil.

Adolph, M. (2009). *Mobile Applications.* International Telecommunication Union.

Agarwal, A. (2012). *Security update & new features.* Retrieved June 25, 2017, from https://blog.dropbox.com/bre/07/security-update-new-features/

Agresti. (2002). *Categorical Data Analysis* (2nd ed.). New York: John Wiley & Sons.

Aira, C. (2001). *Diccionario de autores latinoamericanos.* Buenos Aires: Emecé & Ada Korn.

ALA/ACRL. (2000). *Information literacy competency standards for higher education.* Chicago, IL: ACRL.

Alavi, M., & Leidner, D. E. (2001). Knowledge management and knowledge management systems: Conceptual foundations and research issues. *Management Information Systems Quarterly, 25*(1), 107–136. doi:10.2307/3250961

Alden, J. (2013). Accomodating mobile learning in college programs. *Journal of Asynchronous Learning Networks, 17*(1), 109–122.

Alonso Berrocal, J. L., Gómez Díaz, R., García Figuerola, C., Zazo Rodríguez, A., & Cordón García, J. A. (2011). Propuesta de estudio del campo semántico de los libros electrónicos en Twitter. Zaragoza: *Scire: representación y organización del conocimiento, 17*(2), 87-97. Retrieved from http://ibersid. eu/ojs/index.php/scire/article/view/3939/3710

Alrasheedi, M., Capretz, L. F., & Raza, A. (2015). A Systematic Review of the Critical Factors for Success of Mobile Learning in Higher Education (University Students' Perspective). *Journal of Educational Computing, 52*(2), 252–276. doi:10.1177/0735633115571928

Álvarez Cuervo, R., Roces, J., & Alonso, J. (2006). *Sistema tutorado interactivo para el dibujo de vistas SDV®. XVIII Congreso Internacional Ingeniería Gráfica*, Sitges, Barcelona.

Álvarez Peñín, P., Del Canto, A., Charro, M., Rubio, R., García, R., & Suárez, J. (2001). *Propuesta de un sistema multimedia para minimizar el efecto de la reducción de docencia tradicional en Expresión Gráfica. XIII Congreso Internacional de Ingeniería Gráfica*, Badajoz.

Álvarez Peñín, P., Pando, P., García, R., & Pérez, M. (2006). *Entorno multimedia para uso docente en materias de Expresión Gráfica. XVIII Congreso Internacional de Ingeniería Gráfica*, Sitges, Barcelona.

American Library Association. (1989). *American Library Association Presidential Committee on Information Literacy: final report*. Chicago: American Library Association.

André, M., & Bueno, E. (2008). *Latin American women writers: an encyclopaedia*. New York: Routledge.

Andretta, S. (2007). Phenomenography: A conceptual framework for information literacy education. *Aslib Proceedings*, *59*(2), 152–168. doi:10.1108/00012530710736663

Angiani, G., Fornacciari, P., Mordonini, M., Tomaiuolo, M., & Iotti, E. (2016). Models of participation in social networks. In Social Media Performance Evaluation and Success Measurements (pp. 196-224). Academic Press.

An, H., Kim, S., & Kim, B. (2008). Teacher perspectives on online collaborative learning: Factors perceived as facilitating and impeding successful online group work. *Contemporary Issues in Technology & Teacher Education*, *8*(1), 65–83.

Applications (AINA). (2012). In *IEEE 26th International Conference on* (pp. 868-873). IEEE.

Aramo-Immonen, H., Jussila, J., & Huhtamäki, J. (2014, October). Visualizing informal learning behavior from conference participants Twitter data. *Proceedings of the Second International Conference on Technological Ecosystems for Enhancing Multiculturality*, 603-610. doi:10.1145/2669711.2669962

Area Moreira, M. (2001). *Educar en la sociedad de la información*. Bilbao: Desclée De Brouwer.

Area Moreira, M. (2010). ¿Por qué formar en competencias informacionales y digitales en la educación superior? *Revista de Universidad y Sociedad del Conocimiento*, *7*(2), 2–5.

Área, M. M. A., San Nicolás, M. B., & Fariña, E. (2010). Buenas prácticas de aulas virtuales en la docencia universitaria semipresencial. *Revista de Teoría de la Educación Sociedad de la Información*, *11*(3), 7–31.

Aronson, E., Wilson, T. D., & Akert, R. M. (2010). *Social Psychology*. Upper Saddle River, NJ: Prentice Hall.

Ashour, R., Alzghool, H., Iyadat, Y., & Abu-Alruz, J. (2012). Mobile phone applications in the university classroom: Perceptions of undergraduate students in Jordan. *E-Learning and Digital Media*, *9*(4), 419–425. doi:10.2304/elea.2012.9.4.419

Attwell, G. (2006). *Personal Learning Environments – a position paper*. Retrieved from http://www.knownet.com/writing/weblogs/Graham_Attwell/entries/6521819364

Aviles, L. (1997). Causes and consequences of cooperation and permanent-sociality in spiders. In J. Choe & B. Crespi (Eds.), *Evolution of social behaviour in insects and arachnids* (pp. 467–498). Cambridge, UK: Cambridge Univ. Press.

Baartman, L., & Ruijs, L. (2011). Comparing students' perceived and actual competence in higher vocational education. *Assessment & Evaluation in Higher Education*, *36*(4), 385–398. doi:10.1080/02602938.2011.553274

Back, T., & Schwefel, H. P. (1996, May). Evolutionary computation: An overview. In *Evolutionary Computation, 1996., Proceedings of IEEE International Conference on* (pp. 20-29). IEEE.

Bakkenes, I., Vermunt, J. D., & Wubbels, T. (2010). Teacher learning in the context of educational innovation: Learning activities and learning outcomes of experienced teachers. *Learning and Instruction*, *20*(6), 533–548. doi:10.1016/j.learninstruc.2009.09.001

Balanskat, A., & Engelhardt, K. (2015). *Computing our future. Computer programming and coding Priorities, school curricula and initiatives across Europe*. Brussels, Belgium: European Schoolnet. Retrieved from https://goo.gl/i5aQiv

Balderston, D., & Gonzalez, M. (Eds.). (2004). *Encyclopaedia of Latin American and Caribbean Literature 1900-2003*. London: Routledge.

Ball, D. L., & Forzani, F. M. (2010). What does it take to make a teacher? *Phi Delta Kappan*, *92*(2), 8–12. doi:10.1177/003172171009200203

Baracaldo, N., & Joshi, J. (2013). An adaptive risk management and access control framework to mitigate insider threats. *Computers & Security*, *39*, 237–254. doi:10.1016/j.cose.2013.08.001

Barcelos, T. S., & Silveira, I. F. (2012). Pensamento computacional e educação matemática: Relações para o ensino de computação na educação básica. In *XX Workshop sobre Educação em Computação, Curitiba. Anais do XXXII CSBC* (*Vol. 2*, p. 23). Academic Press.

Bawden, D. (2001). Information and digital literacies: A review of concepts. *The Journal of Documentation*, *57*(2), 218–259. doi:10.1108/EUM0000000007083

Bell, W. J., Roth, L. M., & Nalepa, C. A. (2007). *Cockroaches: Ecology, behaviour, and natural history*. JHU Press.

Benczur, A. A., Csalogany, K., Sarlos, T., & Uher, M. (2005, May). SpamRank–Fully Automatic Link Spam Detection Work in progress. *Proceedings of the First International Workshop on Adversarial Information Retrieval on the Web.*

Benjamin, J., Kallicharan, L., McDonald, I., & Searwar, L. (Eds.). (1998). *They Came in Ships: An Anthology of Indo-Guyanese Writing*. London: Peepal Tree P.

Bergenti, F., Poggi, A., & Tomaiuolo, M. (2014). An Actor Based Software Framework for Scalable Applications. *Lecture Notes in Computer Science*, *8729*, 26–35. doi:10.1007/978-3-319-11692-1_3

Bergenti, F., Rossi, L., & Tomaiuolo, M. (2009). Towards automated trust negotiation in MAS. *Proceeding 10th Workshop on Objects and Agents, WOA'09.*

Berlanga, A. J., & García-Peñalvo, F. J. (2005a). IMS LD reusable elements for adaptive learning designs. *Journal of Interactive Media in Education*, 11.

Berlanga, A. J., & García-Peñalvo, F. J. (2005b). Learning Technology Specifications: Semantic Objects for Adaptive Learning Environments. *International Journal of Learning Technology*, *1*(4), 458–472. doi:10.1504/IJLT.2005.007155

Bers, M. U., Flannery, L., Kazakoff, E. R., & Sullivan, A. (2014). Computational thinking and tinkering: Exploration of an early childhood robotics curriculum. *Computers & Education*, *72*, 145–157. doi:10.1016/j.compedu.2013.10.020

Besteiro Martínez, B. (2015). *Herramienta para el análisis y visualización de las características principales de un MOOC*. Academic Press.

Bethencourt, J., Sahai, A., & Waters, B. (2007). Ciphertext-policy attribute-based encryption. In *Symposium on Security and Privacy* (pp. 321-334). IEEE.

Bevand, M. (2010). *Whitepixel.* Retrieved June 25, 2017, from http://whitepixel.zorinaq.com/

Biddle, B. J. (1986). Recent Development in Role Theory. *Annual Review of Sociology*, 1267–1292.

Bielba Calvo, M., Martínez Abad, F., & Rodríguez Conde, M. J. (En prensa). Validación psicométrica de un instrumento de evaluación de competencias informacionales en la educación secundaria. *Bordón. Revista de pedagogía.*

Bielba Calvo, M., Martínez Abad, F., Herrera García, M. E., & Rodríguez Conde, M. J. (2015). Diseño de un instrumento de evaluación de competencias informacionales en educación secundaria obligatoria a través de la selección de indicadores clave. *Education in the Knowledge Society*, *16*(3), 124–143. doi:10.14201/eks2015163124143

Bielba Calvo, M., Martínez-Abad, F., & Herrera García, M. E. (2014). Selection of key standards to create an instrument for information literacy assessment in Compulsory Secondary Education. In *Proceedings TEEM' 14. Technological Ecosystems for Enhancing Multiculturality* (pp. 475-482). Salamanca: ACM.

Biggs, J. (1999). What the Student Does: Teaching for enhanced learning. *Higher Education Research & Development*, *18*(1), 57–75. doi:10.1080/0729436990180105

Bikhard, M. H. (1993). Representational Content in Humans and Machines. *Journal of Experimental & Theoretical Artificial Intelligence*, *5*(4), 285–333. doi:10.1080/09528139308953775

Bilge, L., Strufe, T., Balzarotti, D., & Kirda, E. (2009, April). All your contacts are belong to us: automated identity theft attacks on social networks. In *Proceedings 18th International Conference on World Wide Web* (pp. 551--560). ACM. doi:10.1145/1526709.1526784

Birbalsingh, F. (Ed.). (1988). *Jahaji Bhai: An Anthology of Indo-Caribbean Literature*. Toronto: TSAR Publications.

Blanco, Á. F., Lacleta, M. L. S., Gené, O. B., & Peñalvo, F. J. G. (2014). Educación en abierto: Integración de un MOOC con una asignatura académica. Teoría de la Educación. *Educación y Cultura en la Sociedad de la Información*, *15*(3), 233.

Blanco, M. (2006). Estudio del sistema diédrico mediante un tutorial multimedia. *XVIII Congreso Internacional Ingeniería Gráfica*, Sitges, Barcelona.

Bloom, B. S. (1953). Thought-processes in lectures and discussions. *The Journal of General Education*, *7*(3), 160–169.

Bloom, H. (1995). *El canon occidental*. Barcelona: Anagrama.

Bodley, J. H. (1994). An anthropological perspective. *Cultural Anthropology: Tribes, States, and the Global System, 3*.

Bodriagov, O., & Buchegger, S. (2013). Encryption for peer-to-peer social networks. In *Security and Privacy in Social Networks* (pp. 47–65). Springer. doi:10.1007/978-1-4614-4139-7_4

Bogdanova, A. M., & Ackovska, N. (2010). Data Driven Intelligent Systems. In M. Gusev (Ed.), *Proceedings of the ICT Innovations 2010* (pp. 1-9). ICT ACT.

Bonett, P., Abad Faciolince, H., & Collazos, O. (n.d.). *Bogotá39. 39 escritores menores de 39*. Bogotá: Hay Festival. Secretaría Distrital de Cultura, Recreación y Deporte de Bogotá. Retrieved from http://www.hayfestival.com/bogota39/es-index.aspx?skinid=7&localesetting=es-ES&currencysetting=EUR&resetfilters=true

Bonneau, J., Anderson, J., Anderson, R., & Stajano, F. (2009, March). Eight friends are enough: social graph approximation via public listings. In *Proceedings 2nd ACM EuroSys Workshop on Social Network Systems* (pp. 13--18). ACM. doi:10.1145/1578002.1578005

Borrás-Gené, O., Martinez-Nuñez, M., & Blázquez-Sevilla, A. (2016). Gamified open Q&A forum model as a platform for a CMOOCs. *INTED2016 Proceedings*, 4903-4910.

Borrás-Gené, O., Martinez-Nuñez, M., & Fidalgo-Blanco, A. (2015). Gamificación de un MOOC y su comunidad de aprendizaje a través de actividades. *Actas del III Congreso Internacional sobre Aprendizaje, Innovación y Competitividad*. CINAIC 2015.

Borrás-Gené, O., Martínez-Nuñez, M., & Fidalgo-Blanco, Á. (2016). New Challenges for the Motivation and Learning in Engineering Education Using Gamification in MOOC. *International Journal of Engineering Education*, *32*(1B), 501–512.

Bostock, M., Ogievetsky, V., & Heer, J. (2011). D³ data-driven documents. *IEEE Transactions on Visualization and Computer Graphics*, *17*(12), 2301–2309. doi:10.1109/TVCG.2011.185 PMID:22034350

Bouarara, H. A., Hamou, R. M., & Amine, A. (2016a). Artificial haemostasis system for modern information retrieval with 3D result-mining. *Journal of Information Technology Research*, *9*(2), 17–46. doi:10.4018/JITR.2016040102

Bouarara, H. A., Hamou, R. M., & Amine, A. (2016b). New Bio Inspired techniques in the filtering of spam. *Journal of Information Technology Research*, *9*(2), 44–77. doi:10.4018/JITR.2016040103

Bourdieu, P. (1966). *Campo de poder, campo intelectual. Itinerario de un concepto.* Buenos Aires: Montressor.

Bourjot, C., Chevrier, V., & Thomas, V. (2003). A new swarm mechanism based on social spiders colonies: From web weaving to region detection. *Web Intelligence and Agent Systems*, *1*(1), 47–64.

Bovens, M. (2010). Two concepts of accountability: Accountability as a virtue and as a mechanism. *West European Politics*, *33*(5), 946–967. doi:10.1080/01402382.2010.486119

Brackmann, C. P., Couto Barone, D. A., Casali, A., & Hernández, S. (2016). Pensamento Computacional: Panorama nas Américas. In F. J. García-Peñalvo & J. A. Mendes (Eds.), *XVIII Simposio Internacional de Informática Educativa, SIIE 2016* (pp. 197–202). Salamanca: Ediciones Universidad de Salamanca.

Breslow, L., Pritchard, D. E., DeBoer, J., Stump, G. S., Ho, A. D., & Seaton, D. T. (2013). Studying learning in the worldwide classroom: Research into edX's first MOOC. *Research & Practice in Assessment, 8.*

Bridge, M. C. (1997). The rise and fall of an executive information system: A case study. *Information Systems Journal*, *7*(4), 277–288. doi:10.1046/j.1365-2575.1997.00021.x

Briz-Ponce, L. (2016). *Análisis de la efectividad en las Aplicaciones m-health en dispositivos móviles dentro del ámbito de la formación médica.* Salamanca: University of Salamanca.

Briz-Ponce, L., & Juanes-Méndez, J. A. (2015). Mobile Devices and Apps, Characteristics and Current Potential on Learning. *Journal of Information Technology Research*, *8*(4), 26–37. doi:10.4018/JITR.2015100102

Briz-Ponce, L., Juanes-Méndez, J. A., & García-Peñalvo, F. J. (2014a). Analysis of Certificated Mobile Application for Medical Education Purposes. In F. J. García-Peñalvo (Ed.), *Proceedings of the Second International Conference on Technological Ecosystems for Enhancing Multiculturality-TEEM14* (pp. 13–17). New York: ACM. doi:10.1145/2669711.2669871

Briz-Ponce, L., Juanes-Méndez, J. A., & García-Peñalvo, F. J. (2014b). Analysis of mobile devices as a support tool for professional medical education in the university school. In *Proceedings of 6th International Conference on Education and New Learning Technologies-EDULEARN14* (pp. 4653–4658). Valencia: IATED Academy.

Briz-Ponce, L., Juanes-Méndez, J. A., & García-Peñalvo, F. J. (2014c). First Approach of Mobile Applications Study for Medical Education purposes. In *Proceedings of the Second International Conference on Technological Ecosystems for Enhancing Multiculturality-TEEM14* (pp. 647–651). New York: ACM. doi:10.1145/2669711.2669968

Brooks, R. A. (1991). Intelligence without reason. In *Proceedings of the 12th international joint conference on Artificial intelligence.* San Francisco, CA: Morgan Kaufmann Publishers.

Brophy, J. E. (2013). *Motivating students to learn.* Routledge.

Brown, S., & McWatt, M. (Eds.). (2004). *The Oxford Book of Caribbean Verse.* Oxford, UK: OUP.

Bruce, C. (1997). *The seven faces of information literacy.* Adelaide: Auslib Press.

Bruce, G. M. et al.. (2012). Recent advances in applying decision science to managing national forests. *Forest Ecology and Management*, *285*, 123–132. doi:10.1016/j.foreco.2012.08.024

Buckley, P., & Doyle, E. (2016). Gamification and student motivation. *Interactive Learning Environments*, *24*(6), 1–14. doi:10.1080/10494820.2014.964263

Bulgurcu, B., Cavusoglu, H., & Benbasat, I. (2010). Information security policy compliance: An empirical study of rationality-based beliefs and information security awareness. *Management Information Systems Quarterly*, *34*(3), 523–548.

Bundy, A. (1998). Information literacy: the key competency for the 21sr century. In *Annual Conference of the International Association of Technological University Libraries* (p. 16). Pretoria, South Africa: Academic Press.

Bundy, A., & Anziil. (2004). *Australian and New Zealand information literacy framework principles, standards and practice*. Australian and New Zealand Institute for Information Literacy. Retrieved from http://www.anziil.org/resources/Info%20lit%202nd%20edition.pdf

Bunnel, F. L., & Boyland, M. (2003). Decision Support Systems: IT is a question not the model. *Journal for Nature Conservation*, *10*(4), 269–279. doi:10.1078/1617-1381-00027

Byrne, A. (2005). La alfabetización informacional desde una perspectiva global: Eldesastre agudiza nuestras mentes. *Anales de Documentación*, *8*, 7–20.

Cabrero, J. (2006). Las TIC y las inteligencias múltiples. Infobit. *Revista para la difusión y el uso educativo de las TIC, 13*, 8-9.

Cabrero, J. (2007). Comunidades virtuales para el aprendizaje. Su utilización en la enseñanza. Eduweb. *Revista de Tecnología de la Información y Comunicación en Educación, 1*(1), 5-22.

Calder, N. (2010). Using Scratch: An Integrated Problem-Solving Approach to Mathematical Thinking. *Australian Primary Mathematics Classroom*, *15*(4), 9–14.

Canadian Forest Service. (1995). *Criteria and Indicators for the Conservation and Sustainable Management of Temperate and Boreal Forests: Montreal Process*. Hull, Canada: Canadian Forest Service, Natural Resources Canada.

Canetti, R., Garay, J., Itkis, G., Micciancio, D., Naor, M., & Pinkas, B. (1999, March). Multicast security: A taxonomy and some efficient constructions. In *Proceedings 8th Annual Joint Conference of the IEEE Computer and Communications Societies* (Vol. 2, pp. 708--716). IEEE. doi:10.1109/INFCOM.1999.751457

Cantoral, R., Farfán, R. M., Lezama, J., & Martínez-Sierra, G. (2006). Socioepistemología y representación: algunos ejemplos. *Revista Latinoamericana de Investigación en Matemática Educativa*.

Caroli, . (1994). Knowledge based tool to support choice of the right air pollution model. In G. Guariso & B. Page (Eds.), *In Computer Support for Environmental Impact Assessment*. Amsterdam: North Holland.

Casany, M. J., Alier, M., Mayol, E., Piguillem, J., Galanis, N., García-Peñalvo, F. J., & Conde, M. Á. (2012). Moodbile: A Framework to Integrate m-Learning Applications with the LMS. *Journal of Research and Practice in Information Technology*, *44*(2), 129–149.

Cassel, C. M., Hackl, P., & Westlund, A. H. (1999). Robustness of partial least-squares method for estimating latent variable quality structures. *Journal of Applied Statistics*, *26*(4), 435–446. doi:10.1080/02664769922322

Cassity, E., & Ang, I. (2006). Industry partnerships and the «Knowledge Society»: The australian experience. *Minerva*, *44*(1), 47–63. doi:10.1007/s11024-005-5412-9

Castañeda, L., & Adell, J. (2011). El desarrollo profesional de los docentes en entornos personales de aprendizaje (PLE). In R. Roig Vila & C. Laneve (Eds.), *La práctica educativa en la Sociedad de la Información: Innovación a través de la investigación / La pratica educativa nella Società dell'informazione: L'innovazione attraverso la ricerca* (pp. 83–95). Alcoy: Marfil.

Castells, M. (1999). *The information age: economy, society and culture*. Oxford, UK: Blackwell.

Castro, M. E., Olmo, R. M. A., & Castro, M. E. (2012). *Desarrollo del pensamiento matemàtico infantil.* Granada: Depósito legal: GR-1173-2002.

Catts, R., & Lau, J. (2008). *Towards Information Literacy Indicators*. Paris: UNESCO.

Catts, R., & Lau, J. (2009). *Hacia unos indicadores de alfabetización informacional*. Madrid: Ministerio de Cultura.

Caudell, T. P., & Mizell. (1992). Augmented reality: An application of heads-up display technology to manual manufacturing processes. "System Sciences, 1992. *Proceedings of the Twenty-Fifth Hawaii International Conference on*. IEEE. doi:10.1109/HICSS.1992.183317

CAUL. (2001). *Information literacy standards*. Canberra: University of South Australia.

Cebrián Herreros, M. (2009). *La sociedad de la informacion y del conocimiento en los paises nordicos*. Barcelona: Gedisa.

Cella, S. (1998). *Diccionario de literatura latinoamericana*. Buenos Aires: El Ateneo.

Cerezo, J. M. (2008). Hacia un nuevo paradigma: la era de la información fragmentada. *Revista Telos*, (76). Retrieved from http://sociedadinformacion.fundacion.telefonica.com/telos/articulocuaderno.a sp@idarticulo=6&rev=76.htm

CERT-NZ. (2017). *Anti Public and Exploit.in release of credentials*. Retrieved June 25, 2017, from https://www.cert.govt.nz/businesses-and-individuals/recent-threats/anti-public-and-exploit-in-release/

Cervini, R., Dari, N., & Quiroz, S. (2014). Estructura familiar y rendimiento académico en países de América Latina: Los datos del Segundo Estudio Regional Comparativo y Explicativo. *rmie. Revista Mexicana de Investigación Educativa, 19*(61), 569–597.

Ceusters, W., Smith, B., Kumar, A., & Dhaen, C. (2004). Mistakes in medical ontologies: Where do they come from and how can they be detected? *Studies in Health Technology and Informatics*. PMID:15853269

Chajri, M., & Fakir, M. (2014). Application of Data Mining in e-Commerce. *Journal of Information Technology Research, 7*(4), 79–91. doi:10.4018/jitr.2014100106

Chakravarti, I. M., Laha, R. G., & Roy, J. (1967). *Handbook of methods of applied statistics*. New York: John Wiley.

Chen, W., Lim, C., & Tan, A. (2010). Pre-Service Teachers' ICT Experiences and Competencies: New Generation of Teachers in Digital Age. In *Proceedings of the 18th International Conference on Computers in Education* (pp. 631–638). Putrajaya, Malaysia: Asia-Pacific Society for Computers in Education.

Cheng, L., Wang, Z. B., Song, Y. H., & Guo, A. H. (2011, September). Cockroach swarm optimization algorithm for TSP. In *Advanced Engineering Forum* (Vol. 1, pp. 226-229). doi:10.4028/www.scientific.net/AEF.1.226

Chen, W., & Chang, E. (2007). Exploring a Digital Ecosystem Conceptual Model and Its Simulation Prototype. In *Proceedings of IEEE International Symposium on Industrial Electronics, 2007 (ISIE 2007)* (pp. 2933 - 2938). IEEE.

Chickering, A. W. & Gamson, Z. F. (1987, March). Seven Principles for Good Practice in Undergraduate Education. *The American Association for Higher Education Bulletin*.

Chin, W. W. (1998). The partial least squares approach to structural equation modelling. Modern Methods for Business Research, 295, 295–336. doi:10.1016/j.aap.2008.12.010

Chin, W. W. (2010). How to write up and report PLS analyses. Handbook of Partial Least Squares, 665–690. doi:10.1007/978-3-540-32827-8

Chirita, P. A., Diederich, J., & Nejdl, W. (2005, October). MailRank: using ranking for spam detection. In *Proceedings of the 14th ACM international conference on Information and knowledge management* (pp. 373-380). ACM.

Chowdhury, G., & Foo, S. (2012). *Digital libraries and information access: Research perspectives*. London: Facet Publishing.

Cialdini, R. B., & Goldstein, N. J. (2004). Social influence: Compliance and conformity. *Annual Review of Psychology*, *55*(1), 591–621. doi:10.1146/annurev.psych.55.090902.142015 PMID:14744228

Clow, D. (2013). MOOCs and the funnel of participation. In *Proceedings of the Third International Conference on Learning Analytics and Knowledge* (pp. 185–189). ACM. doi:10.1145/2460296.2460332

CNMC. (2017). *Informe Mensual Comunicaciones Móviles*. CNMC.

Cohen, L., & Manion, L. (1989). *Research Methods in Education* (3rd ed.). London: Routledge.

Coleman, J. S. (1966). *Equality of educational opportunity* (Vol. 2). Washington, DC: US Department of Health, Education, and Welfare, Office of Education.

Comaroff, J., & Comaroff, J. (1992). *Of totemism and Etnicity*. Ethnography and the Historical Imagination.

Conde González, M. Á., Muñoz Martín, C., & García-Peñalvo, F. J. (2008). M-learning, towards U-learning. In I. Arnedillo Sánchez, & P. Isaías (Eds.), *Proceedings of the IADIS International Conference Mobile Learning 2008* (pp. 196-200). IADIS Press.

Connolly, T. M., Boyle, E. A., MacArthur, E., Hainey, T., & Boyle, J. M. (2012). A systematic literature review of empirical evidence on computer games and serious games. *Computers & Education*, *59*(2), 661–686. doi:10.1016/j.compedu.2012.03.004

Contreras, P., Arancibia, M., & Cárcamo, L. (2009). Hipermedios y cooperación: la Web 2.0 y su incidencia en los procesos educativos [Hypermedia and cooperation: Web 2.0. and its influence in educational processes]. In M. Segura & B. Onetto (Eds.), Diálogos Culturales II. Interfaces Viciadas, comunicación visual y otras mediaciones (pp. 177-221). San José de Rio Preto: Bluecom.

Cope, B., & Kalantzis, M. (2009). Multiliteracies: New literacies, new learning. *Pedagogies*, *4*(3), 164–195. doi:10.1080/15544800903076044

Corbin, J., & Strauss, A. (1990). Grounded theory research: Procedures, canons, and evaluative criteria. *Qualitative Sociology*, *13*(1), 3–21. doi:10.1007/BF00988593

Cordón García, J. A., & García-Figuerola, C. (2012). Introducción. Aventuras, inventos y mixtificaciones del libro electrónico. In Libros electrónicos y contenidos digitales en la sociedad del conocimiento. Mercado, servicios y derechos. Madrid: Pirámide.

Cordón García, J. A. (2004). La visibilidad en los circuitos de la creación: literatura y traducción. In C. Gonzalo García & V. García Yebra (Eds.), *Manual de documentación y terminología para la traducción especializada* (pp. 127–169). Madrid: Arco/Libros.

Cordón García, J. A., Gómez Díaz, R., Alonso Arévalo, J., & Alonso Berrocal, J. L. (2014). *El ecosistema del libro electrónico universitario*. Salamanca: Ediciones Universidad de Salamanca.

Cormack, G. V., Gómez Hidalgo, J. M., & Puertas Sánz, E. (2007a). Feature engineering for mobile (SMS) spam filtering. In *Proceedings of the 30th Annual international ACM SIGIR Conference on Research and Development in information Retrieval*. ACM. doi:10.1145/1277741.1277951

Cormack, G. V., Gómez Hidalgo, J. M., & Puertas Sánz, E. (2007b). Spam filtering for short messages. In *Proceedings of the Sixteenth ACM Conference on Conference on information and Knowledge Management*. ACM. doi:10.1145/1321440.1321486

Cox, P. G. (1996). Some Issues in the Design of Agricultural Decision Support System. *Agricultural Systems*, *52*(2-3), 355–381. doi:10.1016/0308-521X(96)00063-7

Crasborn, F., Hennissen, P., Brouwer, N., Korthagen, F., & Bergen, T. (2011). Exploring a two-dimensional model of mentor teacher roles in mentoring dialogues. *Teaching and Teacher Education*, *27*(2), 320–331. doi:10.1016/j.tate.2010.08.014

Cronk, M. (2012, June). Using gamification to increase student engagement and participation in class discussion. *World conference on educational multimedia, hypermedia and telecommunications*, 2012(1), 311-315.

Crook, C. (2012). The Digital Native in Context: Tensions Associated with Importing Web 2.0 Practices into the School Setting. *Oxford Review of Education, 38*(1), 63-80.

Cross, J. (2011). *Informal learning: Rediscovering the natural pathways that inspire innovation and performance*. John Wiley & Sons.

CRUE-TIC & REBIUN. (2009). *Competencias informáticas e informacionales en los estudios de grado*. Retrieved from http://crue-tic.uji.es/index.php?option=com_remository&Itemid=28&func=startdown&id=226

Cruz-Benito, J., García-Peñalvo, F. J., & Therón, R. (2014). *Defining generic data collectors for Learning Analytics: Facing up the heterogeneous data from heterogeneous environments*. Paper presented at the International Conference on Advanced Learning Technologies (ICALT) 2014, Athens, Greece. doi:10.1109/ICALT.2014.108

Cruz-Benito, J., Therón, R., & García-Peñalvo, F. J. (2016). Software Architectures Supporting Human-Computer Interaction Analysis: A Literature Review. In P. Zaphiris & A. Ioannou (Eds.), *Learning and Collaboration Technologies: Third International Conference, LCT 2016, Held as Part of HCI International 2016, Toronto, ON, Canada, July 17-22, 2016, Proceedings* (pp. 125-136). Cham: Springer International Publishing. doi:10.1007/978-3-319-39483-1_12

Cruz-Benito, J., Therón, R., García Peñalvo, F. J., Maderuelo, C., Pérez-Blanco, J. S., Zazo, H., & Martin-Suarez, A. (2014). Monitoring and feedback of Learning Processes in Virtual Worlds through analytics architectures: A real case. In Á. Rocha, D. Fonseca, E. Redondo, L. P. Reis, & M. P. Cota (Eds.), Sistemas y Tecnologías de Información. Actas de la 9ª Conferencia Ibérica de Sistemas y Tecnologías de Información (Vol. I Artículos, pp. 1126-1131). Barcelona, España: AISTI (Asociación Ibérica de Sistemas y Tecnologías de Información). doi:10.1109/CISTI.2014.6877097

Cruz-Benito, J., Therón, R., García-Peñalvo, F. J., Sánchez-Prieto, J. C., Vázquez-Ingelmo, A., Martín-González, M., & Martínez, J. M. (2017). Improving Success/Completion Ratio in Large Surveys: A Proposal Based on Usability and Engagement. In P. Zaphiris & A. Ioannou (Eds.), *Learning and Collaboration Technologies. Technology in Education: 4th International Conference, LCT 2017, Held as Part of HCI International 2017, Vancouver, BC, Canada, July 9-14, 2017, Proceedings, Part II* (pp. 352-370). Cham: Springer International Publishing. doi:10.1007/978-3-319-58515-4_28

Cruz-Benito, J., Borrás-Gené, O., García-Peñalvo, F. J., Fidalgo Blanco, Á., & Therón, R. (2015). Extending MOOC ecosystems using web services and software architectures *Proceedings of the XVI International Conference on Human Computer Interaction* (pp. 1-7). Vilanova i la Geltrú, Spain: ACM. doi:10.1145/2829875.2829923

Cruz-Benito, J., Borrás-Gené, O., García-Peñalvo, F. J., Fidalgo Blanco, Á., & Therón, R. (2015a). Extending MOOC ecosystems using web services and software architectures. [*Proceedings of the Interacción, 2015*, XV.

Cruz-Benito, J., Borrás-Gené, O., García-Peñalvo, F. J., Fidalgo-Blanco, A., & Therón, R. (2015b). Detection of non-formal and informal learning in Learning Communities supported by social networks in the context of a cooperative MOOC. In *2015 International Symposium on Computers in Education (SIIE)* (pp. 195-200). IEEE. doi:10.1109/SIIE.2015.7451675

Cruz-Benito, J., Borrás-Gené, O., García-Peñalvo, F. J., Fidalgo-Blanco, A., & Therón, R. (2017). *Learning communities in social networks and their relationship with the MOOCs*. IEEE Revista Iberoamericana de Tecnologias del Aprendizaje.

Cruz-Benito, J., Maderuelo, C., García-Peñalvo, F. J., Therón, R., Pérez-Blanco, J. S., Gómez, H. Z., & Martín-Suárez, A. (2016). Usalpharma: A Software Architecture to Support Learning in Virtual Worlds. *IEEE Revista Iberoamericana de Tecnologias del Aprendizaje*, *11*(3), 194–204. doi:10.1109/RITA.2016.2589719

Cruz-Benito, J., Therón, R., García-Peñalvo, F. J., & Pizarro Lucas, E. (2015). Discovering usage behaviors and engagement in an Educational Virtual World. *Computers in Human Behavior*, *47*(0), 18–25. doi:10.1016/j.chb.2014.11.028

Cubrilovic, N. (2009). *RockYou Hack: From Bad To Worse*. Retrieved June 25, 2017, from http://techcrunch.com/2009/12/14/rockyou-hack-security-myspace-facebook-passwords/

Cutillo, L. A., Molva, R., & Strufe, T. (2009). Safebook: A privacy-preserving online social network leveraging on real-life trust. *Communications Magazine, IEEE*, *47*(12), 94–101. doi:10.1109/MCOM.2009.5350374

Cynosure Prime. (2016). *Myspace hashes, length 10 and beyond*. Retrieved June 25, 2017, from http://cynosureprime.blogspot.it/2016/07/myspace-hashes-length-10-and-beyond.html

D'Abreu, J. V. V., & Condori, K. O. V. (2017). Educación y Robótica Educativa. *Revista de Educación a Distancia*, (54).

D'Arcy, J., & Greene, G. (2014). Security culture and the employment relationship as drivers of employees' security compliance. *Information Management & Computer Security*, *22*(5), 474–489. doi:10.1108/IMCS-08-2013-0057

D'Arcy, J., & Herath, T. (2011). A review and analysis of deterrence theory in the IS security literature: Making sense of the disparate findings. *European Journal of Information Systems*, *20*(6), 643–658. doi:10.1057/ejis.2011.23

Dabbagh, N., & Kitsantas, A. (2012). Personal Learning Environments, social media, and self-regulated learning: A natural formula for connecting formal and informal learning. *The Internet and higher education*, *15*(1), 3–8. doi:10.1016/j.iheduc.2011.06.002

Dai, W. (2009). *Crypto++ 5.6.0 Benchmarks*. Retrieved June 25, 2017, from http://www.cryptopp.com/benchmarks-amd64.html

Dasgupta, D., & Forrest, S. (1999, July). Artificial immune systems in industrial applications. In *Intelligent Processing and Manufacturing of Materials, 1999. IPMM'99. Proceedings of the Second International Conference on* (*Vol. 1*, pp. 257-267). IEEE.

Day, C. (2011). Computational Thinking Is Becoming One Of The Three Rs. *Computing in Science & Engineering*, *13*(1), 88–88. doi:10.1109/MCSE.2011.5

Day, R., & Payne, L. (1987). Computer-managed instruction: An alternative teaching strategy. *The Journal of Nursing Education*, *26*(1), 30–36. doi:10.3928/0148-4834-19870101-08 PMID:3029349

DBIR. (2013). *2013 Data Breach Investigations Report*. Retrieved from http://www.verizonenterprise.com/DBIR/

de Certeau, M. (1996). *La invención de lo cotidiano. Tomos I y II*. Universidad Iberoamericana / ITESO / Centro de Estudios Mexicanos y Centroamericanos.

De la Orden, A., & Jornet, J. M. (2012). La utilidad de las evaluaciones de sistemas educativos: El valor de la consideración del contexto. *Bordón*, *64*(2), 69–88.

De Witte, K., & Kortelainen, M. (2013). What explains the performance of students in a heterogeneous environment? Conditional efficiency estimation with continuous and discrete environmental variables. *Applied Economics*, *45*(17), 2401–2412. doi:10.1080/00036846.2012.665602

Delors, J. (1996). *Learning, the treasure within: report to UNESCO of the International Commission on Education for the Twenty-first Century*. Paris: UNESCO.

Deterding, S., Khaled, R., Nacke, L. E., & Dixon, D. (2011, May). Gamification: Toward a definition. In *CHI 2011 Gamification Workshop Proceedings* (pp. 12-15). ACM.

Deutsch, K. W. (1986). *The nerves of government*. New York, NY: The free press.

Dhungana, D., Groher, I., Schludermann, E., & Biffl, S. (2010). Software ecosystems vs. natural ecosystems: learning from the ingenious mind of nature. In *ECSA '10 Proceedings of the Fourth European Conference on Software Architecture: Companion Volume* (pp. 96-102). New York, NY: ACM.

Di Nucci, S., Recoaro, N. and Grieco y Bavio, A. (2014). *Los chongos de Roa Bastos. Narrativa contemporánea de Paraguay*. Buenos Aires: Santiago Arcos Editor.

Di Nucci, S., Recoaro, N., Grieco, & Bavio, A. (2014). *De la Tricolor a la Wiphala. Narrativa contemporánea de Bolivia*. Buenos Aires: Santiago Arcos Editor.

Diaz-Balteiro, L., & Romero, C. (2008). Marking forestry decisions with multiple criteria: A review and an assessment. *Forest Ecology and Management*, *255*(8-9), 3222–3241. doi:10.1016/j.foreco.2008.01.038

Diehm, R.-A., & Lupton, M. (2012). Approaches to learning information literacy: A phenomenographic study. *Journal of Academic Librarianship*, *38*(4), 217–225. doi:10.1016/j.acalib.2012.05.003

Dillenbourg, P. (2002). Over-scripting CSCL: The Risks of Blending Collaborative Learning with Instructional Design. In P.A. Kirschner (Ed.), Inaugural Address, three Worlds of CSCL. Can We Support CSCL? (pp. 61-91). Heerlen: Open Universiteit Nederland.

Dillenbourg, P. (1999a). *Collaborative Learning: Cognitive and Computational Approaches* (2nd ed.). New York, NY: Elsevier Science.

Dillenbourg, P. (1999b). What do you mean by collaborative learning? In P. Dillenbourg (Ed.), *Collaborative-learning: Cognitive and Computational Approaches*. Oxford, UK: Elsevier.

Dirkx, J. M., & Smith, R. O. (2004). Thinking out of a bowl of spaghetti: Learning to learning in online collaborative groups. In T. S. Roberts (Ed.), *Online collaborative learning: Theory and practice* (pp. 132–159). Hershey, PA: Information Science Publishing. doi:10.4018/978-1-59140-174-2.ch006

diSessa, A. A. (2000). *Changing minds: Computers, learning, and literacy*. Cambridge, MA: MIT Press.

Django Software Foundation. (2015). *Django Web Framework*. Retrieved from https://www.djangoproject.com/

Dobber, M., Akkerman, S. F., Verloop, N., & Vermunt, J. D. (2012). Student teachers' collaborative research: Small-scale research projects during teacher education. *Teaching and Teacher Education*, *28*(4), 609–617. doi:10.1016/j.tate.2012.01.009

Dolan, B. (2011). *Nine medical schools that support mobile learning | mobihealthnews*. Retrieved May 9, 2015, from http://mobihealthnews.com/12346/nine-medical-schools-that-support-mobile-learning/

Domingo Coscollola, M & Marquès, P. (2011). Classroom 2.0 Experiences and building on the use of ICT in teaching. *Comunicar: revista científica iberoamericana de comunicación y educación, 37,* 169-175.

Domínguez, A., Saenz-de-Navarrete, J., De-Marcos, L., Fernández-Sanz, L., Pagés, C., & Martínez-Herráiz, J.-J. (2013). Gamifying learning experiences: Practical implications and outcomes. *Computers & Education, 63*, 380–392. doi:10.1016/j.compedu.2012.12.020

Dondis, D. A. (1974). *A primer of visual literacy*. Mit Press.

Donnell, A., & Bucknor, M. (Eds.). (2011). *The Companion to Anglophone Caribbean Literature*. Routledge.

Donnell, A., & Lawson Welsh, S. (Eds.). (1996). *The Routledge Reader in Caribbean Literature*. London: Routledge. doi:10.4324/9780203431054

Dorigo, M., Maniezzo, V., & Colorni, A. (1996). Ant system: optimization by a colony of cooperating agents. *Systems, Man, and Cybernetics, Part B: Cybernetics. IEEE Transactions on, 26*(1), 29–41.

Downes, S. (2008). *MOOC and Mookies: The Connectivism & Connective Knowledge Online Course*. Retrieved June 02, 2017, from: http://www.slideshare.net/Downes/mooc-and-mookiesthe-connectivism-connective-knowledge-online-course-presentation

Downes, S. (2012). *Connectivism and connective knowledge: Essays on meaning and learning networks*. National Research Council Canada. Retrieved June 10, 2017, from http://www.downes.ca/files/books/Connective_Knowledge-19May2012.pdf

Dragusin, R. (2012). *Data breach at IEEE.org: 100k plaintext passwords*. Retrieved June 25, 2017, from http://ieeelog.com/

Driscoll, M. (1997). Defining Internet-based and Web-based training. *Performance Improvement, 36*(4), 5–9. doi:10.1002/pfi.4140360403

Drucker, H., Wu, D., & Vapnik, V. N. (1999). Support vector machines for spam categorization. *IEEE Transactions on Neural Networks, 10*(5), 1048–1054. doi:10.1109/72.788645 PMID:18252607

Eastman, J. R. (1997). *IDRISI for Windows*. Worcester, MA: User's Guide IDRISI Production.

Eden Dahlstrom. (2012). *ECAR Study of undergraduate Students and Information Technology (Research Report)*. EDUCAUSE Center for Applied Research.

Elssied, N. O. F., Ibrahim, O., & Osman, A. H. (n.d.). Enhancement of spam detection mechanism based on hybrid\ varvec {k}-mean clustering and support vector machine. *Soft Computing*, 1-12.

Engel, A., & Onrubia, J. (2010). Patrones de organización grupal y fases de construcción del conocimiento en entornos virtuales de aprendizaje colaborativo [Group patterns and knowledge construction phases in virtual collaborative learning]. *Infancia y Aprendizaje: Journal for the Study of Education and Development, 33*(4), 515–528. doi:10.1174/021037010793139608

Epper, R., & Bates, A.W. (2004). *Enseñar al profesorado cómo utilizar la tecnología. Buenas prácticas de instituciones líderes*. Barcelona: Editorial UOC. Colección Educación y Sociedad Red.

Eshach, H. (2007). Bridging in-school and out-of-school learning: Formal, non-formal, and informal education. *Journal of Science Education and Technology, 16*(2), 171–190. doi:10.1007/s10956-006-9027-1

Esposito, J., Walker, K., & Ehling, T. (2012). *PDA and the university press. A report prepared for The Andrew W. Mellon Foundation*. Baltimore, MD: The Johns Hopkins University Press/Cornell University Library. Retrieved from https://jscholarship.library.jhu.edu/bitstream/handle/1774.2/36210/PDA%20and%20the%20University%20Press%20FINAL%202012-10-15.pdf

Expósito Ortiz, S. (2013). *La importancia de la alfabetización visual en educación primaria*. Academic Press.

Felbo, B. (2017). *What can we learn from emojis?* Retrieved from https://www.media.mit.edu/posts/what-can-we-learn-from-emojis/

Felt, A., & Evans, D. (2008). Privacy protection for social networking APIs. *2008 Web 2.0 Security and Privacy*.

Ferguson, I. S. (1996). *Sustainable Forest Management*. Melbourne: Oxford University Press.

Fernández, A., Reina, R., Mateo, F., & Sánchez, J. (1998). Docencia de dibujo técnico a través de herramientas de animación asistida por ordenador. *X Congreso Internacional Ingeniería Gráfica*, Málaga

Ferrari, A. (2013). *DIGCOMP: a framework for developing and understanding digital competence in Europe* (Y. Punie & B. N. Brečko, Eds.). Luxembourg: Publications Office of the European Union.

Ferreira, J. B., Klein, A. Z., Freitas, A., & Schlemmer, E. (2013). Mobile learning: Definition, uses and challenges. In L. A. Wankel & P. Blessinger (Eds.), Cutting-edge Technologies in Higher Education (pp. 47–82). Bingley, UK: Emerald Group Publishing Limited. doi:10.1108/S2044-9968(2013)000006D005

Fidalgo Blanco, Á., Sein-Echaluce Lacleta, M. L., Borrás Gené, O., & García Peñalvo, F. J. (2014). Educación en abierto: Integración de un MOOC con una asignatura académica. *Education in the Knowledge Society*, *15*(3), 233–255.

Fidalgo Blanco, Á., Sein-Echaluce Lacleta, M. L., & García-Peñalvo, F. J. (2015). Methodological Approach and Technological Framework to break the current limitations of MOOC model. *Journal of Universal Computer Science*, *21*(5), 712–734.

Fidalgo, A., Sein-Echaluce, M. L., & Garcíapeñalvo, F. J. (2013). Coperative MOOC. An integration between cMOOC and xMOOC. *II Congreso internacional sobre Aprendizaje (CINAIC 2013)*, 481-486.

Fidalgo, A., Sein-Echaluce, M. L., Lerís, D., & Castañeda, O. (2013). Teaching Innova Project: The incorporation of adaptable outcomes in order to grade training adaptability. *Journal of Universal Computer Science*, *19*(11), 1500–1521.

Fidalgo-Blanco, Á., Sein-Echaluce, M. L., & García-Peñalvo, F. J. (2015). Epistemological and ontological spirals: From individual experience in educational innovation to the organisational knowledge in the university sector. *Program: Electronic Library and Information Systems*, *49*(3), 266-288. doi:10.1108/PROG-06-2014-0033

Fidalgo-Blanco, Á., Sein-Echaluce, M. L., & García-Peñalvo, F. J. (2014). Knowledge Spirals in Higher Education Teaching Innovation. *International Journal of Knowledge Management*, *10*(4), 16–37. doi:10.4018/ijkm.2014100102

Fiedler, S. (2010). *Personal Learning Environments: concept or technology?* The PLE Conference, Barcelona, Spain. Retrieved on September 23rd 2015 from: http://pleconference.citilab.eu/wp-content/uploads/2010/07/ple2010_submission_45.pdf

Fielding, R. T., & Taylor, R. N. (2002). Principled design of the modern Web architecture. *ACM Transactions on Internet Technology*, *2*(2), 115–150. doi:10.1145/514183.514185

Firte, L., Lemnaru, C., & Potolea, R. (2010, August). Spam detection filter using KNN algorithm and resampling. In *Intelligent Computer Communication and Processing (ICCP), 2010 IEEE International Conference on* (pp. 27-33). IEEE. doi:10.1109/ICCP.2010.5606466

Flores, Á., & Gómez, A. (2013). *La modelación matemática y la enseñanza de las cónicas*. Academic Press.

Fogel, L. J. (1962). Autonomous automata. *Industrial Research*, *4*, 14–19.

Fonseca, D., Martí, N., Redondo, E., Navarro, I., & Sánchez, A. (2014). Relationship between student profile, tool use, participation, and academic performance with the use of Augmented Reality technology for visualized architecture models. *Computers in Human Behavior*, *31*, 434–445. doi:10.1016/j.chb.2013.03.006

Fonseca, D., Redondo, E., & Villagrasa, S. (2015). Mixed-methods research: A new approach to evaluating the motivation and satisfaction of university students using advanced visual technologies. *Universal Access in the Information Society*, *14*(3), 311–332. doi:10.1007/s10209-014-0361-4

Fonseca, D., Villagrasa, S., Navarro, I., Redondo, E., Valls, F., & Sánchez, A. (2017, April). Urban Gamification in Architecture Education. In *World Conference on Information Systems and Technologies* (pp. 335-341). Springer.

Fornell, C., & Larcker, D. F. (1981). Evaluating structural equation models with unobservable variables and measurements error. *JMR, Journal of Marketing Research*, *18*(1), 39–50. doi:10.2307/3151312

Foster, D. W. (1991). *Latin American Writers on Gay and Lesbian Themes: A Bio-Critical Sourcebook*. Austin, TX: University of Texas Press.

Franceschi-Bicchierai, L. (2016). *Another Day, Another Hack: 117 Million LinkedIn Emails And Passwords*. Retrieved June 25, 2017, from https://motherboard.vice.com/en_us/article/78kk4z/another-day-another-hack-117-million-linkedin-emails-and-password

Franchi, E., & Tomaiuolo, M. (2013). Distributed social platforms for confidentiality and resilience. In Social Network Engineering for Secure Web Data and Services (pp. 114--136). Academic Press. doi:10.4018/978-1-4666-3926-3.ch006

Franchi, E., Poggi, A., & Tomaiuolo, M. (2013). Open social networking for online collaboration. *International Journal of e-Collaboration*, *9*(3), 50–68. doi:10.4018/jec.2013070104

Franchi, E., Poggi, A., & Tomaiuolo, M. (2015). Information and password attacks on social networks: An argument for cryptography. *Journal of Information Technology Research*, *8*(1), 25–42. doi:10.4018/JITR.2015010103

Franchi, E., Poggi, A., & Tomaiuolo, M. (2016). Blogracy: A peer-to-peer social network. *International Journal of Distributed Systems and Technologies*, *7*(2), 37–56. doi:10.4018/IJDST.2016040103

Franchi, E., Poggi, A., & Tomaiuolo, M. (2016). Social media for online collaboration in firms and organizations. *International Journal of Information System Modeling and Design*, *7*(1), 18–31. doi:10.4018/IJISMD.2016010102

Fransecky, R. B., & Debes, J. L. (1972). *Visual Literacy: A Way to Learn--A Way to Teach*. Academic Press.

Fransecky, R. B., & Debes, J. L. (1972). *Visual literacy: A way to learn--a way to teach*. Washington, DC: Association for Educational Communications and Technology.

Frechette, J. (2006). Cyber-censorship or cyber-literacy? Envisioning cyber-learning through media education. In D. Buckingham & R. Willett (Eds.), *Digital Generations: Children, Young People, and the New Media*. New York, NY: Routledge.

Freire, P. (1981). *Literacy and revolution*. New York: Continuum.

Freire, P., & Macedo, D. P. (1987). *Literacy: reading the word & the world*. South Hadley, MA: Bergin & Garvey Publishers.

Freudenthal, H. (2006). *Revisiting mathematics education: China lectures* (Vol. 9). Springer Science & Business Media.

Freud, S. (1970). *El malestar en la cultura*. Ediciones AKAL.

Freud, S. (2012). *El porvenir de una ilusión (Serie Great Ideas 16)*. Taurus.

Fuentes, C. (2011). *La gran novela latinoamericana*. Madrid: Alfaguara.

Fuguet, A., & Gómez, A. (Eds.). (1996). *McOndo (una antología de nueva literatura hispanoamericana)*. Barcelona: Gijalbo-Mondadori.

Fundación Telefónica. (2014). *La Sociedad de la Información en España 2013*. Barcelona, Spain: Ariel, S.A.

Fundación Telefónica. (2017). *La Sociedad de la Información en España 2016*. Barcelona, Spain: Ariel, S.A.

Gallegos, A. (2013). *Medical Schools Embrace Benefits of Tablets, Mobile Devices*. Create New Ways of Learning.

Gamazo, A., Olmos-Migueláñez, S., & Martínez-Abad, F. (2016, November). Multilevel models for the assessment of school effectiveness using PISA scores. In *Proceedings of the Fourth International Conference on Technological Ecosystems for Enhancing Multiculturality* (pp. 1161-1166). ACM. doi:10.1145/3012430.3012663

Gammon, J., & O'Neill. (2011). *Ohio LINK-OCLC Collection and Circulation Analysis Project 2011*. Dublin, OH: OCLC Research, September 2011. Retrieved from http://www.oclc.org/content/dam/research/publications/library/2011/2011-06.pdf

García Canclini, N. (1999). *La globalización imaginada*. Buenos Aires: Paidós.

García Castaño, F. J., & Pulido Moyano, R. A. (1992). *Educación multicultural y antropología de la educación*. Academic Press.

García-Holgado, A., & García-Peñalvo, F. J. (2014a). *Architectural pattern for the definition of eLearning ecosystems based on Open Source developments*. Paper presented at the Computers in Education (SIIE), 2014 International Symposium on. doi:10.1109/SIIE.2014.7017711

García-Holgado, A., & García-Peñalvo, F. J. (2014b). *Patrón arquitectónico para la definición de ecosistemas de eLearning basados en desarrollos open source*. Academic Press.

García-Holgado, A., & García-Peñalvo, F. J. (2017a). A metamodel proposal for developing learning ecosystems. In P. Zaphiris & A. Ioannou (Eds.), *Learning and Collaboration Technologies. Novel Learning Ecosystems. 4th International Conference, LCT 2017. Held as Part of HCI International 2017, Vancouver, BC, Canada, July 9–14, 2017. Proceedings, Part I* (pp. 100-109). Springer International Publishing. doi:10.1007/978-3-319-58509-3_10

García-Holgado, A., Cruz-Benito, J., & García-Peñalvo, F. J. (2015). Análisis comparativo de la gestión del conocimiento en la administración pública española. In *La Sociedad del Aprendizaje. Actas del III Congreso Internacional sobre Aprendizaje, Innovación y Competitividad*. Madrid: Fundación General de la Universidad Politécnica de Madrid. Retrieved from http://www.dmami.upm.es/dmami/documentos/liti/Actas_CINAIC_2015.pdf

García-Holgado, A., Cruz-Benito, J., & García-Peñalvo, F. J. (2015). *Analysis of Knowledge Management Experiences in Spanish Public Administration*. Paper presented at the Third International Conference on Technological Ecosystems for Enhancing Multiculturality (TEEM'15), Porto, Portugal. doi:10.1145/2808580.2808609

García-Holgado, A., García-Peñalvo, F. J., Hernández-García, Á., & Llorens-Largo, F. (2015). *Analysis and improvement of knowledge management processes in organizations using the Business Process Model Notation*. Paper presented at the Annual Conference of the Global Innovation and Knowledge Academy. doi:10.1007/978-3-319-22204-2_9

García-Holgado, A., & García-Peñalvo, F. J. (2016). Architectural pattern to improve the definition and implementation of eLearning ecosystems. *Science of Computer Programming*, *129*, 20–34. doi:10.1016/j.scico.2016.03.010

García-Holgado, A., & García-Peñalvo, F. J. (2017b). Preliminary validation of the metamodel for developing learning ecosystems. In J. M. Dodero, M. S. Ibarra Sáiz, & I. Ruiz Rube (Eds.), *Fifth International Conference on Technological Ecosystems for Enhancing Multiculturality (TEEM'17)*. New York, NY: ACM.

García, I., Gros, B. Y., & Noguera, I. (2010). La relación entre las prestaciones tecnológicas y el diseño de las actividades de aprendizaje para la construcción colaborativa del conocimiento [ICT and learning activities design relation for the construction of collaborative knwoledge]. *Cultura y Educación*, *22*(4), 395–418. doi:10.1174/113564010793351867

García, M. A., Deco, C., & Collazos, C. A. (2017). Robotics Based Strategies to Support Computational Thinking: The Case of the Pascual Bravo Industrial Technical Institute. *Journal of Computer Science and Technology*, 17.

García, M., Martín, G., Suárez, J., Pérez, J., & Suárez, F. (2002). *S.A.D. (Sistema de apoyo al dibujo): Una herramienta integrada en la enseñanza de la expresión gráfica. XIV Congreso Internacional de Ingeniería Gráfica*, Santander.

García-Peñalvo, F. J. (2017). Pensamiento computacional en los estudios preuniversitarios. *El enfoque de TACCLE3*. Retrieved from https://repositorio.grial.eu/handle/grial/801

García-Peñalvo, F. J. (Ed.). (2008b). Advances in E-Learning: Experiences and Methodologies. Hershey, PA: Information Science Reference (formerly Idea Group Reference). doi:10.4018/978-1-59904-756-0

García-Peñalvo, F. J., & García Carrasco, J. (2002). Los espacios virtuales educativos en el ámbito de Internet: Un refuerzo a la formación tradicional. Education in the Knowledge Society, 3.

García-Peñalvo, F. J., Hughes, J., Rees, A., Jormanainen, I., Toivonen, T., Reimann, D., . . . Virnes, M. (2016). *Evaluation of existing resources (study/analysis)*. Belgium: TACCLE3 Consortium. doi: 10.5281/zenodo.163112

García-Peñalvo, F. J., Reimann, D., Tuul, M., Rees, A., & Jormanainen, I. (2016). *An overview of the most relevant literature on coding and computational thinking with emphasis on the relevant issues for teachers*. Belgium: TACCLE3 Consortium. doi:10.5281/zenodo.165123

García-Peñalvo, F. J. (2008a). Docencia. In J. Laviña Orueta & L. Mengual Pavón (Eds.), *Libro Blanco de la Universidad Digital 2010* (pp. 29–61). Barcelona: Ariel.

García-Peñalvo, F. J. (2014). Managing the Knowledge Society Construction. *International Journal of Knowledge Management*, *10*(4), iv–vii.

García-Peñalvo, F. J. (2015). Information Technology Research. *Journal of Information Technology Research*, *8*(1), iv–v.

García-Peñalvo, F. J. (2015a). Cómo entender el concepto de presencialidad en los procesos educativos en el siglo XXI. *Education in the Knowledge Society*, *16*(2), 6–12. doi:10.14201/eks2015162612

García-Peñalvo, F. J. (2015b). Engineering contributions to a Knowledge Society multicultural perspective. *IEEE Revista Iberoamericana de Tecnologías del Aprendizaje*, *10*(1), 17–18. doi:10.1109/RITA.2015.2391371

García-Peñalvo, F. J. (2015d). Massive Open Online Courses as Data Sources for Making Decisions in Learning Processes. *Journal of Information Technology Research*, *8*(4), iv–vii.

García-Peñalvo, F. J. (2016). A brief introduction to TACCLE 3 – Coding European Project. In F. J. García-Peñalvo, & J. A. Mendes (Eds.), *2016 International Symposium on Computers in Education (SIIE 16)*. IEEE. doi:10.1109/SIIE.2016.7751876

García-Peñalvo, F. J. (2016). Digital Humanities Data Processing. *Journal of Information Technology Research*, *9*(1), v–viii.

García-Peñalvo, F. J. (2016). Proyecto TACCLE3 – Coding. In F. J. García-Peñalvo & J. A. Mendes (Eds.), *XVIII Simposio Internacional de Informática Educativa, SIIE 2016* (pp. 187–189). Salamanca: Ediciones Universidad de Salamanca.

García-Peñalvo, F. J. (2016). What Computational Thinking Is. *Journal of Information Technology Research*, *9*(3), v–viii.

García-Peñalvo, F. J. (2016b). Technological Ecosystems. *IEEE Revista Iberoamericana de Tecnologias del Aprendizaje, 11*(1), 31–32. doi:10.1109/RITA.2016.2518458

García-Peñalvo, F. J. (2017). Multidisciplinary and interdisciplinary approaches in information technology research. *Journal of Information Technology Research, 10*(1), vi–viii.

García-Peñalvo, F. J. (2018a). Technological ecosystems for enhancing the interoperability and data flows. *Journal of Information Technology Research, 11*(1).

García-Peñalvo, F. J. (Ed.). (2013). *Multiculturalism in Technology-Based Education: Case Studies on ICT-Supported Approaches*. Hershey, PA: IGI Global. doi:10.4018/978-1-4666-2101-5

García-Peñalvo, F. J., & Cruz-Benito, J. (2016). Computational thinking in pre-university education. In F. J. García-Peñalvo (Ed.), *Proceedings of the Fourth International Conference on Technological Ecosystems for Enhancing Multiculturality (TEEM'16)* (pp. 13-17). New York, NY: ACM. doi:10.1145/3012430.3012490

García-Peñalvo, F. J., Cruz-Benito, J., Borrás-Gené, O., & Blanco, Á. F. (2015). Evolution of the Conversation and Knowledge Acquisition in Social Networks related to a MOOC Course. In *Learning and Collaboration Technologies* (pp. 470–481). Springer International Publishing. doi:10.1007/978-3-319-20609-7_44

García-Peñalvo, F. J., Cruz-Benito, J., Maderuelo, C., Pérez-Blanco, J. S., & Martín-Suárez, A. (2014). Usalpharma: A Cloud-Based Architecture to Support Quality Assurance Training Processes in Health Area Using Virtual Worlds. *The Scientific World Journal, 10*. doi:10.1155/2014/659364 PMID:24778593

García-Peñalvo, F. J., Fidalgo-Blanco, Á., & Sein-Echaluce, M. L. (2018). An adaptive hybrid MOOC model: Disrupting the MOOC concept in higher education. *Telematics and Informatics*. doi:10.1016/j.tele.2017.09.012

García-Peñalvo, F. J., & García Carrasco, J. (2005). Educational hypermedia resources facilitator. *Computers & Education, 44*(3), 301–325. doi:10.1016/j.compedu.2004.02.004

Garcia-Peñalvo, F. J., & Garcia-Holgado, A. (2016). *Open Source Solutions for Knowledge Management and Technological Ecosystems*. IGI Global.

García-Peñalvo, F. J., Hernández-García, Á., Conde-González, M. Á., Fidalgo-Blanco, Á., Sein-Echaluce Lacleta, M. L., Alier-Forment, M., & Iglesias-Pradas, S. et al. (2015). Learning services-based technological ecosystems. In G. R. Alves, & M. C. Felgueiras (Eds.), *Proceedings of the Third International Conference on Technological Ecosystems for Enhancing Multiculturality (TEEM'15)* (pp. 467-472). New York: ACM. doi:10.1145/2808580.2808650

García-Peñalvo, F. J., Llorens Largo, F., Molero Prieto, X., & Vendrell Vidal, E. (2017). Educación en Informática sub 18 (EI<18). *ReVisión, 10*(2), 13–18.

García-Peñalvo, F. J., Rees, A. M., Hughes, J., Jormanainen, I., Toivonen, T., & Vermeersch, J. (2016). A survey of resources for introducing coding into schools. In F. J. García-Peñalvo (Ed.), *Proceedings of the Fourth International Conference on Technological Ecosystems for Enhancing Multiculturality (TEEM'16)* (pp. 19-26). New York, NY: ACM. doi:10.1145/3012430.3012491

García-Peñalvo, F. J., & Seoane-Pardo, A. M. (2015). An updated review of the concept of eLearning. Tenth anniversary. *Education in the Knowledge Society, 16*(1), 119–144. doi:10.14201/eks2015161119144

García, R. (1994). *Documentación general (sistemas, redes, centros).Guía del alumno*. Madrid: Síntesis.

García-Sánchez, F., Cruz-Benito, J., Therón, R., & Gómez-Isla, J. (2015). *Designing and building systems and tools to analyze visual communications on social networks*. Paper presented at the Third International Conference on Technological Ecosystems for Enhancing Multiculturality (TEEM'15), Porto, Portugal. doi:10.1145/2808580.2808629

García-Sánchez, F., Therón, R., & Isla, J. G. (2014). Semiotic and technological analysis of photography: a visual literacy study in the educative area. *Proceedings of the Second International Conference on Technological Ecosystems for Enhancing Multiculturality*. doi:10.1145/2669711.2669978

García-Valcárcel Muñoz-Repiso, A., & Mena Marcos, J. J. (2016). Information Technology As a Way To Support Collaborative Learning: What In-service Teachers Think, Know and Do. *Journal of Information Technology Research*, *9*(1), 1–17. doi:10.4018/JITR.2016010101

García-Valcárcel, A., & Mena, J. (2013). *Recursos tecnológicos para la enseñanza e innovación educativa* [Technological resources for teaching and innovation]. Madrid: Síntesis.

García-Valcárcel, A., & Mena, J. (2014). Las TIC en el aprendizaje colaborativo en el aula de Primaria y Secundaria [ICT for collaborative learning in Primary and Secondary Education]. *Comunicar, 42*, 65-74

Garmendia, M., Gisasola, J., & Gorozika, J. (2004). *Enseñanza de la visualización de piezas como resolución de problemas. XVI*. Zaragoza, Huesca: Congreso Internacional de Ingeniería Gráfica.

Garrison, D. R. (2006). Online Collaboration Principles. *Journal of Asynchronous Learning Networks*, *10*(1), 25–34.

Geertz, C. (1987). El desarrollo de la cultura y la evolución de la mente. *La interpretación de las culturas*.

Geisser, S. (1975). The predictive sample reuse method with applications. *Journal of the American Statistical Association*, *70*(350), 320–328. doi:10.1080/01621459.1975.10479865

Ghenname, M., Abik, M., Ajhoun, R., Subercaze, J., Gravier, C., & Laforest, F. (2013). Personalized recommendation based hashtags on e-learning systems. *ISKO-Maghreb. 3rd International Symposium*, 8-9. doi:10.1109/ISKO-Maghreb.2013.6728109

Godino, J. D., Aké, L. P., Gonzato, M., & Wilhelmi, M. R. (2014). Niveles de algebrización de la actividad matemática escolar. Implicaciones para la formación de maestros. *Enseñanza de las ciencias: revista de investigación y experiencias didácticas, 32*(1), 199-219.

Goffman, E. (1959). *The Presentation of Self in Everyday Life*. Anchor Books.

Gómez Cruz, E. (2012). *De la cultura Kodak a la imagen en red: una etnografía sobre fotografía digital*. Academic Press.

Gómez Hidalgo, J. M., Cajigas Bringas, G., Puertas Sanz, E., & Carrero García, F. (2006). Content Based SMS Spam Filtering. In *Proceedings of the 2006 ACM Symposium on Document Engineering*. Amsterdam, The Netherlands: ACM Press.

Gonzales, M. R. (1998). Psicología educacional de las Matemáticas. *Revista de Investigación en Psicología*, 17.

González Barberá, C., Caso Niebla, J., Díaz López, K., & López Ortega, M. (2012). Rendimiento académico y factores asociados: aportaciones de algunas evaluaciones a gran escala. *Bordón. Revista de pedagogía, 64*(2), 51-68.

González, J. A. (1987). Los frentes culturales. Culturas, mapas, poderes y lucha por las definiciones legítimas de los sentidos sociales de la vida. *Estudios sobre las Culturas Contemporaneas*, *1*(3), 5–44.

González, T., & Rodríguez, L. (2010). El valor añadido de las buenas prácticas con TIC en los centros educativos. *Revista de Teoría de la Educación Sociedad de la Información*, *11*(3), 262–282.

González-Talaván, G. (2006). A simple, configurable SMTP anti-spam filter: Greylists. *Computers & Security*, *25*(3), 229–236. doi:10.1016/j.cose.2006.02.005

Goodyear, P. (2005). Educational design and networked learning: Patterns, pattern languages and design practice. *Australasian Journal of Educational Technology*, *21*(1), 82–101. doi:10.14742/ajet.1344

Gorghiu, G., Lindfors, E., Gorghiu, L. M., & And Hämäläinen, T. (2011). Acting as tutors in the ECSUT on-line course - how to promote interaction in a computer supported collaborative learning environment? *Procedia*, *3*, 579–583.

Götz, O., Liehr-Gobbers, K., & Krafft, M. (2010). Evaluation of structural equation models using the partial least squares (PLS) approach. In Handbook of Partial Least Squares (pp. 691-711). Academic Press.doi:10.1007/978-3-540-32827-8_30

Gould, H. W. (1961). The $ q $-Stirling numbers of first and second kinds. *Duke Mathematical Journal*, *28*(2), 281–289. doi:10.1215/S0012-7094-61-02826-5

Gress, C., Fior, M., Hadwin, A., & Winne, P. (2010). Measurement and assessment in computer-supported collaborative learning. *Computers in Human Behavior*, *26*(5), 806–814. doi:10.1016/j.chb.2007.05.012

Grgurina, N., Barendsen, E., Zwaneveld, B., Van Veen, K., & Stoker, I. (2014). *Computational thinking skills in Dutch secondary education: Exploring pedagogical content knowledge.* Paper presented at the ACM International Conference Proceeding Series. doi:10.1145/2674683.2674704

Grier, C., Thomas, K., Paxson, V., & Zhang, M. (2010, October). @ spam: the underground on 140 characters or less. In *Proceedings of the 17th ACM conference on Computer and communications security* (pp. 27-37). ACM. doi:10.1145/1866307.1866311

Grizzle, A., Wilson, C., & UNESCO. (2011). *Media and information literacy curriculum for teachers.* Paris: UNESCO. Retrieved from http://www.unesco.org/new/fileadmin/MULTIMEDIA/HQ/CI/CI/pdf/publications/Media%20and%20information%20literacy%20curriculum%20for%20teachers.pdf

Groh, F. (2012). *Gamification: State of the art definition and utilization.* Institute of Media Informatics Ulm University.

Grosan, C., & Abraham, A. (2011). *Intelligent Systems: A modern Approach* (Vol. 17). Berlin: Springer. doi:10.1007/978-3-642-21004-4_1

Gros, B., & García-Peñalvo, F. J. (2016). Future trends in the design strategies and technological affordances of e-learning. In M. Spector, B. B. Lockee, & M. D. Childress (Eds.), *Learning, Design, and Technology. An International Compendium of Theory, Research, Practice, and Policy* (pp. 1–23). Springer International Publishing. doi:10.1007/978-3-319-17727-4_67-1

Gross, S., Kim, M., Schlosser, J., Lluch, D., Mohtadi, C., Schneider, D., & Ieee. (2014). Fostering Computational Thinking in Engineering Education Challenges, Examples, and Best Practices. *2014 IEEE Global Engineering Education Conference*, 450-459.

Grubor, G., & Barac, I. (2014). Integrated proactive forensics model in network information security. *Sinteza*, 693-699. doi: 10.15308/SInteZa-2014-693-699

Guariso, G., & Werthner, H. (1989). *Environmental Decision Support System.* New York: E. Horwood, Halsted Press.

Gutierrez, A., & Tiner, K. (2012). Educación para los medios, alfabetizaciónmediática y competencia digital. *Comunicar*, *38*(9), 31–39.

Hair, J. F. J., Hult, G. T. M., Ringle, C., & Sarstedt, M. (2014). A primer on partial least squares structural equation modeling (PLS-SEM). *Long Range Planning*, *46*, 328. doi:10.1016/j.lrp.2013.01.002

Hair, J. F., Ringle, C. M., & Sarstedt, M. (2011). PLS-SEM: Indeed a silver bullet. *Journal of Marketing Theory and Practice*, *19*(2), 139–152. doi:10.2753/MTP1069-6679190202

Halász, G., & Michel, A. (2011). Key Competences in Europe: Interpretation, policy formulation and implementation. *European Journal of Education*, *46*(3), 289–306. doi:10.1111/j.1465-3435.2011.01491.x

Hall, M., Frank, E., Holmes, G., Pfahringer, B., Reutemann, P., & Witten, I. H. (2009). The WEKA data mining software: an update. *ACM SIGKDD Explorations Newsletter, 11*(1), 10-18.

Halloy, J., Sempo, G., Caprari, G., Rivault, C., Asadpour, M., Tâche, F., & Deneubourg, J. L. et al. (2007). Social integration of robots into groups of cockroaches to control self-organized choices. *Science*, *318*(5853), 1155–1158. doi:10.1126/science.1144259 PMID:18006751

Hamari, J., Koivisto, J., & Sarsa, H. (2014, January). Does gamification work?--a literature review of empirical studies on gamification. In *System Sciences (HICSS), 2014 47th Hawaii International Conference on* (pp. 3025-3034). IEEE.

Hamari, J., Koivisto, J., & Sarsa, H. (2014, January). Does gamification work?--a literature review of empirical studies on gamification. In *System Sciences (HICSS), 2014 47th Hawaii International Conference on*. IEEE.

Hamou, R. M., Amine, A., & Boudia, A. (2013). A New Meta-Heuristic Based on Social Bees for Detection and Filtering of Spam. *International Journal of Applied Metaheuristic Computing*, *4*(3), 15–33. doi:10.4018/ijamc.2013070102

Handal, B., Macnish, J., & Petocz, P. (2013). Academics adopting mobile devices : The zone of free movement. In H. Carter, M. Gosper, & J. Hedberg (Eds.), *Proceedings 30th ascilite Conference 2013* (pp. 350–361). Sidney, Australia: Macquarie University.

Hanus, M. D., & Fox, J. (2015). Assessing the effects of gamification in the classroom: A longitudinal study on intrinsic motivation, social comparison, satisfaction, effort, and academic performance. *Computers & Education*, *80*, 152–161. doi:10.1016/j.compedu.2014.08.019

Harlen, W., & Deakin Crick, R. (2003). Testing and motivation for learning. *Assessment in Education: Principles, Policy & Practice*, *10*(2), 169–207. doi:10.1080/0969594032000121270

Harnad, S. (1991). Other Bodies Other Minds: A Machine Incarnation of an Old Philosophical Problem. *Minds and Machines*, *1*(1), 43–54.

Havens, T. C., Spain, C. J., Salmon, N. G., & Keller, J. M. (2008, September). Roach infestation optimization. In *Swarm Intelligence Symposium, 2008. SIS 2008. IEEE* (pp. 1-7). IEEE. doi:10.1109/SIS.2008.4668317

Hedström, K., Karlsson, F., & Kolkowska, E. (2013). Social action theory for understanding information security non-compliance in hospitals. *Information Management & Computer Security*, *21*(4), 266–287. doi:10.1108/IMCS-08-2012-0043

Helder, D., Henderson, D., & Orge, B. (2014). *1.000 millones: poesía en lengua española del siglo XXI*. Rosario: Editorial Municipal de Rosario.

Helsinki Process. (1995). *European criteria and indicators for sustainable forest management adopted by the expert level follow-up meetings of Helsinki Conference in Geneva (24 June 1994) and in Antalya (23 June 1995)*. Author.

Henseler, J., Ringle, C. M., & Sinkovics, R. (2009). The use of partial least squares path modeling in international marketing. *Advances in International Marketing*, *20*, 277–319. doi:10.1108/S1474-7979(2009)0000020014

Henseler, J., & Sarstedt, M. (2012). Goodness-of-fit indices for partial least squares path modeling. *Computational Statistics*, *28*(2), 565–580. doi:10.1007/s00180-012-0317-1

Hernández Serrano, M. J., & González Sánchez, M. (2011). *La generación Google. Evolución en las predisposiciones y comportamientos informativos de los jóvenes*. Academic Press.

Hernández, N., González, M., & Muñoz, P. (2014). La planificación del aprendizaje colaborativo en entornos virtuales [Collaborative learning design in virtual envoronments]. *Comunicar*, *42*(21), 25–32.

Hernández-Ramos, J. P., Martínez-Abad, F., García-Peñalvo, F. J., Herrera García, M. E., & Rodríguez-Conde, M. J. (2014). Teachers' attitude regarding the use of ICT. A factor reliability and validity study. *Computers in Human Behavior*, *31*, 509–516. doi:10.1016/j.chb.2013.04.039

Herold, B. (2016). Technology in Education: An Overview. *Education Week*. Retrieved from https://goo.gl/9CxpnL

Herrington, J., Herrington, A., Mantei, J., Olney, I. W., & Ferry, B. (2009). Using mobile tecnologies to develop new ways of teaching and learning. In J. Herrington, J. Mantei, I. W. Olney, B. Ferry, & A. Herrington (Eds.), *New techonologies, new pedagogies: Mobile learning in higher education* (pp. 1–14). Wollongong: University of Wollongong.

Hill, J. R. (2006). Flexible Learning Environments: Leveraging the Affordances of Flexible Delivery and Flexible Learning. *Innovative Higher Education*, *31*(3), 187–197. doi:10.1007/s10755-006-9016-6

Hiltz, S. R., & Turoff, M. (2002). What makes learning effective? *Communications of the ACM*, *45*(4), 56–59. doi:10.1145/505248.505273 PMID:12238525

Hofstede, G. (2017). *Why is culture so important?* Retrieved from http://www.geerthofstede.nl/

Hofstede, G. (2011). Dimensionalizing cultures: The Hofstede model in context. *Online Readings in Psychology and Culture*, *2*(1), 8. doi:10.9707/2307-0919.1014

Holland, J. (1962). Outline for a logical theory of adaptive systems. *Journal of the Association for Computing Machinery*, 3.

Holland, J. H. (1973). Genetic algorithms and the optimal allocation of trials. *SIAM Journal on Computing*, *2*(2), 88–105. doi:10.1137/0202009

Holovaty, A., & Kaplan-Moss, J. (2009). *The definitive guide to Django: Web development done right*. Apress. doi:10.1007/978-1-4302-1937-8

Honan, M. (2012). *How Apple and Amazon Security Flaws Led to My Epic Hacking*. Retrieved June 25, 2017, from http://www.wired.com/gadgetlab/2012/08/apple-amazon-mat-honan-hacking

Huang, J., Dasgupta, A., Ghosh, A., Manning, J., & Sanders, M. (2014, March). Superposter behavior in MOOC forums. In *Proceedings of the first ACM conference on Learning@ scale conference* (pp. 117-126). ACM. doi:10.1145/2556325.2566249

Hug, T. (2013). Competencia mediática y alfabetización visual. Hacia consideraciones más allá de las alfabetizaciones. *Razón y Palabra*, *18*(83).

Hunicke, R., LeBlanc, M., & Zubek, R. (2004, July). MDA: A formal approach to game design and game research. *Proceedings of the AAAI Workshop on Challenges in Game AI*, 4.

Ifinedo, P. (2016). Critical times for organizations: What should be done to curb workers' noncompliance with is security policy guidelines? *Information Systems Management*, *33*(1), 30–41. doi:10.1080/10580530.2015.1117868

Instituto Nacional de Tecnologías Educativas y de Formación del Profesorado (INTEF). (2017). *Marco Común de Competencia Digital Docente*. Author.

International Visual Literacy Association. (2003). *What is visual literacy?* Retrieved from https://web.archive.org/web/20031204235851/http://ivla.org:80/org_what_vis_lit.htm

Irani, D., Webb, S., Pu, C., & Li, K. (2011). Modeling unintended personal-information leakage from multiple online social networks. *IEEE Internet Computing*, *15*(3), 13–19. doi:10.1109/MIC.2011.25

ISO/IEC 27002:2013. (2013). *Information technology security techniques - Code of practice for information security controls*. Retrieved from http://www.iso.org

ISO/IEC. (2009). Information technology – Security techniques – Information security management measurements. ISO/IEC 27004.

ITGI. (2008). *Information security governance: Guidance for information security managers*. Retrieved from www.itgi.org

ITTO. (1992). *Criteria for the Management of Sustainable Tropical Forest Management*. Yokohama: International Tropical Organization.

Jackson, L. D. (2013). Is mobile technology in the classroom a helpful tool or a distraction?: A report of university students' attitudes, usage practices, and suggestions for policies. *International Journal of Technology, Knowledge and Society*, *8*(5), 129–140. doi:10.18848/1832-3669/CGP/v08i05/56335

Jain, K., & Agrawal, S. (2014, August). A hybrid approach for spam filtering using support vector machine and artificial immune system. In *Networks & Soft Computing (ICNSC), 2014 First International Conference on* (pp. 5-9). IEEE. doi:10.1109/CNSC.2014.6906699

James, K., Humpherey, G., & Vilis, T. (2002). "Active" and "Passive" Learning of Three-Dimensional Object Structure Within an Immersive Virtual Reality Environment. *Behavior Research Methods, Instruments, & Computers*, *34*(3), 383–390. doi:10.3758/BF03195466 PMID:12395554

James, R. (2013). ICT's participatory potential in higher education collaborations: Reality or just talk. *British Journal of Educational Technology*, 1–14. doi:10.1111/bjet.12060

Januszewski, A., & Molenda, M. (2008). *Educational technology: A definition with commentary*. Mahwah, NJ: Lawrence Erlbaum Associates.

Jarson, J., & Taub-Pervizpour, L. (2015). Infoglut or Information Literacy? Using Blended Learning for Student Research in «New Information Technologies». *Blended Learning in the Liberal Arts Conference*. Retrieved from http://repository.brynmawr.edu/blended_learning/2015/2015/3

Jeanson, R., Rivault, C., Deneubourg, J. L., Blanco, S., Fournier, R., Jost, C., & Theraulaz, G. (2005). Self-organized aggregation in cockroaches. *Animal Behaviour*, *69*(1), 169–180. doi:10.1016/j.anbehav.2004.02.009

Jindal, N., & Liu, B. (2007, May). Review spam detection. In *Proceedings of the 16th international conference on World Wide Web* (pp. 1189-1190). ACM. doi:10.1145/1242572.1242759

Jirotka, M., Procter, R., Rodden, T., & Bowker, G. (2006). Special issue: Collaboration in e-research. *Computer Supported Cooperative Work*, *15*(4), 251–255. doi:10.1007/s10606-006-9028-x

JISC. (2007). *A report on the JISC CETIS PLE project*. Available online at http://wiki.cetis.ac.uk/Ple/Report

Joachims, T. (1998). *Text categorization with support vector machines: Learning with many relevant features*. Springer Berlin Heidelberg.

João, Z., Dulce, D., Mário, J. S., & Carlos, S. (2014). Group-Based Discretionary Access Control in Health Related Repositories. *Journal of Information Technology Research*, *7*(1), 1–17. doi:10.4018/jitr.2014010106

Johnson, L., Levine, A., & Smith, R. (2009). *The 2009 Horizon Report*. Austin, TX: The New Media Consortium.

Johnson, R. B., & Onwuegbuzie, A. J. (2004). Mixed Methods Research: A Research Paradigm Whose Time Has Come. *Educational Researcher*, *33*(7), 14–26. doi:10.3102/0013189X033007014

Jones, C., Dirckinck-Holmfeld, L., & Lindstrom, B. (2006). A relational, indirect, meso-level approach to CSCL design in the next decade. *Computer-Supported Collaborative Learning*, *1*(1), 35–56. doi:10.1007/s11412-006-6841-7

Jones, C., & Healing, G. (2010). Net generation students: Agency and choice and the new technologies. *Journal of Computer Assisted Learning*, *26*(5), 344–356. doi:10.1111/j.1365-2729.2010.00370.x

Jones, C., Ramanau, R., Cross, S., & Healing, G. (2010). Net generation or Digital Natives: Is there a distinct new generation entering university? *Computers & Education*, *54*(3), 722–732. doi:10.1016/j.compedu.2009.09.022

Joo-Nagata, J., Martínez Abad, F., García-Bermejo Giner, J., & García-Peñalvo, F. J. (2017). Augmented reality and pedestrian navigation through its implementation in m-learning and e-learning: Evaluation of an educational program in Chile. *Computers & Education*, *111*, 1–17. doi:10.1016/j.compedu.2017.04.003

Kafai, Y. B. (1996). Learning design by making games. *Constructionism in Practice: Designing, Thinking and Learning in a Digital World*, 71–96.

Kamii. (1981). *La teoría de Piaget y la educación preescolar.* Madrid: Visor.

Kanowski, P. J., & Williams, K. J. H. (2009). The reality of imagination: Integrating the material and cultural values of old forests. *Forest Ecology and Management*, *258*(4), 341–346. doi:10.1016/j.foreco.2009.01.011

Kapp, K. M. (2012). *The gamification of learning and instruction: game-based methods and strategies for training and education.* John Wiley & Sons.

Karaboga, D., & Basturk, B. (2007). A powerful and efficient algorithm for numerical function optimization: Artificial bee colony (ABC) algorithm. *Journal of Global Optimization*, *39*(3), 459–471. doi:10.1007/s10898-007-9149-x

Katriel, J. (2000). Bell numbers and coherent states. *Physics Letters. [Part A]*, *273*(3), 159–161. doi:10.1016/S0375-9601(00)00488-6

Kaufmann, H., Steinbügl, K., Dünser, A., & Glück, J. (2003). Improving Spatial Abilities by Geometry Education in Augmented Reality - Application and Evaluation Design. *Proceeding of the 2005 VRIC Laval Virtual*, 25-34.

Kazakoff, E., & Bers, M. (2012). Programming in a robotics context in the kindergarten classroom: The impact on sequencing skills. *Journal of Educational Multimedia and Hypermedia*, *21*(4), 371–391.

Keim, D., Andrienko, G., Fekete, J.-D., Görg, C., Kohlhammer, J., & Melançon, G. (2008). Visual analytics: Definition, process, and challenges. *Information Visualization*, 154–175.

Keim, D., Kohlhammer, J., Ellis, G., & Mansmann, F. (2010). *Mastering the information age solving problems with visual analytics.* Eurographics Association.

Kennedy, J., & Eberhart, R. (1995). Particle Swarm Optimization. *Proceedings of IEEE International Conference on Neural Networks, 4,* 1942–1948. doi:10.1109/ICNN.1995.488968

Keppell, M., Suddaby, G., & Hard, N. (2015). Assuring best practice in technology-enhanced learning environments. *Research in Learning* Technology, *23*, 1-DOI:.10.3402/rlt.v23.25728

Kerlinger, F., & Lee, H. (1999). Foundations of behavioral research (4th ed.). Wadsworth Publishing.

Khatoon, B., Hill, K. B., & Walmsley, A. D. (2013). Can we learn, teach and practise dentistry anywhere, anytime? *British Dental Journal*, *215*(7), 345–347. doi:10.1038/sj.bdj.2013.957 PMID:24113956

Kieseberg, P., Schrittwieser, S., Morgan, L., Mulazzani, M., Huber, M., & Weippl, E. (2013). Using the structure of B+-trees for enhancing logging mechanisms of databases. *International Journal of Web Information Systems*, *9*(1), 53–68. doi:10.1108/17440081311316389

Killion, J. (2015). High-quality collaboration benefits teachers and students. *The Learning Professional, 36*(5), 62-64.

Kirkwood, A., & Price, L. (2014). Technology-enhanced learning and teaching in higher education: What is 'enhanced' and how do we know? A critical literature review. *Learning, Media and Technology*, *39*(1), 6–36. doi:10.1080/17439884.2013.770404

Kirschner, P. A. (2002). *Three Worlds of CSCL. Can We Support CSCL.* Open University of the Netherlands.

Klibanski, M. (2006). *Grandes bibliotecas digitales y virtuales del mundo*. Retrieved from http://portal.educ.ar/debates/educacionytic/super-sitios/grandes-bibliotecas-digitales-y-virtuales-del-mundo.php

Kluyver, T., Ragan-Kelley, B., Pérez, F., Granger, B., Bussonnier, M., Frederic, J., . . . Corlay, S. (2016). Jupyter Notebooks—a publishing format for reproducible computational workflows. *Positioning and Power in Academic Publishing: Players, Agents and Agendas*, 87.

Kock, N. (2013). *WarpPLS 4.0 User Manual*. Laredo, TX: ScriptWarp Systems.

Koehler, M., & Mishra, P. (2009). What is technological pedagogical content knowledge (TPACK)? *Contemporary Issues in Technology & Teacher Education*, *9*(1), 60–70.

Kolkowska, E., & Dhillon, G. (2013). Organizational power and information security rule compliance. *Computers & Security*, *33*, 3–11. doi:10.1016/j.cose.2012.07.001

Koza, J. R. (1992). *Genetic Programming: On the Programming of Computers by Means of Natural Selection*. Cambridge, MA: The MIT Press.

Krasser, S., Tang, Y., Gould, J., Alperovitch, D., & Judge, P. (2007, June). Identifying image spam based on header and file properties using C4. 5 decision trees and support vector machine learning. In *Information Assurance and Security Workshop, 2007. IAW'07. IEEE SMC* (pp. 255-261). IEEE.

Kreijns, K., Kirschner, P. A., & Jochems, W. (2003). Identifying the pitfalls for social interactions in computer-supported collaborative learning environments: A review of the research. *Computers in Human Behavior*, *19*(3), 335–353. doi:10.1016/S0747-5632(02)00057-2

Kulik, C. C., & Kulik, J. A. (1991). Effectiveness of computer-based instruction: An updated analysis. *Computers in Human Behavior*, *7*(1-2), 75–94. doi:10.1016/0747-5632(91)90030-5

Labrador, E., & Villegas, E. (2014). Sistema Fun Experience Design (FED) aplicado en el aula. *ReVisión, 7*(2).

Lai, C. C., & Wu, C. H. (2007). *Particle swarm optimization-aided feature selection for spam email classification*. Kumamoto: IEEE. doi:10.1109/ICICIC.2007.442

Lankshear, C., & Knobel, A. (2008). *Nuevos alfabetismos su práctica cotidiana y el aprendizaje en el aula*. Madrid: Morata.

Laroche, M., Habibi, M. R., Richard, M. O., & Sankaranarayanan, R. (2012). The effects of social media based brand communities on brand community markers, value creation practices, brand trust and brand loyalty. *Computers in Human Behavior*, *28*(5), 1755–1767. doi:10.1016/j.chb.2012.04.016

LaValle, S., Lesser, E., Shockley, R., Hopkins, M. S., & Kruschwitz, N. (2011). Big data, analytics and the path from insights to value. *MIT Sloan Management Review*, *52*(2), 21.

Lave, J., & Wenger, E. (1991). *Situated Learning: Legitimate Peripheral Participation.* Cambridge, UK: Cambridge University Press. doi:10.1017/CBO9780511815355

Lavonen, J., Lattu, M., Juuti, K., & Meisalo, V. (2006). Strategy-based development of teacher educators' ICT competence through a co-operative staff development project. *European Journal of Teacher Education, 29*(2), 241–265. doi:10.1080/02619760600617433

Lazaro, H. (2014). *What is EdTech an why should it matter to you?* Retrieved from https://goo.gl/7sPm62

Lea, S., & Callaghan, L. (2011). Enhancing health and social care placement learning through mobile technology. *Journal of Educational Technology & Society, 14*(1), 135–145.

Lee, A. Y. L., & So, C. Y. K. (2014). Media Literacy and Information Literacy: Similarities and Differences. *Comunicar, 21*(42), 137–145. doi:10.3916/C42-2014-13

Lee, E. T., & Wang, J. W. (2003). *Statistical Methods for survival Data Analysis.* John Wiley & Sons, Inc. doi:10.1002/0471458546.fmatter

Lee, J., & Shute, V. J. (2010). Personal and Social-Contextual Factors in K-12 Academic Performance: An Integrative Perspective on Student Learning. *Educational Psychologist, 45*(3), 185–202. doi:10.1080/00461520.2010.493471

Lee, L., Petter, S., Fayard, D., & Robinson, S. (2011). On the use of partial least squares path modeling in accounting research. *International Journal of Accounting Information Systems, 12*(4), 305–328. doi:10.1016/j.accinf.2011.05.002

Levine, T., & Marcus, A. S. (2010). How the structure and focus of teachers' collaborative activities facilitate and constrain teacher learning. *Teaching and Teacher Education, 26*(3), 389–398. doi:10.1016/j.tate.2009.03.001

Levitin, D. J. (2014). *The Organized Mind: Thinking Straight in the Age of Information Overload.* Penguin.

Ley Orgánica 15/1999. (1999). Ley Orgánica 15/1999, de 13 de diciembre, de Protección de Datos de Carácter Personal. *Boletín Oficial Del Estado, 298*(2), 43088-43099.

Li, N., Li, T., & Venkatasubramanian, S. (2007). t-closeness: Privacy beyond k-anonymity and l-diversity. In *IEEE 23rd International Conference on Data Engineering* (pp. 106-115). IEEE.

Liddo, A., Buckingham Shum, S., McAndrew, P., & Farrow, R. (2012). The open education evidence hub: A collective intelligence tool for evidence based policy. In *Joint OER12 and OpenCourseWare Consortium Global 2012.* Cambridge, UK: Conference.

Li, H., Sarathy, R., Zhang, J., & Luo, X. (2014). Exploring the effects of organizational justice, personal ethics and sanction on internet use policy compliance. *Information Systems Journal, 24*(6), 479–502. doi:10.1111/isj.12037

Lihoreau, M., Costa, J. T., & Rivault, C. (2012). The social biology of domiciliary cockroaches: Colony structure, kin recognition and collective decisions. *Insectes Sociaux, 59*(4), 445–452. doi:10.1007/s00040-012-0234-x

Lihoreau, M., & Rivault, C. (2009). Kin recognition via cuticular hydrocarbons shapes cockroach social life. *Behavioural Ecology, 20*(1), 46–53. doi:10.1093/beheco/arn113

Little, J. W. (1990). The persistence of privacy: Autonomy and initiative in teachers' professional relations. *Teachers College Record, 91*(4), 509–536.

Livingstone, S. (2012). Critical reflections on the benefits of ICT in education. *Oxford Review of Education, 38*(1), 9–24. doi:10.1080/03054985.2011.577938

Li, X., Shao, Z., & Qian, J. (2002). Anoptimizing method base on autonomous animates: Fish- swarm algorithm. *Systems Engineering Theory and Practice*, *22*, 32–38.

Li, Y., Liu, Y., & Shu, P. (2015). *Teaching Research and Practice of Blended Leaning Model Based on Computational Thinking Frontiers in Education Conference*. IEEE.

Llorens Largo, F., García-Peñalvo, F. J., Molero Prieto, X., & Vendrell Vidal, E. (2017). La enseñanza de la informática, la programación y el pensamiento computacional en los estudios preuniversitarios. *Education in the Knowledge Society*, *18*(2), 7–17. doi:10.14201/eks2017182717

Llorens, F., & Capdeferro, N. (2011). Facebook's Potential for Collaborative e-Learning. *Revista de Universidad y Sociedad del Conocimiento (RUSC)*, *8*(2), 31–45. Retrieved from http://rusc.uoc.edu/ojs/index.php/rusc/article/view/v8n2-llorens-capdeferro/v8n2-llorens-capdeferro

Llorens-Largo, F. (2015). Dicen por ahí ... que la nueva alfabetización pasa por la programación. *ReVisión*, *8*(2), 11–14.

Llorens-Largo, F. (2015). Dicen por ahí. . . . que la nueva alfabetización pasa por la programación. *ReVisión*, *8*(2), 11–14.

López Meneses, E., Vázquez-Cano, E., & Román Graván, P. (2015). Análisis e implicaciones del impacto del movimiento MOOC en la comunidad científica: JCR y Scopus (2010-13). *Comunicar*, *44*, 73–80. doi:10.3916/C44-2015-08

López-Hernández, F. A., & Silva-Pérez, M. M. (2014). M-learning patters in the virtual classroom. *Universities and Knowledge Society Journal*, *11*(1), 208–221. doi:10.7238/rusc.v11i1.1902

Lotman, I., & Uspenski, B. (1979). Sobre el mecanismo semiótico de la cultura. In *Lotman y Escuela de Tartu: Semiótica de la cultura* (pp. 67–92). Madrid: Cátedra.

Lowenthal, P. R., & Wilson, B. G. (2010). Labels do matter! A critique of AECT's redefinition of the field. *TechTrends*, *51*(1), 38–46. doi:10.1007/s11528-009-0362-y

Lowry, P. B., Posey, C., Bennett, R. J., & Roberts, T. L. (2015). Leveraging fairness and reactance theories to deter reactive computer abuse following enhanced organisational information security policies: An empirical study of the influence of counterfactual reasoning and organisational trust. *Information Systems Journal*, *25*(3), 193–273. doi:10.1111/isj.12063

Luceño C., J. (2012). *Enseñanza, aprendizaje de las competencias matemáticas*. libros en Red.

Lye, S. Y., & Koh, J. H. L. (2014). Review on teaching and learning of computational thinking through programming: What is next for K-12? *Computers in Human Behavior*, *41*, 51–61. doi:10.1016/j.chb.2014.09.012

Mahan, C. G., Young, J. A., Miller, B. J., & Saunders, M. C. (2015). Using Ecological Indicators and a Decision Support System for Integrated Ecological Assessment at Two National Park Units in the Mid-Atlantic Region, USA. *Environmental Management*, *55*(2), 508–522. doi:10.1007/s00267-014-0391-y PMID:25371192

Manovich, L. (2015). Data Science and Digital Art History. *International Journal for Digital Art History*, (1).

Mao, J. (2014). Social media for learning: A mixed methods study on high school students' technology affordances and perspectives. *Computers in Human Behavior*, *33*, 213–223. doi:10.1016/j.chb.2014.01.002

Markoff, J. (2011). Virtual and artificial, but 58,000 want course. *The New York Times*, p. 15.

Marsh, H. W. (1982). SEEQ: A reliable valid and useful instrument for collecting students evaluations of university teaching. *The British Journal of Educational Psychology*, *52*(1), 77–95. doi:10.1111/j.2044-8279.1982.tb02505.x

Martínez Abad, F., Rodríguez Conde, M. J., & García-Peñalvo, F. J. (2014). Evaluación del impacto del término "MOOC" vs "eLearning" en la literatura científica y de divulgación. *Profesorado. Revista de currículum y formación del profesorado, 18*(1), 185-201.

Martínez Abad, F., Olmos Migueláñez, S., & Rodríguez Conde, M. J. (2015). Evaluación de un programa de formación en competencias informacionales para el futuro profesorado de E.S.O. *Review of Education*, (370): 45–70.

Martínez -Núñez, M., Borrás-Gené, O., & Fidalgo- Blanco, Á. (2014). Social community in MOOCs: practical implications and outcomes. In *Proceedings of the Second International Conference on Technological Ecosystems for Enhancing Multiculturality* (pp. 147-154). ACM.

Martínez-Abad, F., Torrijos-Fincias, P., & Rodríguez-Conde, M. J. (2016). The eAssesment of key competences and their relationship performance. *Journal of Information Technology Research, 9*(4), 16–27. doi:10.4018/JITR.2016100102

Martínez-Núñez, M., Borrás-Gene, O., & Fidalgo-Blanco, Á. (2016). Virtual Learning Communities in Google Plus, implications and sustainability in MOOCs. *Journal of Information Technology Research, 9*(3), 18–36. doi:10.4018/JITR.2016070102

Martinez-Perez, B., de la Torre-Diez, I., & Lopez-Coronado, M. (2013). Mobile Health Applications for the Most Prevalent Conditions by the World Health Organization: Review and Analysis. *Journal of Medical Internet Research, 15*(6), e120. doi:10.2196/jmir.2600 PMID:23770578

Martínez-Pérez, B., de la Torre-Díez, I., & López-Coronado, M. (2015). Experiences and Results of Applying Tools for Assessing the Quality of a mHealth App Named Heartkeeper. *Journal of Medical Systems, 39*(11), 1–6. doi:10.1007/s10916-015-0303-6 PMID:26345452

Martín-Gutiérrez, J. (2010). AR_Dehaes: an educational toolkit based on augmented reality technology for learning engineering graphics. *2010 10th IEEE International Conference on Advanced Learning Technologies.*

McAuley, A., Stewart, B., Siemens, G., & Cormier, D. (2010). *The MOOC Model for Digital Practice.* SSHRC Knowledge Synthesis Grant on the Digital Economy.

McInnerney, J., & Robert, T. S. (2004). Collaborative or cooperative learning? In T. S. Roberts (Ed.), *Online collaborative learning: Theory and practice* (pp. 203–214). Hershey, PA: Information Science Publishing. doi:10.4018/978-1-59140-174-2.ch009

McKinney, W. (2011). pandas: a foundational Python library for data analysis and statistics. *Python for High Performance and Scientific Computing*, 1-9.

McLaughin, M. W., & Talbert, J. E. (2006). *Building school-based teacher learning communities: Professional strategies to improve student achievement.* New York: Teachers College Press.

McLuhan, M., & Nevitt, B. (1972). Take today; the executive as dropout. Don Mills.

Mei-Yu, W., & Ming-Hsien, Y. (2013). Enterprise Information Security Management Based on Context-Aware RBAC and Communication Monitoring Technology. *Mathematical Problems in Engineering*, 1–11. doi:10.1155/2013/569562

Mendes, D. (2011). A Semantic Web pragmatic approach to develop Clinical ontolologies, and thus Semantic Interoperability, based in HL7 v2.xml messaging. In *HCist 2011 - Proceedings of the International Workshop on Health and Social Care Information Systems and Technologies*. Paris: Springer-Verlag.

Mendes, D., & Rodrigues, I. P. (2011a). Automatic Ontology Population extracted from SAM Healthcare Texts in Portuguese. Jornadas de Informática da Universidade de Évora.

Mendes, D., & Rodrigues, I. P. (2011b). Well Formed Clinical Practice Ontology Selection. Jornadas de Informática da Universidade de Évora.

Mendes, D., Rodrigues, I. P., Rodriguez-Solano, C., & Baeta, C. (2012). Enrichment/Population of Customized CPR (Computer-based Patient Record) Ontology from Free-text Reports for CSI (Computer Semantic Interoperability). Procedia Technology, 5, 753–762). doi:10.1016/j.protcy.2012.09.083

Mens, T., Claes, M., Grosjean, P., & Serebrenik, A. (2014). Studying evolving software ecosystems based on ecological models. In T. Mens, A. Serebrenik, & A. Cleve (Eds.), *Evolving Software Systems* (pp. 297–326). Berlin: Springer. doi:10.1007/978-3-642-45398-4_10

Merrill, M., & Beck, B. (2012). *League of Legends Account Security Alert.* Retrieved June 25, 2017, from http://euw.leagueoflegends.com/news/league-legends-account-security-alert

Meystre, S. M., Friedlin, F. J., South, B. R., Shen, S., & Samore, M. H. (2010). Automatic de-identification of textual documents in the electronic health record: A review of recent research. *BMC Medical Research Methodology, 10*(1), 70. doi:10.1186/1471-2288-10-70 PMID:20678228

Michavila, F., Martínez, J. M., Martín-González, M., García-Peñalvo, F. J., & Cruz-Benito, J. (2016). *Barómetro de Empleabilidad y Empleo de los Universitarios en España, 2015 (Primer informe de resultados).* Academic Press.

Michavila, F., Martínez, J. M., Martín-González, M., García-Peñalvo, F. J., & Cruz-Benito, J. (2016). *Barómetro de Empleabilidad y Empleo de los Universitarios en España, 2015 (Primer informe de resultados).* Madrid: Observatorio de Empleabilidad y Empleo Universitarios.

Michavila, F., Martín-González, M., Martínez, J. M., García-Peñalvo, F. J., & Cruz-Benito, J. (2015). Analyzing the employability and employment factors of graduate students in Spain: The OEEU Information System. *Proceedings of the 3rd International Conference on Technological Ecosystems for Enhancing Multiculturality.* doi:10.1145/2808580.2808622

Milliot, J. (2011). The Latest E-Book Buying Trends. *Publisher's Weekly.* Retrieved: April 5, 2015, from http://www.publishersweekly.com/pw/by-topic/digital/content-and-e-books/article/48418-the-latest-e-book-buying-trends.html

Mills, E. K., Jr. (1941). *U.S. Patent No. 2,243,973.* Washington, DC: U.S. Patent and Trademark Office.

Minović, M., García-Peñalvo, F. J., & Kearney, N. A. (2016). Gamification in Engineering Education. *International Journal of Engineering Education, 32*(1B), 308–309.

Minsky, M. (1979). The Society Theory of Thinking. In P. Winston (Ed.), *Artificial Intelligence: An MIT Perspective* (Vol. 1). Cambridge, MA: MIT Press.

Mobile Marketing Association. (2017). *Estudio tendencias Mobile 2017.* Author.

Moeller, S., Joseph, A., Lau, J., & Carbo, T. (2011). *Towards Media and Information Literacy Indicators.* Paris: UNESCO.

Mohammad, A. H., & Zitar, R. A. (2011). Application of genetic optimized artificial immune system and neural networks in spam detection. *Applied Soft Computing, 11*(4), 3827-3845.

Molenda, M. (1997). Historical and philosophical foundations of instructional design: A North American view. In R. D. Tennyson, F. Schott, N. M. Seel, & S. Dijkstra (Eds.), *Instructional design. International perspectives: Theory, Research, and Models* (Vol. 1, pp. 41–54). Mahwah, NJ: Lawrence Erlbaum Associates.

Monereo Font, C. Y., & Durán, G. (2003). Entramados métodos de aprendizaje cooperativo y colaborativo [Coopertative and collaborative learning methods]. Barcelona España: Edebe.

Moore, A. J., Ciccone, W. J., & Breed, M. D. (1988). The influence of social experience on the behaviour of male cockroaches, Nauphoeta cinerea. *Journal of Insect Behavior*, *1*(2), 157–168. doi:10.1007/BF01052235

Mora, P., García, & Marín. (2001). *Módulo de AutoCAD orientado a la enseñanza de piezas*. XIII Congreso Internacional de Expresión Gráfica en la Ingeniería, Badajoz.

Moreno-León, J., Robles, G., & Román-González, M. (2016). Comparing Computational Thinking Development Assessment Scores with Software Complexity Metrics. *Proceedings of the IEEE Global Engineering Education Conference.*

Morris, R., & Thompson, K. (1979). Password security: A case history. *Communications of the ACM*, *22*(11), 594–597. doi:10.1145/359168.359172

Mowery, D., Wiebe, J., Visweswaran, S., Harkema, H., & Chapman, W. W. (2012). Building an automated SOAP classifier for emergency department reports. *Journal of Biomedical Informatics*, *45*(0), 71–81. doi:10.1016/j.jbi.2011.08.020 PMID:21925286

Mullis, I. V., Martin, M. O., Kennedy, A. M., Trong, K. L., & Sainsbury, M. (2009). *PIRLS 2011 Assessment Framework.* International Association for the Evaluation of Educational Achievement. Retrieved from http://www.iea.nl

Muntean, C. I. (2011, October). Raising engagement in e-learning through gamification. In *Proc. 6th International Conference on Virtual Learning ICVL* (pp. 323-329). Academic Press.

Murillo, F. J. (2003). El movimiento de investigación de Eficacia Escolar. In *La investigación sobre Eficacia Escolar en Iberoamérica. Revisión internacional del estado del arte.* Bogotá: Convenio Andrés Bello.

Murillo, F. J. (2005). *La Investigación sobre Eficacia Escolar.* Barcelona: Octaedro.

Murteira Mendes, D. J., Pimenta Rodrigues, I., Rodríguez-Solano, C., & Fernandes Baeta, C. (2014). Enrichment/Population of Customized CPR (Computer-Based Patient Record) Ontology from Free-Text Reports for CSI (Computer Semantic Interoperability). *Journal of Information Technology Research*, *7*(1), 1–11. doi:10.4018/jitr.2014010101

Myler, E., & Broadbent, G. (2006). ISO 17799: Standard for security. *Information Management Journal*, *40*(6), 43–52.

Mysiak, J., Giupponi, C., & Rosato, P. (2005). Towards the development of decision support system for water resource management. *Environmental Modelling & Software*, *20*(2), 203–214. doi:10.1016/j.envsoft.2003.12.019

Nath, S. R. (2012). Factors Influencing Primary Students' Learning Achievement in Bangladesh. *Research in Education*, *88*(1), 50–63. doi:10.7227/RIE.88.1.5

National Research Council Committee on Information Technology Literacy. (1999). *Being Fluent with Information Technology*. Washington, DC: National Academy Press.

Navarro, I., & Fonseca, D. (2017). Nuevas tecnologías de visualización para mejorar la representación de arquitectura en la educación. *ACE: architecture, city and environment, 34*, 218-237. DOI:10.5821/ace.12.34.5290

Negnevitsky, M. (2005). *Artificial Intelligence. Intelligent Systems* (2nd ed.). Harlow, UK: Addison Wesley.

Newell, A. (1981). The Knowledge Level, Presidential Address, AAI80, Stanford University, 19 Aug 1980. *AI Magazine*, *2*(2).

Newell, A., & Simon, H. A. (1976). Computer Science as Empirical Enquiry: Symbols and Search. *Communications of the ACM*, *19*(3), 113–126. doi:10.1145/360018.360022

Newman, S., Lynch, T., & Plummer, A. A. (1999). Success and failure of decision support systems: Learning as we go. *Journal of Animal Science*, *77*(E-Suppl), 1–12. doi:10.2527/jas2000.77E-Suppl1e

Nicholson, B., & McClennen, S. (Eds.). (2013). The Generation of '72. Raleigh, NC: Editorial Contracorriente.

Nicholson, S. (2012). A user-centered theoretical framework for meaningful gamification. *Games+ Learning+ Society, 8*(1).

Nielsen, J. (2014). The Digital. *Consumer Reports*. PMID:25350310

Nonaka, I., & Takeuchi, H. (1995). *The knowledge-creating company: How Japanese companies create the dynamics of innovation*. Oxford University Press.

Nunnally, J., & Bernstein, I. (1994). *Psychometric Theory* (3rd ed.). New York: McGraw-Hill; Retrieved from http://scholar.google.com

Nuss, M. A., Hill, J. R., Cervero, R. M., Gaines, J. K., & Middendorf, B. F. (2014). Real-time use of the iPad by third-year medical students for clinical decision support and learning: A mixed methods study. *Journal of Community Hospital Internal Medicine Perspectives, 1*(4), 1–7. doi:10.3402/jchimp.v4.25184 PMID:25317266

O'Farrill, R. (2008). Information Literacy and knowledge management: Preparations for an arranged marriage. *Libri, 58*(3), 155–171. doi:10.1515/libr.2008.017

Oakley, B. R., Felder, M., Brent, R., & Elhajj, I. (2004). Turning student groups into effective teams. *Journal of Student Centered Learning, 2*(1), 9–34.

OASIG. (1996). *The performance of information technology and the role of human and organizational factors*. Organizational Aspect of Information technology Special Interest Group.

Obeidat, B., Al-Sarayrah, S., Tarhini, A., Al-Dmour, R. H., Al-Salti, Z., & Sweis, R. (2016). Cultural influence on strategic human resource management practices: A Jordanian case study. *International Business Research, 9*(10), 94. doi:10.5539/ibr.v9n10p94

Oblinger, D. G., & Maruyama, M. K. (1996). *Distributed learning*. Boulder, CO: CAUSE. Retrieved from https://goo.gl/5Qw24y

Oda, T., & White, T. (2005). Immunity from spam: An analysis of an artificial immune system for junk email detection. In *Artificial Immune Systems* (pp. 276–289). Springer Berlin Heidelberg. doi:10.1007/11536444_21

OECD. (2012). *PISA 2009 technical report*. Paris: OECD Publishing. Retrieved from http://proxy.library.carleton.ca/login?url=http://dx.doi.org/10.1787/9789264167872-en

OECD. (2017). *PISA: Programme for International Student Assessment, OECD Education Statistics (database)*. DOI: 10.1787/data-00365-en

Oechslin, P. (2003). Making a faster cryptanalytic time-memory trade-off. In *Advances in Cryptology-CRYPTO 2003* (pp. 617–630). Springer. doi:10.1007/978-3-540-45146-4_36

Official Journal of the European Union. (2006). *Key competences for Lifelong Learning. European Reference Framework*. Brussels: European Parliament and of the Council. Retrieved from http://eur-lex.europa.eu/LexUriServ/site/es/oj/2006/l_394/l_ 39420061230es00100018.pdf

Ozdalga, E., Ozdalga, A., & Ahuja, N. (2012). The Smartphone in Medicine: A Review of Current and Potential Use Among Physicians and Students. *Journal of Medical Internet Research, 14*(5), e128. doi:10.2196/jmir.1994 PMID:23017375

Paine, K. D. (2011). *Measure what matters: Online tools for understanding customers, social media, engagement, and key relationships*. John Wiley & Sons.

Palloff, R., & Pratt, K. (1999). *Building Learning Communities in Cyberspace: Effective Strategies for the Online Classroom*. San Francisco: Joseey-Bass.

Palou, P. A., Urroz, E., Padilla, I., Chávez-Castañeda, R., & Volpi, J. (2002). Manifiesto Crack I, II, III, IV, V. México: Lateral, 70.

Parsons, K. M., Young, E., Butavicius, M. A., McCormac, A., Pattinson, M. R., & Jerram, C. (2015). The Influence of Organizational Information Security Culture on Information Security Decision Making. *Journal of Cognitive Engineering and Decision Making*, *9*(2), 117–129. doi:10.1177/1555343415575152

Paterson, M., & Pentland, W. (2008). The Use of Hermeneutics in a Mixed Methods Design. *Qualitative Report*, *13*(1), 116–134.

Patil, D. (2012). *Data Jujitsu*. O'Reilly Media, Inc.

Patil, D., & Mason, H. (2015). *Data Driven*. O'Reilly Media, Inc.

Pauli, D. (2016). *Security! experts! slam! Yahoo! management! for! using! old! Crypto!* Retrieved June 25, 2017, from https://www.theregister.co.uk/2016/12/15/yahoos_password_hash/

Paulus, T. M., Woodside, M., & Ziegler, M. F. (2010). 'I tell you, it's a journey, isn't it?' Understanding collaborative meaning making in qualitative research. *Qualitative Inquiry*, *16*(10), 852–862. doi:10.1177/1077800410383124

Peñalvo, F. J. G., Conde, M. J. R., Pardo, A. M. S., González, M. Á. C., Zangrando, V., & Holgado, A. G. (2012). GRIAL (GRupo de investigación en InterAcción y eLearning), USAL. *IE Comunicaciones: Revista Iberoamericana de Informática Educativa,* (15), 85-94.

Pérez, M. A., & Delgado, Á. (2012). De la competencia digital y audiovisual a la competencia mediática: Dimensiones e indicadores [From digital and audiovisual competence to media competence: dimensions and indicators]. *Comunicar*, *39*, 25–34. doi:10.3916/C39-2012-02-02

Pérez-Mateo, M., Romero, M., & Romey, T. (2014). La construcción colaborativa de proyectos como metodología para adquirir competencias digitales [Collaborative projects elaboration as methodology to acquire digital competences]. *Comunicar*, *42*(21), 15–24.

Pfeifer, R., & Verscure, P. (1992). Beyond Rationalism: Symbols, Patterns, and Behaviour. *Connection Science*, *4*(3-4), 313–325. doi:10.1080/09540099208946621

Philips, M. S. (1994). *Measuring Trees and Forests* (2nd ed.). Wallingford, UK: CAB International.

Piaget, J. (1952). The origins of intelligence in children. New York: International Universities Press. doi:10.1037/11494-000

Piaget, J., Beth, E. W., & de Zavala, V. S. (1980). *Epistemología matemática y psicología: relaciones entre la lógica formal y el pensamiento real*. Editorial Critica.

Pifarré, M., & Tomico, O. (2007). Bipolar laddering (BLA): A participatory subjective exploration method on user experience. In *Proc. Conference on designing for user eXperiences*. ACM, doi:10.1145/1389908.1389911

Pinto Molina, M., Cordón, J. A., & Díaz, R. G. (2010). Thirty years of information literacy (1977-2007): A terminological, conceptual and statistical analysis. *Journal of Librarianship and Information Science*, 17.

Pinto-Llorente, A. M., Casillas-Martín, S., Cabezas-González, M., & García-Peñalvo, F. J. (2017). Building, coding and programming 3D models via a visual programming environment. *Quality & Quantity*. doi:10.1007/s11135-017-0509-4

Pizarro, A. (1987). *Hacia una historia de la literatura latinoamericana.* México: El Colegio de México/Universidad Simón Bolívar.

Poggi, A., & Tomaiuolo, M. (2010). Integrating peer-to-peer and multi-agent technologies for the realization of content sharing applications. *Studies in Computational Intelligence*, *324*, 93–107.

Poggi, A., & Tomaiuolo, M. (2013). A DHT-based multi-agent system for semantic information sharing. *Studies in Computational Intelligence*, *439*, 197–213.

Poggi, A., Tomaiuolo, M., & Vitaglione, G. (2003). Security and trust in agent-oriented middleware. *Lecture Notes in Computer Science*, *2889*, 989–1003. doi:10.1007/978-3-540-39962-9_95

Polya, G. (2014). *How to solve it: A new aspect of mathematical method.* Princeton university press.

Pontecorvo, C. (2007). On the conditions for generative collaboration: Learning through collaborative research. *Integrative Psychological & Behavioral Science*, *41*(2), 178–186. doi:10.1007/s12124-007-9017-8 PMID:18193519

Prensky, M. (2008). *Programming Is the New Literacy.* Retrieved from http://www.edutopia.org/literacy-computer-programming

Prensky, M. (2001). Digital Natives, Digital Immigrants Part 1. *On the Horizon*, *9*(5), 1–6. doi:10.1108/10748120110424816

Price, R., Becker, K., Clark, L., & Collins, S. (2011). Embedding information literacy in a first-year business undergraduate course. *Studies in Higher Education*, *36*(6), 705–718. doi:10.1080/03075071003725350

Provos, N., & Mazieres, D. (1999, June). A Future-Adaptable Password Scheme. In *USENIX Annual Technical Conference, FREENIX Track* (pp. 81-91). USENIX.

Puhakainen, P., & Siponen, M. (2010). Improving employees' compliance through information systems security training: An action research study. *Management Information Systems Quarterly*, *34*(4), 767–A4.

Pylyshyn, Z. W. (1978). Computational Model and Empirical Constraints. *Behavioral and Brain Sciences*, *1*(01), 93–127. doi:10.1017/S0140525X00059793

Pylyshyn, Z. W. (1980). Computation and Cognition: Issues in the Foundation of Cognitive Science. *Behavioral and Brain Sciences*, *3*(01), 111–169. doi:10.1017/S0140525X00002053

Pylyshyn, Z. W. (1981). Complexity and the Study of Artificial and Human Intelligence. In J. Haugeland (Ed.), *Mind Design.* Cambridge, MA: MIT Press.

Pylyshyn, Z. W. (1984). *Computation and Cognition, Towards a Foundation for Cognitive Science*. Cambridge, MA: MIT Press.

Qing, H., Zhengchuan, X., Tamara, D., & Hong, L. (2011). Does deterrence work in reducing information security policy abuse by employees? *Communications of the ACM*, *54*(6), 54–60. doi:10.1145/1953122.1953142

Quicios-García, M. del P., Sevillano García, M. L., & Ortega Sánchez, I. (2013). Educational uses of mobile phones by university students in Spain. *The New Educational Review*, *34*(4), 151–163.

Rajdeep, A., Joshi, N.R.C., & Pilli, E.S. (2010). Generic framework for network forensics. *International Journal of Computer Applications, 1*(11), 0975 – 8887.

Ramchand, K. (Ed.). (1966). *West Indian Narrative: An Introductory Anthology.* London: Nelson.

Ramírez-Montoya, M. S., & García-Peñalvo, F. J. (2017). La integración efectiva del dispositivo móvil en la educación y en el aprendizaje. *Revista Iberoamericana de Educación a Distancia*, *20*(2), 29–47. doi:10.5944/ried.20.2.18884

Ramos, G., Baños, & Melgosa. (2003). *Aprendizaje innovador en la visualización de piezas y dispositivos, en la formación de dibujo técnico mediante aplicación hipermedia.* XV Congreso Internacional de Ingeniería Gráfica, Nápoles, Italia.

Rapetti, E., Picco, A., & Vannini, S. (2011). Is Mobile Learning a Resource in Higher Education? Data Evidence From an Empirical Research in Ticino. *Journal of E-Learning and Knowledge Society*, *7*(2), 47–57.

Reimann, D., & Maday, C. (2017). Enseñanza y aprendizaje del modelado computacional en procesos creativos y contextos estéticos. *Education in the Knowledge Society*, *18*(3), 87–97. doi:10.14201/eks20171838797

Report, P. W. C. (2015). *The Global State of Information Security Survey 2015*. Retrieved from http://www.pwc.com

Rey, B. (2006). *Les compétences à l'école: apprentissage et évaluation* (2nd ed.). De Boeck. Retrieved from http://ariane.ulaval.ca/cgi-bin/recherche.cgi?qu=i2804150046

Richey, R. C. (2008). Reflections on the 2008 AECT Definitions of the Field. *TechTrends*, *52*(1), 24–25. doi:10.1007/s11528-008-0108-2

Rigelman, N. M., & Ruben, B. (2012). Creating foundations for collaboration in schools: Utilizing professional learning communities to support teacher candidate learning and visions of teaching. *Teaching and Teacher Education*, *28*(7), 979–989. doi:10.1016/j.tate.2012.05.004

Rizo García, M., & Romeu Aldaya, V. (2006). Hacia una propuesta teórica para el análisis de las fronteras simbólicas en situaciones de comunicación intercultural. *Estudios sobre las Culturas Contemporaneas*, *12*(24).

Rizolli, A. E., & Young, W. J. (1997). Article. *Environmental Modelling & Software*, *12*(2-3), 237–249. doi:10.1016/S1364-8152(97)00016-9

Robertson, A. (2012). *LulzSec hackers post data on 8,000 Twitter accounts, but your passwords are safe*. Retrieved June 25, 2017, from http://www.theverge.com/2012/6/12/3080534/lulzsec-reborn-twitter-tweetgif-hack

Robinson, R., Molenda, M., & Rezabek, L. (2008). Facilitating Learning. In A. Januszewski & M. Molenda (Eds.), *Educational Technology: A Definition with Commentary* (pp. 15–48). New York: Routledge.

Rodríguez Conde, M. J., Olmos Migueláñez, S., & Martínez Abad, F. (2013). Evaluación de competencias informacionales en educación secundaria: Un modelo causal. *Cultura y Educación*, *25*(3), 361–373. doi:10.1174/113564013807749687

Rodríguez-de-Dios, I., & Igartua, J. J. (2016). Skills of Digital Literacy to Address the Risks of Interactive Communication. *Journal of Information Technology Research*, *9*(1), 54–64. doi:10.4018/JITR.2016010104

Rogers, D. L. (2000). A paradigm shift: Technology Integration for Higher Education in the New Millennium. *Educational Technology Review*, 19–33.

Rojas López, A., & García-Peñalvo, F. J. (2016). Relationship of knowledge to learn in programming methodology and evaluation of computational thinking. In F. J. García-Peñalvo (Ed.), *Proceedings of the Fourth International Conference on Technological Ecosystems for Enhancing Multiculturality (TEEM'16)* (pp. 73-77). New York, NY: ACM. doi:10.1145/3012430.3012499

Román-González, M. (2015). Computational Thinking Test: Design guidelines and content validation. *Proceedings of EDULEARN15 Conference*, 2436-2444.

Román-González, M. (2014). Aprender a programar 'apps' como enriquecimiento curricular en alumnado de alta capacidad. *Bordón. Revista de Pedagogía*, *66*(4), 135–155. doi:10.13042/Bordon.2014.66401

Román-González, M., Pérez-González, J.-C., & Jiménez-Fernández, C. (2017). Which cognitive abilities underlie computational thinking? Criterion validity of the Computational Thinking Test. *Computers in Human Behavior*, *72*, 678–691. doi:10.1016/j.chb.2016.08.047

Romero, M., Usart, M., & Ott, M. (2015). Can serious games contribute to developing and sustaining 21st century skills? *Games and Culture*, *10*(2), 148–177. doi:10.1177/1555412014548919

Rovai, A. P. (2002). Sense of community, perceived cognitive learning, and persistence in asynchronous learning networks. *The Internet and Higher Education*, *5*(4), 319–332. doi:10.1016/S1096-7516(02)00130-6

Rubia, B. (2010). La implicación de las nuevas tecnologías en el aprendizaje colaborativo [ICT implications in collaborative learning]. *Tendencias Pedagógicas*, *16*, 89–106.

Sahami, M., Dumais, S., Heckerman, D., & Horvitz, E. (1998, July). A Bayesian approach to filtering junk e-mail. In *Learning for Text Categorization: Papers from the 1998 workshop* (*Vol. 62*, pp. 98-105). Academic Press.

Saint-Germain, R. (2005). Information security management best practice based on ISO/IEC 17799. *Information Management Journal*, *39*(4), 60–65.

Salkind, N. J. (1998). *Métodos de investigación*. D.F, México: Prentice Hall.

Salmon, G. (2004). *E-tivities: The key to active online learning*. London, UK: Routledge.

Sánchez i Peris, F. J. (2015). Gamificación. *Education in the Knowledge Society*, *16*(2), 13–15. doi:10.14201/eks20151621315

Sánchez Prieto, J. C., Olmos Migueláñez, S., & García-Peñalvo, F. J. (2017). ¿Utilizarán los futuros docentes las tecnologías móviles? Validación de una propuesta de modelo TAM extendido. *RED. Revista de Educación a Distancia*(52).

Sánchez-Prieto, J. C., Huang, F., Teo, T., Olmos-Migueláñez, S., & García-Peñalvo, F. J. (2017). *Versión en español de una herramienta para el análisis de la influencia de los factores socioculturales en la aceptación de TIC de los docentes universitarios*. doi:10.5281/zenodo.582401

Sánchez-Prieto, J. C., Olmos-Migueláñez, S., & García-Peñalvo, F. J. (2014). Understanding mobile learning: Devices, pedagogical implications and research lines. *Education in the Knowledge Society*, *15*(1), 20–42.

Sánchez-Prieto, J. C., Olmos-Migueláñez, S., & García-Peñalvo, F. J. (2016). Informal tools in formal contexts: Development of a model to assess the acceptance of mobile technologies among teachers. *Computers in Human Behavior*, *55*, 519–528. doi:10.1016/j.chb.2015.07.002

Sánchez-Prieto, J. C., Olmos-Migueláñez, S., & García-Peñalvo, F. J. (2017). MLearning and pre-service teachers: An assessment of the behavioral intention using an expanded TAM model. *Computers in Human Behavior*, *72*, 644–654. doi:10.1016/j.chb.2016.09.061

Sarrab, M., Alzahrani, A., Al Awan, N., & Alfarraj, O. (2014). From traditional learning into mobile learning in education at the university level: Undergraduate students perspective. *International Journal of Mobile Learning and Organisation*, *8*(3–4), 167–186. doi:10.1504/IJMLO.2014.067014

Sarstedt, M., Ringle, C. M., Smith, D., Reams, R., & Hair, J. F. Jr. (2014). Partial least squares structural equation modeling (PLS-SEM): A useful tool for family business researchers. *Journal of Family Business Strategy*, *5*(1), 105–115. doi:10.1016/j.jfbs.2014.01.002

Sasaki, M., & Shinnou, H. (2005, November). Spam detection using text clustering. In *Cyberworlds, 2005. International Conference on* (pp. 4-pp). IEEE. doi:10.1109/CW.2005.83

Scardamalia, M., & Bereiter, C. (1994). Computer support for knowledge-building communities. *Journal of the Learning Sciences*, *3*(3), 265–283. doi:10.1207/s15327809jls0303_3

Scharff, C., & Brown, H. (2004). Thinking through computing: The power of Learning Communities. *Computer Science Education*, *14*(4), 297–320. doi:10.1080/0899340042000303456

Schell, J. (2014). *The Art of Game Design: A book of lenses*. CRC Press. doi:10.1201/b17723

Scheuermann, R. H., Ceusters, W., & Smith, B. (2009). *Toward an Ontological Treatment of Disease and Diagnosis. In 2009* (pp. 116–120). San Francisco, CA: AMIA Summit on Translational Bioinformatics.

Schmidt-Atzert, L. (1985). *Psicología de las emociones*. Herder.

Schoor, C., & Bannert, M. (2011). Motivation in a computer-supported collaborative learning scenario and its impact on learning activities and knowledge acquisition. *Learning and Instruction*, *21*(4), 560–573. doi:10.1016/j.learninstruc.2010.11.002

Seaborn, K., & Fels, D. I. (2015). Gamification in theory and action: A survey. *International Journal of Human-Computer Studies*, *74*, 14–31. doi:10.1016/j.ijhcs.2014.09.006

Sein-Echaluce, M. L., Fidalgo-Blanco, Á., & García-Peñalvo, F. J. (2016). Students' Knowledge Sharing to improve Learning in Engineering Academic Courses. *International Journal of Engineering Education.*

Selwyn, N. (2011). *Education and Technology: Key Issues and Debates*. London: Continuum International Publishing Group.

Seoane Pardo, A. M. (2016). Computational thinking beyond STEM: an introduction to "moral machines" and programming decision making in Ethics classroom. In F. J. García-Peñalvo (Ed.), *Proceedings of the Fourth International Conference on Technological Ecosystems for Enhancing Multiculturality (TEEM'16)* (pp. 37-44). New York, NY: ACM. doi:10.1145/3012430.3012494

Seoane Pardo, A. M., & García-Peñalvo, F. J. (2014). Pedagogical Patterns and Online Teaching. In F. J. García-Peñalvo & A. M. Seoane Pardo (Eds.), *Online Tutor 2.0: Methodologies and Case Studies for Successful Learning* (pp. 298–316). Hershey, PA: IGI Global. doi:10.4018/978-1-4666-5832-5.ch015

Shah, D. (2015). *MOOCs in 2015: Breaking Down the Numbers*. EdSurge.

Shah, D. (2016). *Monetization Over Massiveness: Breaking Down MOOCs by the Numbers in 2016*. EdSurge.

Shon, H. (2013). *All-In-One CISSP Exam Guide* (6th ed.). McGraw Hill.

Silva, T. P., Santos, I., Almeida, T. A., & Gómez Hidalgo, J. M. (2014). Normalização Textual e Indexacção Semântica Aplicadas na Filtragem de SMS Spam. *Proceedings of the Encontro Nacional de Inteligência Artificial e Computacional (ENIAC).*

Simon, B., Anderson, R., Hoyer, C., & Su, J. (2004). Preliminary Experiences with a Tablet PC Based System to Support Active Learning in Computer Science Courses. *ITiCSE 2004: Proceedings of the 9th Annual Conference on Innovation and Technology in Computer Science Education*, 213–217. doi:10.1145/1007996.1008053

Skemp, R. (1978). Relational understanding and instrumental understanding. *ArithmeticTeacher, 26*, 9-15.

Skinner, B. F. (2011). *About behaviorism*. Vintage.

Slattery, B. (2011). *LastPass, Online Password Manager, May Have Been Hacked.* Retrieved June 25, 2017, from http://www.pcworld.com/article/227223/LastPass_Online_Password_Manager_May_Have_Been_Hacked.html

Smith, B., Ashburner, M., Rosse, C., Bard, J., Bug, W., Ceusters, W., & Lewis, S. et al. (2007). The OBO Foundry: Coordinated evolution of ontologies to support biomedical data integration. *Nature Biotechnology*, *25*(11), 1251–1255. doi:10.1038/nbt1346 PMID:17989687

Snalune, P. (2015). The Benefits of Computational Thinking. *ITNOW*, *57*(4), 58–59. doi:10.1093/itnow/bwv111

Sommestad, T., Hallberg, J., Lundholm, K., & Bengtsson, J. (2014). Variables influencing information security policy compliance. *Information Management & Computer Security*, *22*(1), 42–75. doi:10.1108/IMCS-08-2012-0045

Sommestad, T., Karlzén, H., & Hallberg, J. (2015). The sufficiency of the theory of planned behavior for explaining information security policy compliance. *Information & Computer Security*, *23*(2), 200–217. doi:10.1108/ICS-04-2014-0025

Spector, J. M. (2015). *Foundations of educational technology: Integrative approaches and interdisciplinary perspectives*. New York: Routledge.

Statista. (2017a). *Google Play: number of available apps 2009-2017 Number of available applications in the Google Play Store from December 2009 to March 2017*. Author.

Statista. (2017b). *Number of available apps in the Apple App Store from July 2008 to January 2017*. Author.

Steil, J. J., Röthling, F., Haschke, R., & Ritter, H. (2004). Situated robot learning for multi-modal instruction and imitation of grasping. *Robotics and Autonomous Systems*, *47*(2), 129–141. doi:10.1016/j.robot.2004.03.007

Stieger, S., & Reips, U.-D. (2010). What are participants doing while filling in an online questionnaire: A paradata collection tool and an empirical study. *Computers in Human Behavior*, *26*(6), 1488–1495. doi:10.1016/j.chb.2010.05.013

Stone, M. (1974). Cross-validatory choice and assessment of statistical predictions. *Journal of the Royal Statistical Society. Series B. Methodological*, *36*, 111–147. doi:10.2307/2984809

Strijbos, J. W., & Fischer, F. (2007). Methodological challenges for collaborative learning research. *Learning and Instruction*, *17*(4), 389–393. doi:10.1016/j.learninstruc.2007.03.004

Stroud, D. (2008). Social networking: An age-neutral commodity—Social networking becomes a mature web application. *Journal of Direct Data and Digital Marketing Practice*, *9*(3), 278–292. doi:10.1057/palgrave.dddmp.4350099

Suppes, P., & Morningstar, M. (1969). Computer-Assisted Instruction. *Science*, *166*(3903), 343–350. doi:10.1126/science.166.3903.343 PMID:5812034

Suykens, J. A., & Vandewalle, J. (1999). Least squares support vector machine classifiers. *Neural Processing Letters*, *9*(3), 293–300. doi:10.1023/A:1018628609742

Tabaa, Y., & Medouri, A. (2013). LASyM: A learning analytics system for MOOCs. *International Journal of Advanced Computer Science and Applications*, *4*(5), 113–119. doi:10.14569/IJACSA.2013.040516

TACCLE 3 Consortium. (2017). *TACCLE 3: Coding Erasmus + Project website*. Retrieved from http://www.taccle3.eu/

Tang, L. M., & Kay, J. (2014). Gamification: Metacognitive scaffolding towards long term goals? *Personalization Approaches in Learning Environments, 63*.

Tardón, C. G., & de la Vega, R. A. (2014). #dametuits: Programa de formación y comunicación corporativa gamificada. *Obra digital: revista de comunicación*, (7), 47-68.

Tarhini, A. (2016). *The Effects of Cultural dimensions and Demographic Characteristics on E-learning acceptance.* arXiv preprint arXiv:1607.01492

Tarhini, A., Elyas, T., Akour, M. A., & Al-Salti, Z. (2016). Technology, Demographic Characteristics and E-Learning Acceptance: A Conceptual Model Based on Extended Technology Acceptance Model. *Higher Education Studies*, *6*(3), 72–89. doi:10.5539/hes.v6n3p72

Tarhini, A., Hone, K., & Liu, X. (2015). A cross-cultural examination of the impact of social, organisational and individual factors on educational technology acceptance between British and Lebanese university students. *British Journal of Educational Technology*, *46*(4), 739–755. doi:10.1111/bjet.12169

Tarhini, A., Teo, T., & Tarhini, T. (2016). A cross-cultural validity of the E-learning Acceptance Measure (ElAM) in Lebanon and England: A confirmatory factor analysis. *Education and Information Technologies*, *21*(5), 1269–1282. doi:10.1007/s10639-015-9381-9

Taweesiriwate, A., Manaskasemsak, B., & Rungsawang, A. (2012, March). Web spam detection using link-based ant colony optimization. Advanced Information Networking. doi:10.1109/AINA.2012.118

Tenenhaus, M., Vinzi, V. E., Chatelin, Y. M., & Lauro, C. (2005). PLS path modelling. *Computational Statistics & Data Analysis*, *48*(1), 159–205. doi:10.1016/j.csda.2004.03.005

Tennina, L. (Ed.). (2014). Antología Saraus. Movimiento - Literatura - Periferia - São Paulo. Buenos Aires: Tinta Limón.

Teri, S., Acai, A., Griffith, D., Mahmoud, Q., Ma, D. W. L., & Newton, G. (2013). Student use and pedagogical impact of a mobile learning application. *Biochemistry and Molecular Biology Education*, *42*(2), 121–135. doi:10.1002/bmb.20771 PMID:24375862

Terzo, O., Ruiu, P., Bucci, E., & Xhafa, F. (2013). *Data as a service (DaaS) for sharing and processing of large data collections in the cloud.* Paper presented at the Complex, Intelligent, and Software Intensive Systems (CISIS), 2013 Seventh International Conference on.

Toktarova, V. I., Blagova, A. D., Filatova, A. V., & Kuzmin, N. V. (2015). Design and Implementation of Mobile Learning Tools and Resources in the Modern Educational Environment of University. *Review of European Studies*, *7*(8), 318–324. doi:10.5539/res.v7n8p318

Tolmie, A., & Boyle, J. (2000). Factors influencing the success of computer-mediated communication (CMC) environments in university teaching: A review and case study. *Computers & Education*, *34*(2), 119–140. doi:10.1016/S0360-1315(00)00008-7

Tomaiuolo, M. (2013). dDelega: Trust management for web services. *International Journal of Information Security and Privacy*, *7*(3), 53–67. doi:10.4018/jisp.2013070104

Tomaiuolo, M. (2013). *Trust management and delegation for the administration of web services.* Organizational, Legal, and Technological Dimensions of Information System Administration.

Tomaiuolo, M., Poggi, A., & Franchi, E. (2013). Supporting Social Networks With Agent-Based Services. *International Journal of Virtual Communities and Social Networking*, *5*(1), 62–74. doi:10.4018/jvcsn.2013010104

Toman, M. A., & Ashton, P. M. S. (1996). Sustainable forest ecosystems and management: A review article. *Forest Science*, *42*(3), 366–377.

Torres, R., & Edirisingha, P. (2010). Flight of the PELICANS. *Proceedings of The PLE Conference 2010.* Retrieved from http://pleconference.citilab.eu

Torres, R., Martín, R., & Ortiz, A. (2011). *New Technologies, New Environments: developing Personal Learning Environments in Secondary Education.* Presented at McLuhan Galaxy Conference, Barcelona, Spain.

Torres, R., Edirisingha, P., & Mobbs, R. (2009). Putting the pieces together: Conceptual frameworks for building PLEs with Web 2.0 Tools. In *Distance and e-learning in transition*. Wiley-ISTE.

Touya, K., & Fakir, M. (2014). Mining Students' Learning Behavior in Moodle System. *Journal of Information Technology Research, 7*(4), 12–26. doi:10.4018/jitr.2014100102

Tramullas Saz, J. (2002). Propuestas de concepto y definición de la biblioteca digital. In *e-LIS, 2002*. Retrieved: April 4, 2015, from http://eprints.rclis.org/15118/1/04_2002.pdf

Tsang, J. (2013). Doble personalidad. *Entrevista a Alberto Fuguet.Buen Salvaje, 5*, 20-23. Retrieved: April 2, 2015, from http://buensalvaje.com/tag/mcondo/

Turiel, J. (2010). La manía de las listas: 100 bibliotecas digitales y 'recolectores' de libros electrónicos. *bid - textos universitaris de biblioteconomia i documentació, 24*, 1-10. Retrieved: April 4, 2015, from en: http://bid.ub.edu/24/turiel2.htm

UNESCO. (2009). La nueva dinámica de la educación superior y la investigación para el cambio social y el desarrollo. In Conferencia Mundial sobre la Educación Superior (pp. 1–9). París: United Nations Educational, Scientific and Cultural Organization.

UNESCO. (2011). *UNESCO Mobile Learning Week Report*. Paris.

UNESCO. (2012). *2012 Paris OER Declaration*. Retrieved April 20, 2016, from: http://www.unesco.org/new/fileadmin/MULTIMEDIA/HQ/CI/CI/pdf/Events/English_Paris_OER_Declaration.pdf

UNESCO. (2015). *Keystones to foster inclusive knowledge societies: Access to information and knowledge, freedom of expression, privacy, and ethics on a global internet*. París: UNESCO. Retrieved from http://apo.org.au/node/55441

Ungerman, M. (2005). Creating and enforcing an effective information security policy. *Information Systems Control Journal, 6*. Retrieved from http://www.isaca.org

University of Wisconsin Digital Collections. (n.d.). *The Art Collection. Base de Datos de Editoriales Cartoneras*. Madison, WI: University of Wisconsin. Retrieved: May 5, 2015 from http://digicoll.library.wisc.edu/WebZ/SearchOrBrowse?sessionid=01-53965-684142241

Urbach, N., & Ahlemann, F. (2010). Structural equation modelling in information systems research using partial least squares. *Journal of Information Technology Theory and Application, 11*, 5–40. Retrieved from http://iris.ebs.edu

US GAO. (2012). *Information security: Better implementation of controls for mobile devices should be encouraged*. Retrieved from http://www.gao.gov/products/GAO-12-757

Vaibhav, A., & Gupta, P. (2014). Gamification of MOOCs for increasing user engagement. *MOOC, Innovation and Technology in Education (MITE). 2014 IEEE International Conference*, 19-20. doi:10.1109/MITE.2014.7020290

Valimaa, J., & Hoffman, D. (2008). Knowledge society discourse and higher education. *Higher Education: The International Journal of Higher Education and Educational Planning, 56*(3), 265–285. doi:10.1007/s10734-008-9123-7

Vallerand, R. J., Pelletier, L. G., Blais, M. R., Briere, N. M., Senecal, C., & Vallieres, E. F. (1992). The academic motivation scale: A measure of intrinsic, extrinsic, and amotivation in education. *Educational and Psychological Measurement, 52*(4), 1003–1017. doi:10.1177/0013164492052004025

Van Harmelen, M. (2008). Design trajectories: Four experiments in PLE implementation. *Interactive Learning Environments, 16*(1), 35–46. doi:10.1080/10494820701772686

Vance, A., Lowry, P. B., & Eggett, D. (2013). Using Accountability to Reduce Access Policy Violations in Information Systems. *Journal of Management Information Systems, 29*(4), 263–289. doi:10.2753/MIS0742-1222290410

Varma, K. V., Ferguson, I., & Wild, I. (2000). Decision support system for the sustainable forest management. *Forest Ecology and Management*, *128*(1-2), 49–55. doi:10.1016/S0378-1127(99)00271-6

Vázquez-Ingelmo, A., Cruz-Benito, J., & García-Peñalvo, F. J. (2017). *Improving the OEEU's data-driven technological ecosystem's interoperability with GraphQL.* Paper presented at the 5th International Conference on Technological Ecosystems for Enhancing Multiculturality, Cádiz, Spain.

Vázquez-Ingelmo, A., Cruz-Benito, J., & García-Peñalvo, F. J. (2017). Improving the OEEU's data-driven technological ecosystem's interoperability with GraphQL. In J. M. Dodero, M. S. Ibarra Sáiz, & I. Ruiz Rube (Eds.), *Fifth International Conference on Technological Ecosystems for Enhancing Multiculturality (TEEM'17)*. New York, NY: ACM.

Vee, A. (2013). Understanding Computer Programming as a Literacy. *LiCS. Literacy in Composition Studies*, *1*(2), 42–64. doi:10.21623/1.1.2.4

Vicente García, R. (2013). *Los libros electrónicos en las colecciones de las bibliotecas públicas.* Barcelona: Revista bid, textos universitarios de biblioteconomia documentació nº 30, June 2013. Retrieved: May 10, 2015, from http://bid.ub.edu/es/30/devicente.htm#sthash.UPe1oIq8.dpuf

Vijayan, J. (2007). *TJX data breach: At 45.6M card numbers, it's the biggest ever.* Retrieved June 25, 2017, from http://www.computerworld.com/s/article/9014782

Vila, A. R. (2016). Latin American and Caribbean Literature Transposed into Digital. Journal of Information Technology Research, 9(1). doi:10.4018/JITR.2016010103

Villagrasa, S., Fonseca, D., Redondo, E., & Duran, J. (2014). Teaching case of gamification and visual technologies for education. *Journal of Cases on Information Technology*, *16*(4), 38–57. doi:10.4018/jcit.2014100104

Vinacua, B. V. (2007). *Análisis estadístico con SPSS 14: estadística básica (3rd ed.).* McGraw-Hill Interamericana de España S.L.

Visvanathan, A., Hamilton, A., & Brady, R. R. W. (2012). Smartphone apps in microbiology-is better regulation required? *Clinical Microbiology and Infection*, *18*(7), 218–220. doi:10.1111/j.1469-0691.2012.03892.x PMID:22563840

Vygotsky, L. S. (1978). *Mind in society: The development of higher psychological processes.* Cambridge, MA: Harvard University Press.

Wagner, D. A. (1998). *Alfabetización: construir el futuro.* Paris: UNESCO.

Walters, W. H. (2012). Patron-Driven Acquisition and the Educational Mission of the Academic Library. *University Libraries, Library Resources & Technical Services Vol. 56.* Piscataway, NJ: Rutgers. Retrieved: May 7, 2015, fromhttp://journals.ala.org/lrts/article/view/5528/6796

Walther, J. B. (1996). Computer-Mediated Communication: Impersonal, Interpersonal, and Hyperpersonal Interaction. *Communication Research*, *23*(1), 3–43. doi:10.1177/009365096023001001

Walwyn Michael, A. (Ed.). (2001). *Sugar Is All: Caribbean Short Stories and Poems from St. Kitts and Nevis.* Morris Publisher.

Wang, H. B., Yu, Y., & Liu, Z. (2005). SVM classifier incorporating feature selection using GA for spam detection. In *Embedded and Ubiquitous Computing–EUC 2005* (pp. 1147–1154). Springer Berlin Heidelberg. doi:10.1007/11596356_113

Wang, J., Gupta, M., & Raghav Rao, H. R. (2015). Insider threats in a financial institution: Analysis of attack-proneness of information systems applications. *Management Information Systems Quarterly*, *39*(1), 91–112. doi:10.25300/MISQ/2015/39.1.05

Watzlawick, P., & Watzlawick, P. (1976). *Teoría de la comunicación humana; interacciones, patologías y paradojas.* Barcelona: Herder.

Weber, G. (1971). *Inner-city children can be taught to read: four successful schools.* Washington, DC: Council for Basic Education.

Wegerif, R. (2006). A dialogic understanding of the relationship between CSCL and teaching thinking skills. *Computer-Supported Collaborative Learning, 1*(1), 143–157. doi:10.1007/s11412-006-6840-8

Wells, G. (2001). The case for dialogic inquiry. In G. Wells (Ed.), *Action talk and text: Learning and teaching through inquiry* (pp. 171–194). New York: Teachers College Press.

Wenger, E. (1998). *Communities of Practice: Learning, Meaning, and Identity.* Cambridge, UK: Cambridge University Press. doi:10.1017/CBO9780511803932

Wenger, E., Mcdermott, R. A., & Snyder, W. M. (Eds.). (2002). *Cultivating communities of practice: A guide to managing knowledge.* Boston: Harvard Business School Publishing.

Werbach, K., & Hunter, D. (2012). *For the win: How game thinking can revolutionize your business.* Wharton Digital Press.

White, D., & Le Cornu, A. (2011). Visitors and Residents: A new typology for online engagement. *First Monday, 16*(9).

Wickham, H. (2014). Tidy data. *Journal of Statistical Software, 59*(10), 1–23. doi:10.18637/jss.v059.i10 PMID:26917999

Wikipedia. (2011). *Translation memory.* Retrieved from http://en.wikipedia.org/wiki/Translation_memory/

Williams, T. C., & Zahed, H. (1996). Computer-based training versus traditional lecture: Effect on learning and retention. *Journal of Business and Psychology, 11*(2), 297–310. doi:10.1007/BF02193865

Willison, R., & Warkentin, M. (2013). Beyond deterrence: An expanded view of employee computer abuse. *Management Information Systems Quarterly, 37*(1), 1–20. doi:10.25300/MISQ/2013/37.1.01

Wills, G. (2008). Linked data views. Handbook of data visualization, 217-241.

Wilson, S., Liber, O., Johnson, M., Beauvoir, P., Sharples, P., & Milligan, C. (2007). Personal Learning Environments: Challenging the dominant design of educational systems *Journal of e-Learning and Knowledge Society, 3*(3), 27-38.

Wing, J. M. (2008). Computational thinking and thinking about computing. *Philosophical Transactions of the Royal Society a-Mathematical Physical and Engineering Sciences, 366*(1881), 3717-3725. doi:10.1098/rsta.2008.0118

Wing, J. M. (2006). Computational Thinking. *Communications of the ACM, 49*(3), 33–35. doi:10.1145/1118178.1118215

Winograd, T., & Flores, F. (1986). *Understanding Computers and Cognition.* Norwood, NJ: Ablex Publishing Company.

Wischenbart, R., Carrenho, C., Licher, V., Kovac, M. & Vinhuta, M. (2013). *Global eBook: Current Conditions & Future Projections.* London: O'Reilly.

Yadav, A., Mayfield, C., Zhou, N. E., Hambrusch, S., & Korb, J. T. (2014). Computational Thinking in Elementary and Secondary Teacher Education. *Acm Transactions on Computing Education, 14*(1), 1–16. doi:10.1145/2576872

Yang, S. J. H. (2006). Context Aware Ubiquitous Learning Environments for Peer-to-Peer Collaborative Learning. *Journal of Educational Technology & Society, 9*(1), 188–201.

Yan, J., Blackwell, A., Anderson, R., & Grant, A. (2004). Password memorability and security: Empirical results. *Security & Privacy, IEEE, 2*(5), 25–31. doi:10.1109/MSP.2004.81

Yaokumah, W., Brown, S., & Dawson, A. A. (2016). Towards Modelling the Impact of Security Policy on Compliance. *Journal of Information Technology Research*, *9*(2), 1–16. doi:10.4018/JITR.2016040101

Yıldırım, S. (2012). Teacher Support, Motivation, Learning Strategy Use, and Achievement: A Multilevel Mediation Model. *Journal of Experimental Education*, *80*(2), 150–172. doi:10.1080/00220973.2011.596855

Yin, S. (2012). *Last.FM Joins eHarmony, LinkedIn to Celebrate Breach Week*. Retrieved June 25, 2017, from http://www.pcmag.com/article.aspx/curl/2405492

Young, W. J. (1997). An Expert System for estimating diffuse source nutrient export rates. *AI Applications*, *11*(2), 11–18.

Yu, E., & Deng, S. (2011). Understanding Software Ecosystems: A Strategic Modeling Approach. In S. Jansen, J. Bosch, P. Campbell, & F. Ahmed (Eds.), *IWSECO-2011 Software Ecosystems 2011. Proceedings of the Third International Workshop on Software Ecosystems* (pp. 65-76). Aachen, Germany: CEUR Workshop Proceedings.

Zapata-Ros, M. (2015). Pensamiento computacional: Una nueva alfabetización digital. *Revista de Educación a Distancia*, (46).

Zelder, N. (2009). Infoglut: Its the disease of the new millennium. How do we treat it? *IEEE Spectrum*, *46*(10), 30-32, 52.

Zeleny, M. (1987). Management support systems: Towards integrated knowledge management. *Human Systems Management*, *7*(1), 59–70.

Zhang, H., Chen, G., Ooi, B. C., Tan, K.-L., & Zhang, M. (2015). In-memory big data management and processing: A survey. *IEEE Transactions on Knowledge and Data Engineering*, *27*(7), 1920–1948. doi:10.1109/TKDE.2015.2427795

Zhang, J., Reithel, B. J., & Li, H. (2009). Impact of perceived technical protection on security behaviors. *Information Management & Computer Security*, *17*(4), 330–340. doi:10.1108/09685220910993980

Zheleva, E., & Getoor, L. (2009, April). To join or not to join: the illusion of privacy in social networks with mixed public and private user profiles. In *Proceedings 18th international conference on World Wide Web* (pp. 531--540). ACM. doi:10.1145/1526709.1526781

Zhou, B., Pei, J., & Luk, W. (2008). A brief survey on anonymization techniques for privacy preserving publishing of social network data. *ACM SIGKDD Explorations Newsletter*, *10*(2), 12–22. doi:10.1145/1540276.1540279

Zichermann, G., & Cunningham, C. (2011). *Gamification by design: Implementing game mechanics in web and mobile apps*. O'Reilly Media, Inc.

About the Contributors

Francisco José García-Peñalvo received his bachelor's degree in computing from the University of Valladolid (Spain), and his PhD. degree from the University of Salamanca, where he is currently the Head of the Research Group in Interaction and e-Learning (GRIAL). His main research interests focus on eLearning, computers and education and digital ecosystems. He is the Editor in Chief of the Education in the Knowledge Society journal and the Journal of Information Technology Research. He coordinates the Doctoral Program in Education in the Knowledge Society.

* * *

Benaouda Amri originates from Algeria. After having his Bsc in natural science from the University of Tiaret (Algeria), he started studying engineering in soil science. He has gotten his diploma with excellent grade. He worked on the subject: Modeling of organic matter dynamics and nitrification in soils. After he started studying Magister (Doctorate of third cycle, French system) for three years. In his thesis he has gotten very honorable mark. The subject was: The modeling and Application of Artificial Intelligence on the organic matter dynamics, oxygen diffusion and temperature transfer in soils. He worked in the National Agency for the Territorial Planning, after he taught at the university as an Assistant Professor. His studies and work were in French language. Now, he is a PhD candidate at Salamanca University, working on the subject: An Intelligent System for the Territorial Planning.

Oriol Borras-Gené is a PhD in Telecommunications and Telecommunications engineering and technician in e-learning in the Tele-Education Cabinet (GATE) from Technical University of Madrid (Spain). His research lines are Social Networks, e-Learning, Collaborative Learning and Gamification.

Hadj Ahmed Bouarara received a licence degree in computer Science and Master diploma in computer modeling of knowledge and reasoning from the Computer Science department of Tahar Moulay University of Saida Algeria. Now Hadj Ahmed BOUARARA is a phd student in Web and Knowledge Engineering in Dr. Tahar Moulay University. His research interests Data Mining, Knowledge Discovery, Metaheuristic, Bio-inspired techniques, Retrieval Information, Cloud Computing and images processing.

Laura Briz-Ponce obtained his PhD from University of Salamanca and has a Bachelor Degree of Engineering of Telecommunications from Technical University of Madrid. She also holds a Master Degree of Business Administration from IE Business School. She has developed her professional life in telecommunications sector, working as Project Manager and Product Manager. Her main research interests are m-learning, TIC, mobile technologies, Computers & Education, Medical Education, Higher Education and medical imaging technologies. In addition, she has participated in different congress and she is a reviewer of several international journals.

Juan Cruz-Benito received the M.Sc. degree in intelligent systems from the University of Salamanca, Spain, in 2013, where he is currently pursuing the Ph.D. degree in computer sciences. He is one of the youngest members of the Research Group Interaction and eLearning (GRIAL), where he specializes in software solutions based on technology ecosystems. He is an expert in educational virtual worlds, human-computer interaction analysis, and technologies for educational purposes, disciplines that he has developed in many innovation and research projects. He has participated in several European and national projects, such as Tagging, Recognition and Acknowledgment of Informal Learning Experiences, Virtual Alliances for Learning Society, USALSIM Virtual Campus, or the Spanish Observatory for University Employability and Employment, where he participated as a Software Engineer, Researcher, and Developer.

Angel Fidalgo-Blanco is PhD in Computer Science from the University of Las Palmas de Gran Canaria, professor at the Technical University of Madrid (UPM, Spain), director of the Laboratory for Innovation in Information Technologies and member of the research group "Computer Science in Education and Knowledge Management". His main research lines are cooperative work, knowledge management and e-learning. Highlights of software products include search engines on ontologies, which currently offers the official service at two universities and two Spanish Government Ministries. He was founding partner of spin-off on software solutions and author of registered software "CSORA: Search by Ontology" He has been the principal researcher of the project "Searcher of Educational Innovation Good Practices" (funded by Spanish Government) and over 40 R+D+i projects. He is chairman and member of various scientific committees of international conferences and publishes in different scientific journals.

Adriana Gamazo is a PhD student in the Department of Didactics, Organisation and Research Methods at the Faculty of Education of the University of Salamanca. She conducted her undergraduate studies in the fields of psychology and pedagogy. Her research lines revolve around educational assessment, factors related to student academic achievement, school effectiveness and multilevel analysis. She also collaborates with the National Center for Educational Innovation and Research (Ministry of Education) and the Eurydice European network as an external expert.

Felicidad García-Sánchez holds a Bachelor of Fine Arts (2012, University of Salamanca, Spain) and Master Degree (MsC) in Architecture and Interior Design (2013, University of Salamanca, Spain) and is a PhD Candidate in the Education of Knowledge Society programme of the University of Salamanca. She is member of the GRIAL Research Group, where participates as the art director and head of design. She has participated in several national and international projects like VALS (Virtual Alliances for Learning Society), the Spanish Observatory for University Employability and Employment (OEEU), where she participated as visual designer and creative and associations like TRAILS (Talking and Reflecting About Informal Learning Spaces). She focuses her research in the usage of images as a communication language, specially in the context of social networks and the Internet, as can be observed in her Google

Scholar public profile, available at https://scholar.google.es/citations?user=_JS4VMQAAAAJ Apart of her contribution in scientific publications, she has been organizer, technical committee and reviewer for several conferences like Technological Ecosystems for Enhancing Multiculturality (TEEM), "Congreso Internacional sobre Aprendizaje, Innovación y Competitividad" (CINAIC), etc.

Juan Antonio Juanes Méndez is a permanent lecturer at the University of Salamanca, where he obtained his PhD in Medicine and Surgery, and a Software Technician by the Pontificia University of Salamanca. He is a member of the Experts' Board in ACADEMIA programme (Health Scienced Field), in the National Programme of Acreditation for Permanent Leacturers in Health Sciences, at the National Agency of Quality Assessment and Screditation (ANECA). He is professor of Human Anatomy and teaches at the Medicine, Psychology and Pharmacy Faculties. He teaches as well in different Master's programmes: Neurosciences, Clinical Neuropsychology and Communication and Information Technologies in Education. Moreover, he is also a teacher of the Universitary Posgraduate Course on Dementia and Alzheimer diseases. Master en Física y Tecnología Médica He has been invited to take part in the Scientific Committee on Information and Communication Technologies (ICTs)in the Spanish Review of Medical Radiology. He coordinates the official reserach group Advanced Medical Visualization Systems, at the University of Salamanca; and collaborated with the research group: Grup d'Anatomia Virtual i de Simulació, del Centre de Recursos per a l'Aprenentatge i la Investigació, at the University of Barcelona. He has participated as director and collaborator in more than 30 funded research and innovation projects. He has been awarded with 14 research and teaching awards from Royal Medicine Academies and other national institutions. He is a co-author of 20 books of teaching kind. He has collaborated in different projects involving diverse medical and university centres of Spain. He has published more than 150 articles in international and national academic journals, and has delivered more than 330 papers at international and international conferences. The moment being, he has supervised 20 doctoral theses. He has delivered more than 60 lectures in courses, days and symposia. He has been a chair in sessions and round tables at different Conferences and Simposia.

Peace Kumah is currently a doctoral student at the SMC University, Switzerland pursuing Business Administration with specialization in Human Resource Management. She holds MBA (Human Resource Management) degree from Wisconsin International University College, Accra, Ghana. She has several years of teaching experience and occupies leadership positions in various educational institutions. Her research interest includes strategic human resource training and development, organizational leadership and motivation, change management, employment relations, human resource and information technology.

Emiliano Labrador, B.Sc. degree in Multimedia by the Universitat Oberta de Catalunya (UOC), and the M.Sc. degree in Creation and Multimedia Engineering by La Salle School of Engineering from Universitat Ramon Llull. He is currently the Director of the Master in Multimedia Creation and Serious Games (MCM) at La Salle-URL and member of the GRETEL. His main research interest are the application of game mechanics to enhance the user experience in various fields, with special focus on education. His work as a media artist has led him to exhibit work at festivals like ArtFutura, Sonar and others. He collaborates with the Entertainment Technology Center at Carnegie Mellon University (US) as a faculty advisor collaborating in gamification and serious games projects with companies such as Telefónica, Roland, Infojobs, Saint Jude Hospital, and others. He has published several articles and book chapters, as well as spoken and organized industry conferences. He is charge of the co-organization of gEducation Workshop that will be hold within the Gamification World Congress (GWC 2014 and 2015). He has been partner in Gamewise, Leonardo European Project (2012-2014) http://game-wise.eu

Luis Martin-Fernandez is a PhD in Systems and Services Engineering for the Information Society from Technical University of Madrid (Spain). His research lines are Knowledge Management, Social Networks and Collaborative Learning.

Martín Martín-González is a researcher at the UNESCO Chair in University Management and Policy of the Technical University of Madrid (UPM). Prior to his work in UPM he has been working as a researcher in the Faculty of Economics in the Autonomous University of Madrid (UAM). He holds a PhD in Economics and a Master's degree in Economic Development and Public Policy from the UAM. He graduated in Economics from the University of La Laguna (ULL). His working fields are: Economics of Education, Economics of Higher Education, Employability, Evaluation of Educational Policies, Public Economics, Applied Economics, Economic Development, Higher Education and Vocational Training.

Fernando Martinez-Abad is currently an Assistant Professor in the Department of Teaching, Organization and Research Methods at the University of Salamanca, and teaches undergraduate and postgraduate courses in the field of Research Methods and Diagnosis in Education. He is a member of the Research Institute 'University Institute of Educational Sciences' and an active researcher in the Consolidated Research Unit GRIAL. He is specialized in the field of quantitative research methods in Social Sciences and has contributed for the past 5 years on various national and international competitive research projects. He has also participated as co-author in recent years in high-impact scientific journal articles in the field of assessment and research methods in education and made contributions to numerous national and international conferences, being awarded for some of the papers presented.

Margarita Martinez-Núnez is Lecturer Professor in the Department of Business Organization, Management and Statistics at the Technical University of Madrid (Spain). Her research interests are designed to promote knowledge focusing on ICT and their role in improving communication and learning, technological innovation and competitive advantage in organizations. She focuses her research on the conceptual change ICT has brought about in education, particularly, in social networking and mobile device.

David José Murteira Mendes, born in Évora (Portugal) in the 4th of July of 1966 achieved a PhD degree in Computer Science in Universidade de Évora (Portugal) in 2014. Previously he owns a MSc in Applied Artificial Intelligence also from the same University, a MBA from Universidad de Extremadura (Badajoz, Spain) a BSc in Computer Science from Universidade Nova de Lisboa (Portugal). He has developed a Software Engineer career from 1989 and developed mainly software for government organizations in particular for Hospitals intended both for management and clinical support. He acted as IT coordinator near the Mozambican Health Ministry after having designed the national Health Information System according to the governments memorandum "Health Strategic Program 2009/14". He has special interest in scuba diving being an active PADI instructor and underwater photographer.

Agostino Poggi is full professor of Computer Engineering at the Department of Engineering and Architecture of the University of Parma. His research focuses on agent, Web and object-oriented technologies and their use to develop distributed and complex systems. He is author of more than two hundreds of technical papers in international scientific journals, books and refereed conference proceedings, and his scientific contribution has been recognized through the "System Research Foundation Outstanding Scholarly Contribution Award" in 1988 and through the "Innovation System Award" in 2001. He is currently editor of the Software Practice & Experience Journal and in the editorial board of the following scientific journals: International Journal of Agent-Oriented Software Engineering, International Journal on Advances in Intelligent Systems, International Journal of Hybrid Intelligent Systems and International Journal of Multiagent and Grid Systems.

María José Rodríguez-Conde is a professor at the University of Salamanca in Teacher Training in Research and Evaluation Methodology. Currently she is the director of the Institute of Educational Sciences of the University of Salamanca / Spain. She has extensive experience in statistical analysis of social data (SPSS software) education assessment methodology and data analysis. Member of GRAIL Group, author of more than fifty research articles on education and social sciences. She directs and develops research activities in various projects of regional, national and international levels. She has directed more than twenty doctoral theses. She teaches at the Faculty of Education and the University of Salamanca.

José Carlos Sánchez-Prieto is a Researcher in Training at the Faculty of Education of the University of Salamanca (Spain), where he conducts his PhD research within the Programme on Education in the Knowledge Society. His area of research is the assessment of attitudes among in-service and pre-service teachers.

Michele Tomaiuolo received a M.Eng. in Computer Engineering and a PhD in Information Technologies from the University of Parma. Currently he is an assistant professor at the Department of Information Engineering, University of Parma. He has given lessons on Foundations of Informatics, Object-Oriented Programming, Software Enigineering, Computer Networks, Mobile Code and Security. He participated in various national and international research projects, including the EU funded @lis TechNet, Agentcities, Collaborator, Comma, and the national project Anemone. His current research activity is focused on peer-to-peer social networking, with attention to security and trust management, multi-agent systems, semantic web, rule-based systems, peer-to-peer networks.

Ricardo Torres Kompen (Santiago de Chile, 1968) holds both a B Sc (1991) and a M Sc (2000) in Chemical Engineering. He has been working as a professor in HE institutions since 1996. In 2001 he started research in the field of technology enhanced learning and multimedia, with a focus on Personal Learning Environments based on Web 2.0 services; he holds a Ph D in Multimedia Engineering, Universitat Politècnica de Catalunya (2016). He currently teaches at La Salle Campus Barcelona, in the areas of Mathematics, Information Systems for Business, and Computer Engineering.

Patricia Torrijos-Fincias is a research personnel and professor in training in the Department of Teaching, Organization and Research Methods at the University of Salamanca. She is a member of Research Institute 'University Institute of Educational Sciences' and an active researcher in the Research Consolidated Unit GRIAL. Her lines of research are above competencies development, teachers' training and emotional education. She is co-author in recent years in medium-impact scientific journal articles in the field of emotional competencies about the no-university staff and she has contributed with different contributions to numerous national and international conferences, that completed her doctoral research. In addition, she is collaborating in national and international competitive research projects.

Eva Villegas holds a Master in Multimedia Creation and Serious Games (La Salle-URL), Master in Accessible Technologies for Information Societies Services (UOC) and degree in Multimedia by the UPC (Polytechnical University of Catalonia). Nowadays as a coordinator and professor of Degree in Multimedia Engineering and coordinator and professor of Master in User eXperience (MUX). Her research is done mainly around new ways to user experience methods and evaluate the gamification in education. She participates as author and reviewer of articles in National and International scientific congress and carries out projects of usability, gamification, accessibility and user experience for organizations. Her work has received several awards including the 2008 Fundació Agrupació Mútua award on research in the field of disabilities.

Winfred Yaokumah is a senior lecturer and the Dean of the Faculty of Engineering, Science and Computing at the Pentecost University College, Accra, Ghana. He obtained his PhD in Information Technology with specialization in Information Assurance and Security at the Capella University, USA. He has published extensively in several international journals, including the Information Management & Computer Security, International Journal of Technology Diffusion, Journal of Information Technology Research, International Journal of Information Systems and Social Change, International Journal of IT/Business Alignment and Governance, and the International Journal of Information Systems in the Service Sector. His research interest includes information security, e-services, IT governance, cloud computing, technology diffusion, information security governance, and IT leadership.

Index

N

O

P

Q

S

T

V

www.ingramcontent.com/pod-product-compliance
Lightning Source LLC
LaVergne TN
LVHW060304011026
849161LV00011B/2768